Achieving sustainable cultivation of bananas

Volume 2: Germplasm and genetic improvement

It is widely recognised that agriculture is a significant contributor to global warming and climate change. Agriculture needs to reduce its environmental impact and adapt to current climate change whilst still feeding a growing population, i.e. become more 'climate-smart'. Burleigh Dodds Science Publishing is playing its part in achieving this by bringing together key research on making the production of the world's most important crops and livestock products more sustainable.

Based on extensive research, our publications specifically target the challenge of climate-smart agriculture. In this way we are using 'smart publishing' to help achieve climate-smart agriculture.

Burleigh Dodds Science Publishing is an independent and innovative publisher delivering high quality customer-focused agricultural science content in both print and online formats for the academic and research communities. Our aim is to build a foundation of knowledge on which researchers can build to meet the challenge of climate-smart agriculture.

For more information about Burleigh Dodds Science Publishing simply call us on +44 (0) 1223 839365, email info@bdspublishing.com or alternatively please visit our website at www.bdspublishing.com.

Related titles:

Achieving sustainable cultivation of bananas Volume 1: Cultivation techniques
Print (ISBN 978-1-78676-156-9); Online (ISBN 978-1-78676-158-3, 978-1-78676-159-0)

Achieving sustainable cultivation of tropical fruits
Print (ISBN 978-1-78676-284-9); Online (ISBN 978-1-78676-286-3, 978-1-78676-287-0)

Achieving sustainable cultivation of mangoes
Print (ISBN 978-1-78676-132-3); Online (ISBN 978-1-78676-134-7, 978-1-78676-135-4)

Critical issues in plant health: 50 years of research in African agriculture
Print (ISBN 978-1-78676-232-0); Online (ISBN 978-1-78676-235-1, 978-1-78676-234-4)

Advances in postharvest management of horticultural produce
Print (ISBN 978-1-78676-288-7); Online (ISBN 978-1-78676-290-0, 978-1-78676-291-7)

Chapters are available individually from our online bookshop:
https://shop.bdspublishing.com

climate•SMART•publishing
IN AGRICULTURAL SCIENCE

BURLEIGH DODDS SERIES IN AGRICULTURAL SCIENCE
NUMBER 86

Achieving sustainable cultivation of bananas

Volume 2: Germplasm and genetic improvement

Edited by Professor Gert Kema, Wageningen University and Research, The Netherlands; and Professor André Drenth, The University of Queensland, Australia

burleigh dodds
SCIENCE PUBLISHING

Published by Burleigh Dodds Science Publishing Limited
82 High Street, Sawston, Cambridge CB22 3HJ, UK
www.bdspublishing.com

Burleigh Dodds Science Publishing, 1518 Walnut Street, Suite 900, Philadelphia, PA 19102-3406, USA

First published 2021 by Burleigh Dodds Science Publishing Limited

Library of Congress Control Number: 2020942304

British Library Cataloguing in Publication Data
A catalogue record for this book is available from the British Library

ISBN 978-1-78676-344-0 (Print)
ISBN 978-1-78676-347-1 (PDF)
ISBN 978-1-78676-346-4 (ePub)
ISSN 2059-6936 (print)
ISSN 2059-6944 (online)

DOI 10.19103/AS.2020.0070

Typeset by Deanta Global Publishing Services, Dublin, Ireland

Contents

Part 3 Genetic improvement through breeding

Series list

Title	Series number
Achieving sustainable cultivation of maize - Vol 1 From improved varieties to local applications Edited by: Dr Dave Watson, CGIAR Maize Research Program Manager, CIMMYT, Mexico	001
Achieving sustainable cultivation of maize - Vol 2 Cultivation techniques, pest and disease control Edited by: Dr Dave Watson, CGIAR Maize Research Program Manager, CIMMYT, Mexico	002
Achieving sustainable cultivation of rice - Vol 1 Breeding for higher yield and quality Edited by: Prof. Takuji Sasaki, Tokyo University of Agriculture, Japan	003
Achieving sustainable cultivation of rice - Vol 2 Cultivation, pest and disease management Edited by: Prof. Takuji Sasaki, Tokyo University of Agriculture, Japan	004
Achieving sustainable cultivation of wheat - Vol 1 Breeding, quality traits, pests and diseases Edited by: Prof. Peter Langridge, The University of Adelaide, Australia	005
Achieving sustainable cultivation of wheat - Vol 2 Cultivation techniques Edited by: Prof. Peter Langridge, The University of Adelaide, Australia	006
Achieving sustainable cultivation of tomatoes Edited by: Dr Autar Mattoo, USDA-ARS, USA & Prof. Avtar Handa, Purdue University, USA	007
Achieving sustainable production of milk - Vol 1 Milk composition, genetics and breeding Edited by: Dr Nico van Belzen, International Dairy Federation (IDF), Belgium	008
Achieving sustainable production of milk - Vol 2 Safety, quality and sustainability Edited by: Dr Nico van Belzen, International Dairy Federation (IDF), Belgium	009
Achieving sustainable production of milk - Vol 3 Dairy herd management and welfare Edited by: Prof. John Webster, University of Bristol, UK	010
Ensuring safety and quality in the production of beef - Vol 1 Safety Edited by: Prof. Gary Acuff, Texas A&M University, USA & Prof. James Dickson, Iowa State University, USA	011
Ensuring safety and quality in the production of beef - Vol 2 Quality Edited by: Prof. Michael Dikeman, Kansas State University, USA	012
Achieving sustainable production of poultry meat - Vol 1 Safety, quality and sustainability Edited by: Prof. Steven C. Ricke, University of Arkansas, USA	013
Achieving sustainable production of poultry meat - Vol 2 Breeding and nutrition Edited by: Prof. Todd Applegate, University of Georgia, USA	014

Acknowledgements

We wish to acknowledge the following for their help in reviewing particular chapters:

- Chapter 2: Prof. Rodomiro Ortiz, Swedish University of Agricultural Sciences, Sweden
- Chapter 4: Dr Christophe Jenny, CIRAD, France
- Chapter 12: Dr Sébastien Ricci, CIRAD, France; and Dr Benjamin Dugdale, Queensland University of Technology, Australia
- Chapter 15: Dr Claudia Fortes Ferreira, Embrapa, Brazil

Introduction

Bananas are the world's most exported and valuable fruit. Banana production faces a number of challenges stemming from an extreme narrow genetic base on which commercial cultivation is based which leaves the crop highly vulnerable to a range of biotic and abiotic stresses. Due to the increasing global spread of banana diseases over the last 50 years there is an urgent need to widen the genetic base and incorporate disease resistance into new varieties while retaining or improving key yield and quality characteristics whilst also improving tolerances to abiotic stresses. This requires a thorough understanding of banana genetics that underlies any breeding effort, irrespective of the strategies that are being followed. These key issues are addressed in this volume

This book is the second in a three-volume series on bananas. Volume 1 published in 2018 focuses on the cultivation of bananas, this second volume is focused on germplasm and genetic improvement, and Volume 3 is devoted to pests and diseases. Each volume stands on its own but together we believe the three books provide a comprehensive review of all key aspects for a process towards sustainable banana production.

Volume 2 begins with an overview of the challenges faced trying in contemporary banana breeding programs. This is followed by an overview of the progress made in identifying and classifying Musa germplasm to broaden the genetic base as a foundation for a more sustainable banana industry in the future. Chapters review ways of classifying banana cultivars and exploiting current germplasm collections. Insights are provided into the wide genetic diversity in wild seeded *Musa* species and its poor representation in gene banks. This is followed by chapters on genetic improvement through breeding comprising a range of conventional breeding techniques, improving fertility, trait identification, new ways for developing hybrid varieties as well as several techniques aimed at achieving genetic improvement without hybridization.

Chapter 1 sets the scene by providing an overview of genetic improvement in bananas over the last century. Bananas were among the first crops cultivated by man and continue to be highly important for the livelihoods and food security of millions of people throughout the tropics and subtropics. Yet almost all banana varieties and landraces grown have been around for a very long time and some were selected in antiquity and clonally propagated ever since. The high level of sterility and parthenocarpy in bananas, producing edible fruit without seeds, has until the middle of the last century excluded genetic recombination as a means to select novel hybrids and achieve genetic gain. The fact that Cavendish and closely related varieties represent nearly half of the world's banana production has given rise to the extremely narrow genetic basis for contemporary banana production as demonstrated by the emergence and subsequent spread of a range of pathogens, which underscores the genetic vulnerability of the global banana industry.

Resistance to diseases and pests is the best form of crop protection and it is imperative to develop new cultivars that perform well in changing growing environments. Currently, banana improvement - with relevant gains in resistance to diseases and pests - is accomplished through either conventional or non-conventional breeding strategies but compared to many other crops banana breeding is in its infancy. Various chapters critically review a variety of genetic improvement strategies that have been employed over the last century such as hybridization approaches, mutation breeding, somaclonal selection, genetic transformation and marker assisted selection. The role of genomics and more recent molecular breeding approaches are also discussed.

Part 1 Classification

The first part of the book begins with fundamental genetics in *Musa* and cultivars. Chapter 2 presents an overview of cytogenetics, genetics and genomics research carried out to elucidate the meiotic chromosome behaviour in the hybrids and the mapping of genes. Special attention is paid to the different origins of chromosomes and DNA sequences in the banana genomes. Finally, the chapter illustrates cytogenetic and linkage mapping of a diploid *M. acuminata* ssp. *malaccensis* using the diploid *M. acuminata* 'Pahang' as a reference genome and highlights the occurrence of structural rearrangements in the genome.

The next chapter focuses on identifying and classifying banana cultivars. Chapter 3 begins with an overview of different banana cultivars, followed by a discussion of the different classification methods and levels such as genera, sections and species. The chapter also highlights the current banana cultivar groups and examines how different cultivars can be identified and concludes with a summary of future research trends in the area.

In Chapter 4 current *Musa* germplasm collections are explored. They are a source of germplasm diversity which is the foundation for genetic improvement. However, the germplasm information systems aimed at increasing their utilisation and impact faced several challenges. The chapter discusses conservation and germplasm integrity, characterization of accessions and information systems that add value to collection documentation. Furthermore, it reviews information systems and their key role in collection management, with a focus on the importance of specific user interfaces and tools in order to simplify and accelerate research and breeding efforts relying on *Musa* germplasm. It considers a range of environmental and experimental data that play an important role in selection of accessions for breeding and other purposes and a strategy is outlined for reducing gaps in collections using *in situ* banana collections.

Part 2 Broadening the genetic base

Part 2 opens with a chapter on the scope of collecting wild *Musa* species germplasm. The value of crop wild relatives as sources of diversity to secure crop production is increasingly recognized. In banana, past collecting and conservation efforts focused mainly on the two primary ancestors of cultivated bananas, *M. acuminata* Colla and *M. balbisiana* Colla. However, more than 70 taxa of wild *Musa* species exist in the wild and contain a vast array of genes which can be deployed in genetic improvement programs. Chapter 5 discusses the hitherto limited exploration of genetic diversity in wild bananas and how the industry focus on clonal propagation of the banana crop has discouraged the efficient collecting and conservation of wild *Musa* species. The change in focus from collecting only single clonal individuals to seed populations is presented and practical issues linked to collecting populations are discussed within the scope of wild *Musa* species reproductive biology.

Chapter 6 examines the collection and evaluation of wild *Musa* species consisting of about 70 different taxa distributed across the Asian tropics and subtropics from India, Southern China and Southern Japan in the north to Northern Australia in the south and the islands in the Pacific Ocean to the west. The chapter begins by highlighting the importance of collecting and characterizing wild *Musa* germplasm and provides an estimation of the genetic diversity missing in the *Musa* Germplasm Transit Center (ITC). It also discusses the importance of conserving cultivars versus wild germplasm, before concluding with a summary on why maintaining and expanding the ITC collection is so important in terms of conserving banana germplasm from all over the world.

Chapter 7 expands on topics touched upon in Chapter 6 and focuses on the collection and evaluation of banana and plantain landraces in Africa which is sometimes considered as a secondary centre of diversity for bananas. The chapter provides an overview of the different banana and plantain landraces present in Africa followed by an examination of landrace diversity and its conservation. It continues with discussing access to and utilization of characterization and evaluation data of banana and plantain landraces before highlighting the challenges of maintaining collected landraces in Africa. The chapter concludes with recommendations for addressing challenges and exploiting opportunities for the sustainable deployment of *Musa* landraces in Africa.

Chapter 8 covers seed germination, preservation and population genetics of wild *Musa* germplasm. Crop wild relatives play an important role in modern plant breeding as they contain important traits for agriculture. Their conservation is therefore essential for future sustainable agriculture and food security. The chapter focuses on the different aspect and techniques that are required to establish seed collections of wild bananas and covers details of the physiology and morphology of banana seeds a well as difficulties encountered during germination of banana seed and putative solutions and possibilities for medium- and long-term maintenance. The chapter concludes with a summary of future research trends.

Chapter 9 reviews the safe dissemination of banana germplasm resources. It begins by examining the general recommendations for international *Musa* germplasm exchange, followed by methods to assure high health status germplasm from field and *in vitro* collections. The chapter also discusses cryopreservation of *Musa* germplasm, before reviewing seed and DNA banking. Due to sterility of bananas and the need to vegetatively propagate planting material, the need for virus indexing to ensure plant health is highlighted. The importance of indexing *Musa* germplasm for existing and novel viruses is discussed, particularly related to Banana Streak Virus, which can be integrated into the banana B genome. The chapter also covers practical aspects related to physical shipment of germplasm and the legal regulations surrounding international germplasm transfer and concludes with highlighting future research trends.

Part 3 Genetic improvement through breeding

The final part of the book starts with a discussion on approaches which can be used to improve banana breeding. A case is made for more genetically diverse bananas which meet specific demands such as market acceptable fresh fruit, cooking bananas or varieties dedicated to processing. Chapter 10 highlights the need for sustainable genetic solutions, followed by an evolutionary analysis of *Musa* species and the events that gave rise to gamete sterility producing the contemporary edible banana fruit. The chapter also discusses different breeding strategies to achieve genetic gain and gives an excellent overview of the difference between 'pedigree breeding' and 'reconstructive breeding' strategies and their perspective towards achieving future breeding goals.

Chapter 11 deals with overcoming the fertility crisis in bananas. Edible bananas are normally parthenocarpic and seedless, a condition which ensures edible fruits, but this infertility limits the use of existing varieties in crosses to generate new combinations of targeted traits for crop improvement. Hybridizations involving edible bananas result in very few hybrid seeds with low or no viability. The inability to effectively use existing varieties with desired traits as parents for new crosses and backcrosses constitutes a fertility crisis which seriously hampers progress in banana breeding. The chapter discusses key processes surrounding effective pollination, fertilization and viable seed production in relation to our hitherto knowledge concerning banana reproductive biology. For instance, it explores possible limitations of seed-set and provides insights to overcome the fertility crisis towards a much-needed renaissance in banana breeding, which would enable gene recombinations through intra- and interspecific crosses.

Chapter 12 focuses on targeted improvement of Cavendish clones. Despite hundreds of different available banana cultivars, growers and the banana export industry prefer cultivars from the Cavendish subgroup. All Cavendish cultivars are triploid and parthenocarpic. Therefore, improvements depend on somaclonal variation which has resulted in a diverse range of morphological phenotypes. Through clonal propagation and more recently the development of *in vitro* culture, preferred clones can be rapidly mass propagated and conveniently offered to growers. The chapter outlines the potential of ongoing selection and genetic improvement technologies that are presently available and how these have been utilized to improve the performance of Cavendish cultivars to overcome today's biological and environmental challenges.

Chapter 13 concentrates on developing hybrid varieties with improved properties, specifically focusing on research undertaken at Embrapa in Brazil. It opens with a discussion concerning the background of the 11 hybrid banana cultivars that have been released by Embrapa, followed by a review of using fertile seeded germplasm as a basis of the breeding strategy. The chapter outlines the development and use of improved diploids - containing multiple desirable traits - in the process of breeding sterile banana cultivars. Resistance breeding to Fusarium wilt race 1 is discussed, followed by a review of Embrapa's breeding cycle and some alternative methods to classical crossbreeding strategies. The chapter concludes with the importance of additional research to support the banana breeding effort.

The great challenge for bananas into the future is to develop cultivars with high agronomic acceptability and yield potential, possessing multiple disease resistances and abiotic stress tolerances, a high level of transportability and enhanced fruit quality. In short, the two major genetic improvement strategies are conventional breeding, delivering new cultivars, and genetic modification, delivering improved current cultivars. Chapter 14 reviews banana improvement through the process of genetic modification and gene editing with a focus on the journey required for these improved varieties to reach farmers' fields. The chapter contains a case study on biofortified East African Highland bananas and concludes with an overview of future research trends.

Chapter 15 addresses the development and use of rapid phenotyping methods, genetics and genomic selection approaches for tolerance to environmental stress factors in banana. This final chapter discusses banana physiology in relation to its agro-ecological environment and the phenotyping of traits such as tolerance to sub-optimal temperature water availability. A workflow is presented to phenotype for these traits in a high throughput-controlled environment. Once interesting traits are identified, strategies are outlined to identify potential molecular markers that are closely linked to genes encoding for these traits, including genome-wide association studies (GWAS), associative transcriptomics and proteomics. The chapter concludes with highlighting how correlations between phenotype and genetic markers will aid future banana breeding.

Taken together, the chapters in this volume show the challenges of genetic improvement of bananas due to the inherent fertility issues and the lack of basic genetic research which underpins breeding. At the same time, all chapters also highlight the urgent need for genetic improvement, specifically for disease resistance, and in general call for broadening the genetic basis of the current banana industry. To achieve this available *Musa* germplasm must be collected, protected *in situ*, moved around the world in a safe manner, safeguarded, and characterised for the presence of useful traits. The underlying genetics needs to be unravelled and enabling technologies for either direct or indirect select for these traits need to be developed. Novel approaches in genetic improvement and breeding as put forward and scrutinized in this volume await deployment to bypass the lack of fertility in edible banana varieties. A truly sustainable future for bananas lies in a coordinated, synergistic and multidisciplinary combination of approaches from a production, an employment, a logistic and consumer perspective. Future banana production entails a smooth integration of germplasm selection, breeding, agronomy and pre- and post-harvest integrated pest and disease management.

Chapter 1

An overview of genetic improvement in bananas over the last century

Mike Smith, Queensland Department of Agriculture and Fisheries, Australia; and Michael Pillay, Vaal University of Technology, South Africa

1 Introduction

Bananas were among the first crops cultivated by man and continue to be important for the livelihoods of millions of people throughout the tropics and subtropics. Yet many cultivars and landraces are susceptible to biotic and abiotic stresses. Genetic resistance to diseases and pests is the best form of crop protection and it is imperative to develop new cultivars that perform well across a variety of changing environments. Currently, the production of improved cultivars with relevant gains in resistance to diseases and pests and other preferred agronomic traits has been accomplished through both conventional and non-conventional breeding. This chapter reviews the various genetic improvement strategies used to develop new banana cultivars and discusses the strategies that have been employed over the last one hundred years. They include hybridization approaches, mutation breeding, genetic transformation and marker-assisted selection. The role of genomics and more recent molecular breeding approaches are also discussed.

Vegetatively propagated crops, such as banana, play an important role as a source of food and income especially for small-scale farmers in many countries. Banana is also a vital export crop for a number of countries and to which it contributes a large part of the gross domestic product (GDP). Like many other

http://dx.doi.org/10.19103/AS.2020.0070.20

crops, bananas are also affected by a number of diseases and pests (Table 1) that are responsible for severe yield losses in both small-scale and large commercial plantations. Therefore, genetic improvement of the crop is essential to mitigate current and envisaged biotic and abiotic stresses. However, almost exclusively, farmers were and are still cultivating a small number of clones selected by their forebearers, some of which extend back thousands of years.

Banana breeding began in 1922 in Trinidad and four years later in Jamaica with the singular objective of developing a Fusarium wilt-resistant 'Gros Michel' (AAA) export dessert banana (Simmonds, 1966; Shepherd, 1974). Initially, banana breeding was considered a paradox since most of the commercially important cultivars are parthenocarpic triploids ($2n = 3x = 33$) and largely sterile. Seedlessness of the fruit is an essential characteristic that consumers

Table 1 Resistance to major banana diseases and pests are objectives of most conventional and non-conventional breeding programmes

Pathogen or pest	Crop effect	Genetic resistance identified
Fusarium wilt (various *Fusarium* spp.) race 1	Wilting and death of susceptible cultivars	Yes*
Fusarium wilt (*Fusarium odoratissimum)* TR 4	Wilting and death of susceptible cultivars	Yes*
Black Sigatoka (*Pseudocercospora fijiensis*)	Loss of green leaf; severely reduced yields; very poor shelf life of fruit	Yes
Yellow Sigatoka (*Pseudocercospora musae*)	As above, but usually not as severe	Yes
Xanthomonas wilt (*Xanthomonas campestris* pathovar *musacearum*)	Internal discolouration and rotting of fruit leading to a 'top-down' infection of plant	Incomplete resistance
Moko; Bugtok (*Ralstonia solanacearum* biovar 1)	Internal discolouration and rotting of fruit leading to a 'top-down' infection of plant	Incomplete resistance
Blood disease (*Ralstonia syzygii* subsp. *celebesensis*)	Internal discolouration and rotting of fruit leading to a 'top-down' infection of plant	Incomplete resistance
Banana Bunchy Top Virus (BBTV)	Severe stunting of plants leading to crop failure	None
Nematodes (*Radopholus similis, Pratylenchus goodeyi, Helicotylenchus multicinctus*)	Damage to roots leading to toppling and reduced yield	Yes*
Banana weevil (*Cosmopolites sordidus, Castniomera licus)*	Boring and damage to rhizome and stems leading to snapping and reduced yield	None

* Complete resistance has been identified as well as levels of partial resistance.

demand and is a trait that must be retained by the breeder. In order to generate new cultivars, breeders need to conduct crosses that produce seeds from which to grow plants and make their selections. Despite the purported sterility of most cultivars, banana breeding turned out to be possible and was initially based on the ability of 'Gros Michel' and its shorter natural mutants to produce seeds at a very low frequency, for example, there was an average of two seeds per bunch when hand pollinated with wild diploids ($2n = 2x = 22$) (Simmonds, 1966). It was discovered that 'Gros Michel' contributed an unreduced triploid gamete that yielded tetraploids ($2n = 4x = 44$) when fertilized with pollen from a diploid.

Although Fusarium wilt-resistant tetraploid hybrids were produced by the Imperial College of Tropical Agriculture (Trinidad), other undesirable attributes such as poor shelf life, short fingers or weak pedicels meant that none of the hybrids was commercially acceptable (Stover and Simmonds, 1987). Cavendish, selected in antiquity by farmers in the centre of origin of banana in Southeast Asia, was reported to be resistant to *Fusarium* spp. (so-called race 1 strains), which caused the Panama disease or Fusarium wilt epidemics in 'Gros Michel' and eventually replaced it in the international trade.

Today there are twelve banana-breeding programmes around the world that differ in their objectives and breeding strategies, while addressing distinct regional needs. Most of these programmes have been reviewed recently (Tenkouano et al., 2011b; Ortiz and Swennen, 2014) and their key characteristics are outlined in the next section. While resistance to Fusarium wilt (especially to the devastating tropical race 4 - TR4) remains a priority, resistance to Sigatoka leaf diseases (*Pseudocercospora fijiensis* and *P. musae*) and a complex of plant-parasitic nematodes (*Radopholus similis, Pratylenchus goodeyi* and *Heliocotylenchus multicinctus*) are also important breeding objectives (Table 1). With this in mind, breeders aim to develop resistant cultivars that retain acceptable organoleptic properties and fruit quality. Recently, fruit quality traits such as increased pro-vitamin A has gained prominence as an objective for banana breeding (Amah et al., 2019a,b,c). Hybrids should also be productive over a wide range of conditions and have a short stature that is easy to manage and harvest.

Recent breakthroughs in banana breeding have overturned earlier attitudes to breeding strategies. Aguilar Morán (2013) showed that it was possible to improve the triploid Cavendish cultivar through conventional breeding as they are not as sterile as previously considered. The rate of seed production was approximately one seed for every 100 pollinated bunches. Advances in genetic marker and sequencing technologies have allowed the development of tools to decipher the complexity of the *Musa* genome and significantly increase the insights into the genetic make-up of our banana cultivars, which will vastly improve the efficiency of banana breeding. Bakry and Horry (2016) have shown that only a few hybridization events have occurred to produce our present-day cultivars from their wild, seedy ancestors. Molecular markers can now be utilized to best select parents that combine fertility with other desirable traits to mimic the sequence of crosses that occurred naturally over several millennia (Bakry and Horry, 2016). This approach is particularly important in a biological context of gamete sterility in desirable but seedless cultivars and difficulties to get large quantities of seeds for which recombination and introgression of desirable genes remains the greatest challenge.

Figure 1 Conventional breeding sometimes involves hand pollinating hundreds of bunches (above) to recover a few seeds (below), which are usually embryo-cultured to obtain plants for further rounds of selection and breeding. (Picture courtesy of International Institute of Tropical Agriculture, Uganda).

Currently, the production of improved cultivars with relevant gains in resistance to diseases and pests and other preferred agronomic traits has been accomplished through both conventional and non-conventional breeding. In conventional breeding hundreds of bunches have often been hand-pollinated to obtain a few seeds (Fig. 1); however, some crosses have been more productive and have accelerated genetic improvement goals. These various strategies are discussed in the next section.

2 Conventional breeding: evolutionary and reconstructive approaches

Tenkouano et al. (2011b) discuss two competing breeding philosophies, which however, are not mutually exclusive. Evolutionary breeding, which many breeding programmes referred to as the conventional approach, attempts to mimic the evolutionary development of the *Musa* species complex. In this process, female fertile triploids (AAA, AAB or ABB genomes) with good agronomic and fruit characteristics are crossed with diploid accessions of *M. acuminata* (AA) or *M. balbisiana* (BB), which often are the sources of resistance genes the breeders seek. The 3*x*/2*x* crosses produce primary hybrids with large phenotypic variation subtended by ploidy and genome polymorphisms, among which diploids and tetraploid progenies that are disease resistant and agronomically desirable are selected and intermated by way of 4*x*/2*x* crosses to produce secondary triploid hybrids.

The second school of thought may be described as reconstructive breeding, whereby breeders attempt to recreate the landraces by first determining their most likely ancestors from the pool of diploids available and subsequently using only such putative or closely related derivatives in crosses aiming at synthesizing improved variants of the susceptible landraces. This approach advocated by Stover and Buddenhagen (1986) involves the use of colchicine to double the chromosome number of natural and improved diploids in order to induce the formation of tetraploids, which are crossed with other diploids leading to the synthesis of sterile triploid individuals.

The earliest banana-breeding programme, set up at the Imperial College of Tropical Agriculture (Trinidad), is no longer operational. The honour of being the oldest continuous banana-breeding programme in operation is that of Fundación Hondureña de Investigación Agrícola (FHIA), which was established in 1959 by the United Fruit Company in La Lima, Honduras, to develop a Fusarium wilt-resistant 'Gros Michel' (Rowe, 1987; Rowe and Rosales, 1994). FHIA's programme favoured an evolutionary approach that initially focussed on developing tetraploid dessert bananas, but gradually expanded to produce plantain and cooking bananas. Rowe and Rosales (1994) stated that the breeding objectives of FHIA by the latter part of the twentieth century were to develop productive dwarf dessert bananas, plantain and cooking banana hybrids with resistance to black Sigatoka, burrowing nematode and race 1 and 4 of Fusarium wilt. A top priority was to safeguard smallholder farmers as it was recognized that these were staple crops used for domestic consumption in many countries throughout the tropics.

Orjeda (2000) concluded from a global International *Musa* testing programme that the four FHIA hybrids included in the field trials (FHIA-01, FHIA-03, FHIA-17 and FHIA-23) were consistently the best-yielding genotypes in comparison to the eleven hybrids from other breeding programmes, as well as local cultivars under evaluation. In fact, FHIA-17 and FHIA-23 were the most stable and best yielding hybrids in all 18 countries involved in the programme. Further in-country evaluation of various FHIA hybrids have demonstrated their improved agronomic performance in a variety of geographic locations, whether it is the lowland tropics of Rwanda (Gaidashova et al., 2008) and Cameroon (Tenkouano et al., 2011a), the wet subtropics of Brazil (Nomura et al., 2013) and Australia (Smith et al., 2014), and the dry subtropics of Mozambique (Uazire et al., 2008). In Cuba, disease-resistant FHIA hybrids are increasingly replacing susceptible banana and plantain cultivars and, as early as 2000, accounted for over 12% of total banana production (Frison and Sharrock, 2000). Importantly, Aguilar Morán (2013) stated that three tetraploid hybrids (FHIA-01, FHIA-18 and FHIA-25) are resistant to TR4, but large-scale comparative field trials in TR4-infected fields are required to further substantiate this.

It is also important to note that FHIA never entirely lost sight of the importance of developing disease-resistant dessert bananas for export (Rowe and Rosales, 1994). Starting in 1992, and on the assumption that Cavendish (AAA) has low fertility, an ambitious crossing programme was initiated to produce second-generation triploid Cavendish hybrids that are resistant to TR4 and black Sigatoka (Aguilar Morán, 2013).

Brazil launched a banana breeding programme in 1976 by initially collecting mainly diploids to be used in crosses with the AAB group, and more specifically the Prata subgroup of dessert bananas (Shepherd et al., 1994; Amorim et al., 2013). The Empresa Brasileira de Pesquisa Agropecuária (Embrapa) programme utilizes an evolutionary breeding approach to produce high-yielding genotypes, with shorter plant stature, shorter cycle and resistance to Fusarium wilt and Sigatoka leaf diseases, and with fruits having the qualities of either Prata or Silk bananas. A number of hybrids have been released to Brazilian farmers, as well as to the international community, with a tetraploid hybrid 'BRS Platina' performing well in commercial field trials (Amorim et al., 2013).

Research on plantain at IITA (International Institute of Tropical Agriculture) in Nigeria began in 1973 when several African governments called for the urgent need to control black Sigatoka as millions of smallholders depended on plantains for their livelihoods (Ortiz and Vuylsteke, 1994; Tenkouano et al., 2011b). A plantain (AAB) evolutionary breeding programme followed in 1987 in Nigeria, eventually including a programme for East African Highland bananas (EAHBs, AAA) in Uganda. A decade later, several improved plantain-derived hybrids with resistance to black Sigatoka had been produced and tested in many countries with 'PITA-14' being one of the first-generation tetraploid hybrids being grown in Nigeria and neighbouring countries (Tenkouano et al., 2011b). In Côte d'Ivoire, plantain growers selected 'PITA-3' and 'FHIA-21' based on their improved agronomic characteristics, which have been mass-propagated and made available to farmers in four West African countries (Tenkouano et al., 2019). With regard to EAHBs, which have low female-fertility rates, 'NARITA-7' was released to farmers in Uganda (Batte et al., 2019). However, there has been some resistance to uptake some of the NARITA hybrids by smallholders who cite astringency, altered organoleptic qualities and the occasional presence of seeds in the pulp as key constraints to acceptability in highland regions (Karamura et al., 2016). This demonstrates the dilemma banana breeders face with new disease-resistant hybrids, whereby smallholders prefer the taste and texture of local landraces (Tenkouano et al., 2011a; Nyine and Pillay, 2011). The Centre Africain de Recherches sur les Bananiers et Agriculture (CARBAP) comprises another African plantain breeding programme located in Cameroon that developed the tetraploid 'CRBP-39' which was later released for evaluation (Tenkouano et al., 2011b). Ongoing funding for the breeding programmes in Africa has always been a serious risk to these programmes and they are currently supported temporarily by external funding from the Bill and Melinda Gates Foundation. This unique project seeks to improve the production and productivity of banana in Tanzania and Uganda, through the development of hybrid banana varieties that are expected to have 30% higher yield compared to the current varieties grown by farmers under the same conditions (https://breedingbetterbananas.org/).

The Centre de Coopération Internationale en Recherche Agronomique pour le Développement (CIRAD) has been a strong proponent of the reconstructive approach of banana breeding (Bakry and Horry, 2016). A hybridisation programme commenced in Guadeloupe in the mid-1980s to improve diploids sourced from Papua New Guinea and Southeast Asia (Tezanas du Montcel, 1986). Bakry and Horry (1994) stated that this reconstructive approach has several advantages: (1) the final product, being triploid, is highly sterile, (2) the synthesis is flexible due to the high diversity available in the diploid germplasm and (3) the use of a doubled diploid (colchicine treatment) leads to the retention, in the triploid, of the genetic structure of the diploid (Fig. 2).

One serious problem that was encountered, and now overcome, was the presence of endogenous banana streak virus (eBSV) sequences present in

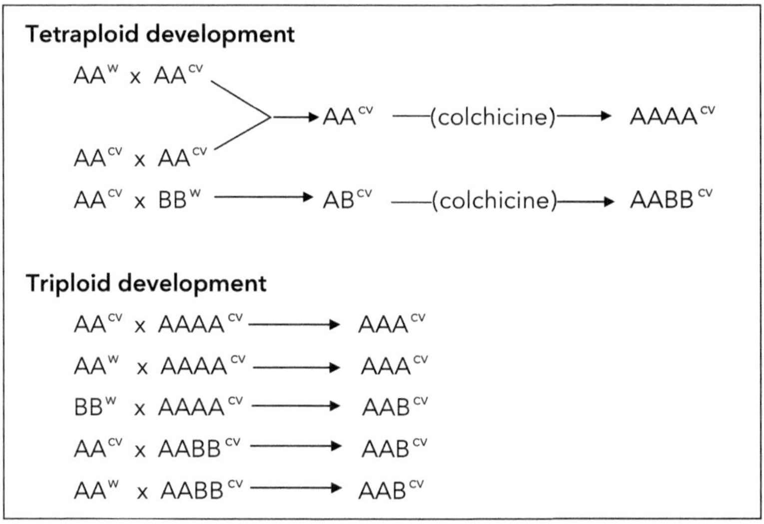

Figure 2 Reconstructive breeding strategy at CIRAD (Bakry and Horry, 1994). cv – cultivar, w – wild species.

the genome of *M. balbisiana* genitors (Umber et al., 2016). These researchers report on the successful marker-assisted breeding of *M. balbisiana* diploid genitors devoid of all infectious eBSVs, paving the way for the synthesis of new AAB and ABB hybrids. A plantain-like hybrid, 'K74', obtained at CARBAP is a good example of this new strategy (Bakry and Horry, 2016).

Other conventional banana-breeding programmes exist in India including the National Research Centre for Banana (NRCB), Tamil Nadu Agricultural University and Kerala Agricultural University (Tenkouano et al., 2011b). Although banana breeding in India developed as early as 1949 in Tamil Nadu, there were no specific breeding objectives other than to understand the genetics and breeding behaviour of banana (Sathiamoorthy, 1994). It was not until 1971 that extensive inter-diploid crosses were made to develop new diploids with the primary objectives of resistance to nematodes, Sigatoka leaf diseases and Fusarium wilt. Uma et al. (2011) summarized the results from 79 diploid cross combinations for their compatibility traits and factors influencing them. In terms of hybrid development, Menon et al. (2011) reported success in developing a black Sigatoka-resistant French plantain tetraploid hybrid in Kerala, while NRCB have recently released 'Dwarf Karpuravalli' which is notable for its resistance to wind damage. The difficulty of breeding for smallholders in India, as for Africa, is that regional and subregional preferences vary considerably, and these determine the choice of cultivars or landraces grown.

The Chinese Academy of Agricultural Sciences, the Guangdong Academy of Agricultural Sciences and the Indonesian Fruits Research Institute also have initiated breeding programmes based on both conventional and molecular genetic approaches.

3 Non-conventional breeding

3.1 Somaclonal variation and mutation breeding

Spontaneous mutations have contributed to the origin of almost all of the edible extant banana cultivars. In vegetatively propagated crops like banana it is still important to generate additional genetic variation to facilitate the further selection of desirable traits. One way of doing this is through mutation breeding and in vitro culture (also referred to as in vitro mutagenesis), the combination ensuring more effective induction and selection of induced somatic mutants. Further cycles of in vitro propagation will ensure the formation of periclinal chimeras or homohistont individuals, thereby increasing the chance of obtaining stable mutants. There are several important reviews that have examined the importance of mutation breeding for banana improvement (Roux, 2004; Chai et al., 2004; Khayat et al., 2004; Jain et al., 2011). The extent of variation can be enhanced by the choice of physical and chemical mutagens administered to micropropagated shoot tips or to embryogenic cell suspensions plated on a semi-solid culture medium. Physical mutagens such as gamma-rays, x-rays and carbon ion-beams have been used for mutation induction (Jain et al., 2011). Gamma irradiation (dosage of 10–60 Gy; radiosensitivity is related to ploidy and genomic constitution) has reasonable reproducibility and has a high and uniform penetration of multicellular systems. Chemical mutagens such as ethyl methanesulphonate (EMS; 15–200 mM) and sodium azide can also be used and produce a high density of predominantly point mutations (Jain et al., 2011). The main problem with mutagenesis approaches is that no favourable genes for disease resistance or quality can be introduced and that existing favourable characteristics are frequently reduced due to random mutations. Other problems involved in using physical mutagens are the high costs of establishing the radiation source and the simultaneous induction of chromosomal and gene mutations. Chemical mutagens are carcinogenic in nature and penetrate multicellular tissues non-uniformly.

Ploidy has also been manipulated chemically by applying colchicine or oryzalin to micropropagated shoot tips to produce autotetraploids, which is widely used in triploid (reconstructive) breeding programmes (Tenkouano et al., 2011b). According to Predieri (2001), banana has benefited more from in vitro-induced mutation than any other fruit crop. This is due in no small part to the resources put into this approach by many institutions after considering the difficulties associated with breeding triploid cultivars.

Somaclonal variation occurs during the normal course of banana in vitro propagation. In a classic paper by Larkin and Scowcroft (1981), somaclonal variation was defined as genetic variation generated during tissue culture. This variation can involve point mutations at one end of the spectrum to gross ploidy changes at the other end. Variation is most likely the result of both intrinsic factors, such as the genetic stability of the genotype under investigation, and genetic alterations induced during the process of tissue culture. Smith (1988) and Damasco et al. (1998) summarized the following known factors that influence the level of somaclonal variation observed in banana: (1) a greater variation is observed if the callus or adventitious budding is a significant phase of the propagation cycle, and this can be influenced by the nature and concentration of plant-growth regulators used in the medium, (2) the frequency of somaclonal variants increases with time in culture and number of propagation cycles, (3) some genotypes are more prone to genetic instability than others and the types of variants generated also varies between genotypes and (4) characters based on multiple genes each having an additive effect, such as, for example, height of plants.

Somaclonal variation was exploited in the selection of new cultivars with partial resistance to Fusarium wilt. This was achieved by micropropagating the cultivar 'Giant Cavendish' (Hwang and Ko, 2004). Initially, the selected mutants had inferior horticultural characteristics; however, further cycles of micropropagation and recurrent selection have produced agronomically acceptable selections while retaining reduced susceptibility to the pathogen (Lee et al., 2011; Molina et al., 2016). The use of gamma irradiation and EMS has also been used to develop Fusarium wilt-resistant banana cultivars with improved performance to Fusarium wilt or productivity.

Many of these improved selections have subsequently been released to industry and grown commercially, and examples include 'Novaria' (Novak et al., 1990), 'Mutaria' (Ho et al., 2001), 'Klue Hom Thong KU1' (Roux, 2004), 'DPM-25' (Smith et al., 2006), 'Tai-Chiao No. 5' (Lee et al., 2011), as well as 'Adi' and 'Jaffa' (Khayat and Ortiz, 2011).

Roux (2004) stated that one of the obstacles of mutation breeding comes from the random nature of mutation induction. Hence as many plants as possible need to be screened and with bananas, field screening is expensive, laborious and site-specific. Consequently, selection efficiency could be increased by the development of mass screening methods in the screenhouse or at the in vitro level (Roux et al., 2009). Despite these difficulties, mutation breeding continues to have an important role in banana improvement.

3.2 Genetic transformation

Genetic transformation offers the potential of introducing desirable genes into already established cultivars. Efforts to transform bananas and plantains have largely focussed on pest and disease resistance and pro-vitamin A. Genes, once identified, isolated and cloned into a suitable plasmid vector, may be introduced to a banana plant from any organism. Many useful genes for banana breeding may be found in wild bananas or early landraces. Transformation protocols have been developed for an ever-expanding list of banana cultivars and are the subject of several reviews (Tripathi et al., 2008; Tripathi, 2011; Khanna and Deo, 2016).

Direct DNA transfer via protoplast electroporation has been reported (Sági et al., 1995a), but has been superseded by both biolistic and *Agrobacterium*-mediated transformation approaches, with the latter procedure finding the most favour. Most transformation protocols use embryogenic cell suspensions which offer high transformation efficiencies. For example, Khanna et al. (2004) described a centrifugation-assisted *Agrobacterium*-mediated protocol that resulted in 25–65 plants per 50 mg of settled cell volume of embryogenic suspension cells that gave rise to morphologically normal, transgenic plants. May et al. (1995) and Tripathi et al. (2005) have successfully used meristems from micropropagated apical shoots. This technique has wider applications across many cultivars but may result in chimeric plants. It also necessitates more cycles of selection and regeneration to obtain stably transformed plants expressing the gene of interest.

The most widely used selection agents in the recovery of transgenic banana cells and tissues following the incorporation of a suitable marker gene are kanamycin, geneticin and hygromycin (May et al., 1995; Sági et al., 1995b; Khanna et al., 2004). Herbicides such as glufosinate, chlorsulfuron, imidazolinone and sulfonylurea have also been used as selection agents (Becker et al., 2000; Ganapathi et al., 2001; Matsumoto et al., 2002). The use of marker genes under the control of constitutive promoters such as the 35S promoter from CaMV, or the ubiquitin promoter from maize, works as efficiently for selection of *Agrobacterium*-transformed cells as for biolistic-mediated transformation and have been the most commonly used to drive gene expression to date. Viral promoters derived from banana streak virus and banana bunchy top virus have also been used in banana transformation (Schenk et al., 2001; Hermann et al., 2001).

Once transgenic embryos and shoot tips have survived the selection process they are multiplied in vitro and must begin the long process of screening and selection in the glasshouse and field to ensure the target gene is stably incorporated and expressed in a plant(s) that retains all the other desirable characteristics of the cultivar chosen for transformation.

One of the first reports of a transgenic banana showing host-plant resistance against a major pest was done with Cavendish, which was transformed using *Agrobacterium tumefaciens* to express an engineered rice cystatin (OcIΔD86) that provided host-plant resistance to the burrowing nematode (*Radopholus similis*) (Atkinson et al., 2004). Since then transgenic approaches show potential for the genetic improvement of bananas and plantains with a wide range of available genes that may confer resistance to many fungi, bacteria, nematodes, viruses and even abiotic stresses that threaten production (Tripathi et al., 2008; Tripathi, 2011; Khanna and Deo, 2016). In addition, various research groups have targeted banana for the development of 'edible vaccines' conferring immunity to hepatitis B (Mason and Arntzen, 1995; Kumar et al., 2005), as well as for pro-vitamin A and iron biofortification (Khanna and Deo, 2016; Paul et al., 2017, 2018). Banana was thought to be a good candidate for edible vaccines and biofortification as they are eaten raw, appeal to children, are relatively inexpensive to produce and are grown in many developing countries (Hassler, 1995). However, the commercial release of transgenic plants has faced many regulatory hurdles and consumer acceptance continues to be an issue. As a result, conventional breeding and non-GMO approaches continue to be a major focus for banana improvement.

4 Marker-assisted selection and mapping

The identification and functional characterization of agronomical important genes is a major challenge (Biswas, 2016). Disease-related traits are considered the most important characteristics by the *Musa* research and breeding community, followed by abiotic stress and quality traits. One of the first markers to be used by breeders was that differentiating A and B genomes (Fig. 3).

Emediato et al. (2011) identified over 50 distinct NBS-LRR-type resistance gene analogues (RGAs) in the resistant-wild diploid *M. acuminata* ssp. *burmannicoides* 'Calcutta 4'. The *Musa* non-expressor of pathogenesis-related genes 1A (MNPR1A) was greatly expressed after infections with a *Fusarium* strain with higher and earlier expression in the Cavendish somaclone GCTCV-218 than in the susceptible cultivar 'Grande Naine' (Endah et al., 2008). Peraza-Echeverria et al. (2009) isolated and characterized the entire cDNA sequences of RGA2 and RGA5 from the root transcriptome of *M. acuminata* ssp. *malaccensis*. The RGA2 was used to produce a transgenic Cavendish line that was resistant to TR4 (Dale et al., 2017). This is the first report of Fusarium wilt resistance in transgenic bananas in the field. Dale et al. (2017) have shown that Cavendish encodes three RGA2 homologues and suggested the prospect of using gene-editing technology to generate TR4 resistance.

Genome-editing techniques, particularly clustered regularly interspersed short-palindromic repeats (CRISPR/Cas9) (Osakabe and Osakabe, 2015) can make significant gains in the genetic improvement of banana. The first three crop species in which CRISPR/Cas9 technology was demonstrated included sorghum (Jiang et al., 2013), rice (Zhang et al., 2014) and wheat (Wang et al., 2014). Since then there has been an explosion in research in genome editing

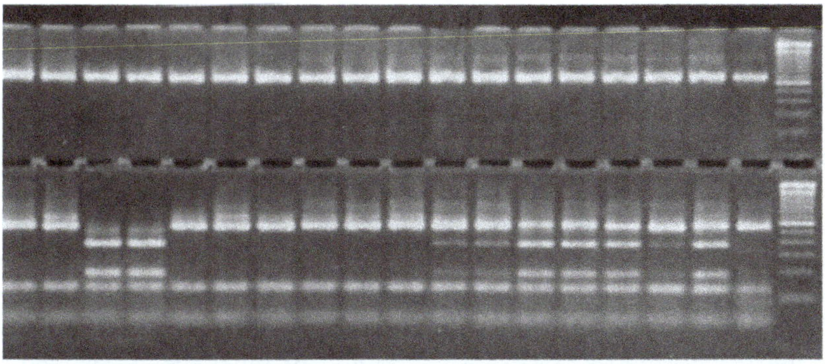

Figure 3 TOP: fragment size of the ITS region in 18 accessions of *Musa*. BOTTOM: *Rsa*I restriction endonuclease fragment patterns of the internal transcribed spacer (ITS) sequence in *Musa* accessions. The top band represents the diagnostic fragment of the A genome while the next two bands represent the B genome. Accessions from left to right: *M. acuminata* ssp. *malaccensis*, *M. acuminata* ssp. *zebrina* (AA, 1–2); *M. balbisiana* 'Singapuri' and 'Los Banos' (BB, 3–4); 'Pisang lilin', 'Tjau lagada' (AAcv, 5–6); 'Lacatan', 'Gros Michel', 'Dwarf Cavendish', 'Yangambi km5' (AAA, 7–10); 'Baka', 'Agbagba' (AAB, 11–12); 'Bluggoe', 'Cardaba', 'Lep Chang Kut' (ABB, 13–15); '6939-1 (AAAB, 16); '612-74' (AABB, 17); 'Mbwazirume' (AAA, 18).

of plants using the CRISPR/Cas9 and its modifications. Current research using CRISPR/Cas9 technology has already shown that disease resistance can be modified, as well as other areas in crop improvement including yield and nutrition. The precise modification of genes in elite cultivars will save the time-consuming task of backcrossing that is usually involved in conventional breeding schemes. Another advantage of CRISPR is that it can edit multiple target genes simultaneously with a single molecular construct (Liu et al., 2017b; Borrelli et al., 2018), but in banana we still need to know which genes to edit. CRISPR/Cas9 is still at an early stage of its development and it is difficult to predict the licensing agreements needed to commercialise this technology and the regulations that will apply to CRISPR/Cas9 and even whether the plants altered with this technology will be considered as GMOs.

Recently, Tripathi et al. (2019) used CRISPR/Cas9 technology to irreversibly silence the latent eBSV in the B genome. The study showed that 75% of the edited events tested remained asymptomatic in comparison to the control non-edited plants under water-stress conditions. This study confirmed that the inhibition of eBSV and reversal of its ability to be converted into infectious viral particles is now a reality. CRISPR/Cas9 technology is currently being pursued for resistance to banana Xanthomonas wilt at IITA.

Similarity searches using the entire ORF of both genes revealed considerable sequence similarity with the *I2* gene, known to confer resistance to *F. oxysporum* in tomato. By comparing the susceptible cultivar 'BX' and a resistant wild banana *M. yunnanensis*, ('YN') using metabolomics, biochemical and molecular biology methods, Li et al. (2017) revealed that the resistant 'YN' expressed pathogenesis-related (PR) genes, especially PR1, PR4, PR5-1 and PDF2.2, at higher levels than the susceptible 'BX'.

An important characteristic of the export banana is its green fruit life and response to artificial ripening using ethylene. Henry et al. (2011) investigated the structure and regulation of the abscisic acid, stress, ripening (A*sr*) gene family in banana. This A*sr* gene family contained at least four members exhibiting the typical two exons, one intron structure and the 'ABA/WDS' (abscisic acid water-deficit stress) domain characteristic to A*sr* genes. In meristem cultures, mA*sr*1 and mA*sr*3 were induced by osmotic stress and wounding, while mA*sr*3 and mA*sr*4 were induced by exposure to ABA. The mA*sr*3 gene exhibited the most variation both in terms of amino acid sequence and expression pattern, making it the most promising candidate for further functional studies. Cenci et al. (2014) and Cenci and Rouard (2017) studied the set-up of a framework of orthologous groups determined by an expert sequence comparison of NAC and GRAS transcription factors from both monocots and dicots. NAC genes are known for their responses to abiotic and biotic stresses as demonstrated by Shan et al. (2014), Negi et al. (2016) and Tak et al. (2017), and are potential target genes for crop improvement.

Hu et al. (2016) as well as Goel et al. (2016) demonstrated that the bZIP and WRKY gene families are involved in the development, ripening and abiotic response of banana. Banana-ovate family protein MaOFP1 and MADS-Box protein MuMADS1 antagonistically regulate banana-fruit ripening (Liu et al., 2015, 2017a). DREB transcription factors were also identified as associated with fruit ripening in banana (Kuang et al., 2017). Within the six analysed gene families studied by Jourda et al. (2016), eight starch synthesis genes (*Ma*AGP1, *Ma*AGP6, *Ma*AGP7, *Ma*SS3, *Ma*SS8, *Ma*SS10, *Ma*GBSS1 and *Ma*GBSS2), four amylases-encoding genes (*Ma*AMY12, *Ma*BAM4, *Ma*BAM6 and *Ma*BAM7), three SBE- and three DBE-encoding genes are major players involved in the starch metabolism of mature green and/or ripening banana fruits.

Candidate genes for the seedless phenotype in banana were recently detected by performing genome-wide association studies (GWAS) on a panel of 105 accessions that were specifically selected based on available molecular data (Sardos et al., 2016). Genes in 13 candidate genomic regions were potentially linked with the seedless phenotype (i.e. parthenocarpy combined with female sterility). These findings are consistent with the earlier findings of Dodds and Simmonds (1948) who stipulated that parthenocarpy in banana is caused by the combination of at least three genes, among which one would be major and dominant.

A number of gene expression studies based using RNAseq have also been published for stress-related traits such as TR4 (Li et al., 2012, 2017; Wang et al.,

2012; Bai et al., 2013), Black Leaf Streak (Sánchez Timm et al., 2016; Noar and Daub, 2016), drought (Muthusamy et al., 2016; Zorrilla-Fontanesi et al., 2016) and other abiotic stress factors (Lee et al., 2015; Hu et al., 2017), providing lists of candidate genes to be further validated.

Buah et al. (2016) suggested that CCD4 expression may contribute to the low-carotenoid status of Cavendish fruit compared to 'Asupina'. However, the primary difference accounting for the variation between 'Asupina' and the low-carotenoid accumulating triploid (AAA) Cavendish banana is likely due to the enhanced storage sink strength in 'Asupina' and Fe'i banana resulting from the formation of chromoplasts in mature fruit. It is possible that this is due to a homologue of the cauliflower *Or* gene within 'Asupina' that is yet to be identified.

Musa acuminata species are involved in the domestication of a majority of edible bananas. The sequencing of the 523-megabase genome of a *M. acuminata* (A genome) doubled-haploid-Pahang (DH) genotype ($2n$=22) has provided the stepping stone for genetic improvement of bananas (D'Hont et al., 2012). Its wild relative *M. acuminata* ssp. *malaccensis* 'Pahang' is a specimen collected from Malaysia's Pahang province before it was introduced to the Banana Board's Jamaican Gene bank in the late 1940s. Seventy percent of the assembly was anchored along the 11 *Musa* linkage groups of the Pahang genetic map and 36 542 genes have been positioned on the 11 chromosomes. Transposable elements accounted for almost half of the *Musa* sequence (D'Hont et al., 2012). This publication revealed three rounds of whole-genome duplication in the *Musa* lineage and led to the discovery of conserved non-coding sequences predating the monocotyledon-eudicotyledon divergence. In 2016, an improved reference sequence was produced using additional sequencing data and semi-automatic bioinformatics tools that allowed high-quality assemblies, and will serve to further study *Musa* species in greater detail (Martin et al., 2016). This new version of the reference genome and gene annotation for *M. acuminata* is a more effective tool for the functional, evolutionary and structural analysis of banana genomes. It is now available to the scientific community on the Banana Genome Hub (http://banana-genome-hub.southgreen.fr/) (Droc et al., 2013) available through the South Green Platform that supports the banana community with various bioinformatics applications and resources (South Green collaborators, 2016).

5 Conclusion and future trends

Breeding bananas started with the simple objective of developing a Fusarium wilt-resistant 'Gros Michel'. However, breeding objectives changed and proliferated as new programmes started and local farmers' needs were addressed. Objectives are now very diverse and complex, and they differ in various regions. Much has been learned of banana evolution through the molecular sciences and is now being put to use for the genetic improvement of this important crop. Yet breeders still use only a limited pool of parents compared to the great natural diversity that exists. It is clear that much research is still needed to reduce reproductive barriers and streamline the selection process which first starts with the identification of germplasm containing the genes of interest (Fig. 4).

The future will see greater integration of molecular and bioinformatics tools to manage and analyse *Musa* datasets. Breeders have recognized the

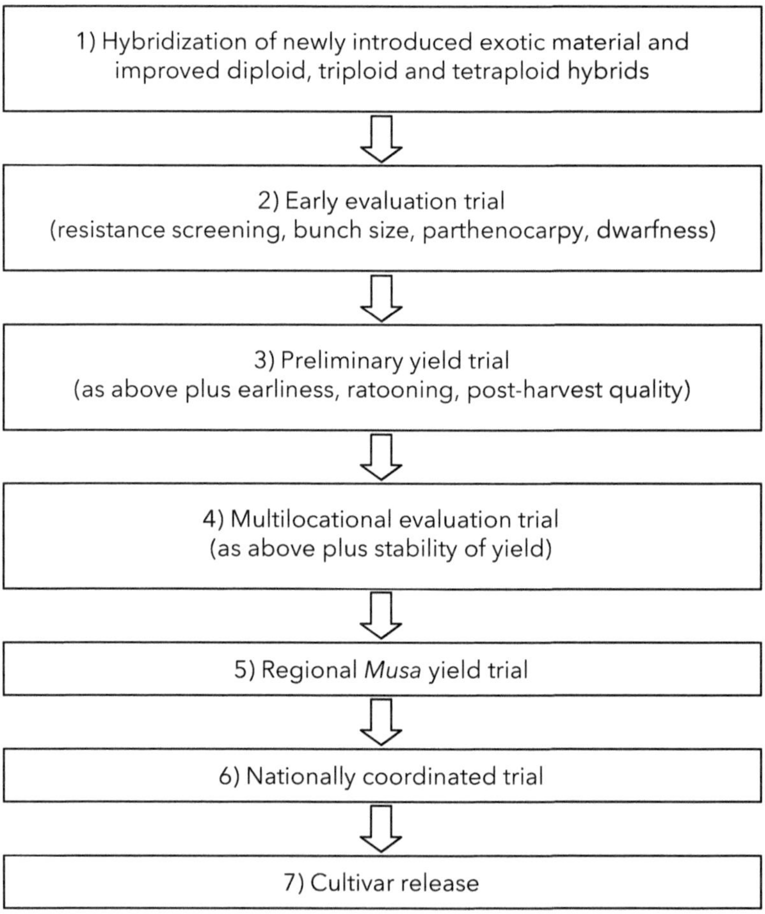

Figure 4 Typical flow diagram of the banana and plantain selection process at the IITA. Each step may take approximately 2 years, so that 15 years may be required from crossing to release of an improved cultivar. Furthermore, each plant occupies about 6 m^2 of field space and less than 1% of the plants in the early evaluation trial typically make it to the next evaluation step.

need to identify genes for important traits such as resistance and tolerance to important biotic and abiotic stresses, respectively, yield potential, fruit quality, parthenocarpy, fruit quality and nutritional value, while also providing information about alleles in diverse germplasm. If this can be combined with molecular information gained on host-pathogen interactions, as well as pathogen diversity, great strides can be made in resistance breeding. An important requirement for disease evaluation is appropriate phenotyping tools and the availability of sites for screening pre-breeding materials and hybrids that do not jeopardise quarantine measures in either the country where breeding takes place or in the countries undertaking regional evaluation. Emphasis needs to be given to screen germplasm for resistance to TR4.

More affordable genotyping and sequencing, readily available software for genomic prediction and a greater understanding of the requirement and limitations of genomic selection are revolutionizing plant breeding. Beyond genomics, there is increasing emphasis on the integration of other -omics information into breeding programmes (Kissel et al., 2016; Zorrilla-Fontanesi et al., 2016; Campos et al., 2018: van Wesemael et al., 2018; Drapal et al., 2019; Price et al., 2019; Bhuiyan et al., 2020).

Recently, an understanding of the genes involved in meiotic crossovers, combined with the ability to inactivate those genes, is opening up new possibilities to enhance recombination which could be used with other state-of-the-art technologies such as genomic selection and genome editing to enhance the pace and efficiency of genetic improvement (Mieulet et al., 2018).

Aside from molecular genetics, further conservation efforts, particularly of wild types and improved diploids, are very important. Better ways to increase seed production during crossing, to rescue embryos from seed and to conserve seed are also needed (Roux et al., 2011). Banana improvement is a continuous process and together, those involved in the genetic improvement of banana, are making the ability to produce desirable, elite cultivars much more likely.

6 Where to look for further information

The reader is encouraged to visit Alliance Bioversity – CIAT's website for a list of the more pertinent publications covering the genetic improvement of *Musa*. In order to facilitate the *Musa* research and breeding community to easily identify and select germplasm from the global gene pool based on a combination of passport, genotypic and phenotypic data, the *Musa* Germplasm Information System (MGIS), the database for global ex situ-held banana genetic resources, has been re-developed to address those needs in a user-friendly way (Ruas et al., 2017) and complement Banana Genome Hub, which already centralized most of the genomic resources on banana.

For those wanting to keep up to date with banana and plantain research developments, upcoming conferences and workshops, news and views, and publications go to the Pro*Musa* website (www.promusa.org) and register your interest in joining a network of people working on banana. The Pro*Musa* Secretariat, together with the International Society for Horticultural Science and host institutes, regularly hold conferences around the globe and publish proceedings of papers and posters presented at these events in *Acta Horticulturae*.

Finally, for those interested in banana breeding it is hard to go past the book of Pillay and Tenkouano (2011).

For further reading:

- Jain, S. M. and Swennen, R. (2004). *Banana Improvement: Cellular, Molecular Biology and Induced Mutations.* Science Publishers, Enfield, NH.
- Jones, D. R. (1994). *The Improvement and Testing of* Musa: *A Global Partnership*. International Network for the Improvement of Banana and Plantain, Montpellier, France.
- Ortiz, R. (2011). *Musa*. In: Kole, C. (Ed.), *Wild Crop Relatives: Genomic and Breeding Resources, Tropical and Subtropical Fruits*. Springer-Verlag, Berlin. pp. 97–128.
- Pillay, M. and Tenkouano, A. (2011). *Banana Breeding: Progress and Challenges*. CRC Press, Boca Raton, FL.
- Pillay, M., Ude, G. and Kole, C. (2012). *Genetics, Genomics and Breeding of Bananas*. Science Publishers, Enfield, NH.
- Roux, N., Baurens, F. C., Doležel, J., Hřibová, E., Heslop-Harrison, P., Town, C., Sasaki, T., Matsumoto, T., Aert, R., Remy, S., Souza, M. and Lagoda, P. (2008). Genomics of banana and plantain (*Musa* spp.), major staple crops in the tropics. In: Moore, P. H. and Ming, R. (Eds), *Genomics of Tropical Crop Plants*. Springer, New York. pp. 83–111.
- Stover, R. H. and Simmonds, N. W. (1987). *Bananas*. Longman Scientific & Technical, London.
- Tripathi, L., Tripathi, J. N., Tenkouano, A. and Bramel, P. (2008). Banana and plantain. In: Kole, C. and Hall, T.C. (Eds), *Transgenic Tropical and Subtropical Fruits and Nuts*. Wiley-Blackwell, Oxford, UK. pp. 77–107.

7 Acknowledgements

Thanks to Matthieu Rouard, Julie Sardos and Nicolas Roux for providing their insights into molecular banana breeding. We also acknowledge the support given by Alliance Bioversity – CIAT (and its previous iterations – Bioversity

International and the International Network for the Improvement of Banana and Plantains) in facilitating networking opportunities and promoting programmes around the world involved in the genetic improvement of *Musa*.

8 References

Aguilar Morán, J. F. (2013). Improvement of Cavendish banana cultivars through conventional breeding. *Acta Horticulturae* 986, 205–208.

Amah, D., Alamu, E., Adesokan, M., van Biljon, A., Maziya-Dixon, B., Swennen, R. and Labuschagne, M. (2019a). Variability of carotenoids in a *Musa* germplasm collection and implications for provitamin A biofortification. *Food Chemistry X* 2, 100024. Available at: http://www.sciencedirect.com/science/article/pii/S2590157519300264.

Amah, D., van Biljon, A., Brown, A., Perkins-Veazie, P., Swennen, R. and Labuschagne, M. (2019b). Recent advances in banana (*Musa* spp.) biofortification to alleviate vitamin A deficiency. *Critical Reviews in Food Science and Nutrition* 59(21), 3498–3510. Available at: https://www.ncbi.nlm.nih.gov/pubmed/29999424.

Amah, D., van Biljon, A., Maziya-Dixon, B., Labuschagne, M. and Swennen, R. (2019c). Effects of *in vitro* polyploidization on agronomic characteristics and fruit carotenoid content: implications for banana genetic improvement. *Frontiers in Plant Science* 10(1450), 1450. Available at: https://www.frontiersin.org/article/10.3389/fpls.2019.01450.

Amorim, E. P., dos Santos-Serejo, J. A., Amorim, V. B. O., Ferreira, C. F. and Silva, S. O. (2013). Banana breeding at Embrapa Cassava and fruits. *Acta Horticulturae* 986, 171–176.

Atkinson, H. J., Grimwood, S., Johnston, K. and Green, J. (2004). Prototype demonstration of transgenic resistance to the nematode *Radopholus similis* conferred on banana by a cystatin. *Transgenic Research* 13(2), 135–142.

Bai, T. T., Xie, W. B., Zhou, P. P., Wu, Z. L., Xiao, W. C., Zhou, L., Sun, J., Ruan, X. L.. and Li, H. P. (2013). Transcriptome and expression profile analysis of highly resistant and susceptible banana roots challenged with *Fusarium oxysporum f. sp. cubense* tropical Race 4. *PLOS ONE* 8(9), e73945.

Bakry, F. and Horry, J. P. (1994). *Musa* breeding at CIRAD-FLHOR. *Musa* breeding at FHIA. In: Jones, D. R. (Ed.). *The Improvement and Testing of Musa: A Global Partnership*. INIBAP, Montpellier, France. pp. 169–175.

Bakry, F. and Horry, J.-P. (2016). Advances in genomics: applications to banana breeding. *Acta Horticulturae* 1114, 171–180.

Batte, M., Swennen, R., Uwimana, B., Akech, V., Brown, A., Tumuhimbise, R., Persson Hovmalm, H., Geleta, M. and Ortiz, R. (2019). Crossbreeding East African highland bananas: lessons learnt relevant to the botany of the crop after 21 years of genetic enhancement. *Frontiers in Plant Science* 10(81), 1–9.

Becker, D. K., Dugdale, B., Smith, M. K., Harding, R. M. and Dale, J. L. (2000). Genetic transformation of Cavendish banana (*Musa* spp., AAA group) cv. 'Grand Nain' via microprojectile bombardment. *Plant Cell Reports* 19(3), 229–234.

Bhuiyan, F., Campos, N. A., Swennen, R. and Carpentier, S. (2020). Characterizing fruit ripening in plantain and Cavendish bananas: a proteomics approach. *Journal of*

Proteomics 214, 103632. Available at: http://www.sciencedirect.com/science/article/pii/S187439191930404X.

Biswas, M. K. (2016). Genes and markers: application in banana crop improvement. In: Mohandas, S. and Ravishankar, K. V. (Eds), *Banana: Genomics and Transgenic Approaches for Genetic Improvement*. Springer, Singapore. pp. 35–50.

Borrelli, V. M. G., Brambilla, V., Rogowsky, P., Marocco, A. and Lanubile, A. (2018). The enhancement of plant disease resistance using CRISPR/Cas9 technology. *Frontiers in Plant Science* 9, 1245.

Buah, S., Mlalazi, B., Khanna, H., Dale, J. L. and Mortimer, C. L. (2016). The quest for golden bananas: investigating carotenoid in a Fe'i group *Musa* cultivar. *Journal of Agricultural and Food Chemistry* 64(16), 3176–3185.

Campos, N. A., Swennen, R. and Carpentier, S. C. (2018). The plantain proteome, a focus on allele specific proteins obtained from plantain fruits. *Proteomics* 18(3–4). doi: 10.1002/pmic.201700227.

Cenci, A., Guignon, V., Roux, N. and Rouard, M. (2014). Genomic analysis of NAC transcription factors in banana (*Musa acuminata*) and definition of NAC orthologous groups for monocots and dicots. *Plant Molecular Biology* 85(1–2), 63–80.

Cenci, A. and Rouard, M. (2017). Large-scale evolutionary analyses of GRAS family in angiosperms. *Frontiers in Plant Science* 8, 273.

Chai, M., Ho, Y. W., Liew, K. W. and Asif, J. M. (2004). Biotechnology and *in vitro* mutagenesis for banana improvement. In: Jain, S. M. and Swennen, R. (Eds), *Banana Improvement: Cellular, Molecular Biology and Induced Mutations*. Science Publishers, Inc., Enfield, NH. pp. 59–77.

Dale, J., James, A., Paul, J. Y., Khanna, H., Smith, M., Peraza-Echeverria, S., Garcia-Bastidas, F., Kema, G., Waterhouse, P., Mengersen, K. and Harding, R. (2017). Transgenic Cavendish bananas with resistance to Fusarium wilt tropical race 4. *Nature Communications* 8(1), 1496.

Damasco, O. P., Adkins, S. W., Godwin, I. D. and Smith, M. K. (1998). Use of a SCAR-based marker for the early detection of dwarf off-types in micropropagated Cavendish bananas. *Acta Horticulturae* 461, 157–164.

D'Hont, A., Denoeud, F., Aury, J. M., Baurens, F. C., Carreel, F., Garsmeur, O., Noel, B., Bocs, S., Droc, G., Rouard, M., Da Silva, C., Jabbari, K., Cardi, C., Poulain, J., Souquet, M., Labadie, K., Jourda, C., Lengellé, J., Rodier-Goud, M., Alberti, A., Bernard, M., Correa, M., Ayyampalayam, S., McKain, M. R., Leebens-Mack, J., Burgess, D., Freeling, M., Mbéguié-A-Mbéguié, D., Chabannes, M., Wicker, T., Panaud, O., Barbosa, J., Hribova, E., Heslop-Harrison, P., Habas, R., Rivallan, R., Francois, P., Poiron, C., Kilian, A., Burthia, D., Jenny, C., Bakry, F., Brown, S., Guignon, V., Kema, G., Dita, M., Waalwijk, C., Joseph, S., Dievart, A., Jaillon, O., Leclercq, J., Argout, X., Lyons, E., Almeida, A., Jeridi, M., Dolezel, J., Roux, N., Risterucci, A. M., Weissenbach, J., Ruiz, M., Glaszmann, J. C., Quétier, F., Yahiaoui, N. and Wincker, P. (2012). The banana (*Musa acuminata*) genome and the evolution of monocotyledonous plants. *Nature* 488(7410), 213–217.

Dodds, K. S.and Simmonds, N. W. (1948). Sterility and parthenocarpy in diploid hybrids of *Musa*. *Heredity* 2(1), 101–117.

Drapal, M., de Carvalho, E. B., Rouard, M., Amah, D., Sardos, J., Van den Houwe, I., Brown, A., Roux, N., Swennen, R. and Fraser, P. D. (2019). Metabolite profiling characterises chemotypes of Musa diploids and triploids at juvenile and pre-flowering growth stages. *Scientific Reports* 9(1), 4657.

Droc, G., Larivière, D., Guignon, V., Yahiaoui, N., This, D., Garsmeur, O., Dereeper, A., Hamelin, C., Argout, X., Dufayard, J. F., Lengellé, J., Baurens, F. C., Cenci, A., Pitollat, B., D'Hont, A., Ruiz, M., Rouard, M. and Bocs, S. (2013). The banana genome hub. *Database*, bat035.

Emediato, F. L., De Camargo Teixeira, C., Pappas Júnior, G. J., Miller, R. N. G. and Passos, M. A. N. (2011). Candidate gene discovery: resistance gene analog characterisation and gene expression analysis in *Musa-Mycosphaerella* host-pathogen interactions. *Acta Horticulturae* 897, 179–185.

Endah, R., Beyene, G., Kiggundu, A., Van Den Berg, N., Schlüter, U., Kunert, K. and Chikwamba, R. (2008). Elicitor and Fusarium-induced expression of NPR1-like genes in banana. *Plant Physiology and Biochemistry* 46(11), 1007–1014.

Frison, E. A. and Sharrock, S. (2000). The potential for use of disease-resistant varieties as organic bananas. In: Holderness, M., Sharrock, S., Frison, E. and Kairo, M. (Eds), *Organic Banana 2000: Towards an Organic Banana Initiative in the Caribbean*. INIBAP, Montpellier, France. pp. 143–150.

Gaidashova, S. V., Karemera, F. and Karamura, E. B. (2008). Agronomic performance of introduced banana varieties in lowlands of Rwanda. *African Crop Science Journal* 16(1), 9–16.

Ganapathi, T. R., Higgs, N. S., Balint-Kurti, P. J., Arntzen, C. J., May, G. D. and Van Eck, J. M. (2001). *Agrobacterium*-mediated transformation of embryogenic cell suspensions of the banana cultivar Rasthali (AAB). *Plant Cell Reports* 20(2), 157–162.

Goel, R., Pandey, A., Trivedi, P. K. and Asif, M. H. (2016). Genome-wide analysis of the *Musa* WRKY gene family: evolution and differential expression during development and stress. *Frontiers in Plant Science* 7(299), 1–13.

Hassler, S. (1995). Bananas and biotech consumers. *Nature Biotechnology* 13(5), 417.

Henry, I. M., Carpentier, S. C., Pampurova, S., Van Hoylandt, A., Panis, B., Swennen, R. and Remy, S. (2011). Structure and regulation of the *Asr* gene family in banana. *Planta* 234(4), 785–798.

Hermann, S. R., Becker, D. K., Harding, R. M. and Dale, J. L. (2001). Promoters derived from banana bunchy top virus-associated components S1 and S2 drive transgene expression in both tobacco and banana. *Plant Cell Reports* 20(7), 642–646.

Ho, Y. W., Mak, C. and Liew, K. W. (2001). Selection of banana cultivars for tolerance to Fusarium wilt race 4 in Malaysia. In: Molina, A. B., Nik Masdek, N. H. and Liew, K. W. (Eds), *Banana Fusarium Wilt Management: Towards Sustainable Cultivation*. International Network for the Improvement of Bananas and Plantains – Asia and the Pacific Network, Los Baños, Laguna, Philippines. pp. 234–242.

Hu, W., Ding, Z., Tie, W., Yan, Y., Liu, Y., Wu, C., Liu, J., Wang, J., Peng, M., Xu, B. and Jin, Z. (2017). Comparative physiological and transcriptomic analyses provide integrated insight into osmotic, cold, and salt stress tolerance mechanisms in banana. *Scientific Reports* 7, 43007.

Hu, W., Wang, L., Tie, W., Yan, Y., Ding, Z., Liu, J., Li, M., Peng, M., Xu, B. and Jin, Z. (2016). Genome-wide analyses of the bZIP family reveal their involvement in the development, ripening and abiotic stress response in banana. *Scientific Reports* 6, 30203.

Hwang, S. C. and Ko, W. H. (2004). Cavendish banana cultivars resistant to *Fusarium* wilt acquired through somaclonal variation in Taiwan. *Plant Disease* 88(6), 580–588.

Jain, S. M., Till, B., Suprasanna, P. and Roux, N. (2011). Mutations and cultivar development of banana. In: Pillay, M. and Tenkouano, A. (Eds), *Banana Breeding, Progress and Challenges*. CRC Press, Boca Raton, FL. pp. 203–217.

Jiang, W., Zhou, H., Bi, H., Fromm, M., Yang, B. and Weeks, D. P. (2013). Demonstration of CRISPR/Cas9/sgRNA-mediated targeted gene modification in *Arabidopsis*, tobacco, sorghum and rice. *Nucleic Acids Research* 41(20), e188.

Jourda, C., Cardi, C., Gibert, O., Giraldo Toro, A., Ricci, J., Mbéguié-A-Mbéguié, D. and Yahiaoui, N. (2016). Lineage-specific evolutionary histories and regulation of major starch metabolism genes during banana ripening. *Frontiers in Plant Science* 7, 1778.

Karamura, E. B., Tinzaara, W., Kikulwe, E., Ochola, D., Ocimata, W. and Karamura, D. (2016). Introduced banana hybrids in Africa: seed systems, farmers' experiences and consumers' perspectives. *Acta Horticulturae* 1114, 239–244.

Khanna, H. K., Becker, D., Kleidon, J. and Dale, J. (2004). Centrifugation assisted *Agrobacterium tumefaciens*-mediated transformation (CAAT) of embryogenic cell suspensions of banana (*Musa* spp. Cavendish AAA and Lady finger AAB). *Molecular Breeding* 14(3), 239–252.

Khanna, H. K. and Deo, P. C. (2016). Novel gene transfer technologies. In: Mohandras, S. and Ravishankar, K. V. (Eds), *Banana: Genomics and Transgenic Approaches for Genetic Improvement*. Springer Nature, Singapore. pp. 127–140.

Khayat, E., Duvdevani, A., Lahav, E. and Ballesteros, B. A. (2004). Somaclonal variation in banana (*Musa acuminata* cv. Grande Naine): genetic mechanism, frequency and application as a tool for clonal selection. In: Jain, S. M. and Swennen, R. (Eds), *Banana Improvement: Cellular, Molecular Biology and Induced Mutations*. Science Publishers, Inc., Enfield, NH. pp. 97–109.

Khayat, E. and Ortiz, R. (2011). Genetics of important traits in *Musa*. In: Pillay, M. and Tenkouano, A. (Eds), *Banana Breeding, Progress and Challenges*. CRC Press, Boca Raton, FL. pp. 71–83.

Kissel, E., Vanhove, A. C., Garcia, S., Panis, B., Rouard, M., Cenci, A., Roux, N., Zorrilla, J., Swennen, R. and Carpentier, S. C. (2016). Abiotic stress research in crops using -omics approaches: drought stress and banana in the spotlight. *Acta Horticulturae* 1114, 81–90.

Kuang, J. F., Chen, J. Y., Liu, X. C., Han, Y. C., Xiao, Y. Y., Shan, W., Tang, Y., Wu, K. Q., He, J. X. and Lu, W. J. (2017). The transcriptional regulatory network mediated by banana (*Musa acuminata*) dehydration-responsive element binding (MaDREB) transcription factors in fruit ripening. *New Phytologist* 214(2), 762–781.

Kumar, G. B., Ganapathi, T. R., Revathi, C. J., Srinivas, L. and Bapat, V. A. (2005). Expression of hepatitis B surface antigen in transgenic banana plants. *Planta* 222(3), 484–493.

Larkin, P. J. and Scowcroft, W. R. (1981). Somaclonal variation – a novel source of variability from cell cultures for plant improvement. *Theoretical and Applied Genetics* 60(4), 197–214.

Lee, S. Y., Su, Y. U., Chou, C. S., Liu, C. C., Chen, C. C. and Chao, C. P. (2011). Selection of a new somaclone cultivar 'Tai-Chiao No. 5' (AAA, Cavendish) with resistance to Fusarium wilt of banana in Chinese Taipei. *Acta Horticulturae* 897, 391–397.

Lee, W. S., Gudimella, R., Wong, G. R., Tammi, M. T., Khalid, N. and Harikrishna, J. A. (2015). Transcripts and microRNAs responding to salt stress in *Musa acuminata* Colla (AAA Group) cv. Berangan roots. *PLoS ONE* 10(5), e0127526.

Li, C. Y., Deng, G. M., Yang, J., Viljoen, A., Jin, Y., Kuang, R. B., Zuo, C. W., Li, Z. C., Yang, Q. S., Sheng, O., Wei, Y. R., Hu, C. H., Dong, T. and Yi, G. J. (2012). Transcriptome profiling of resistant and susceptible Cavendish banana roots following inoculation with *Fusarium oxysporum* f. sp. cubense tropical race 4. *BMC Genomics* 13(1), 374.

Li, W., Berhal, C., Sun, J. and Peng, M. (2017). Metabolomic, biochemical, and gene expression analyses reveal the underlying responses of resistant and susceptible banana species during early infection with *Fusarium oxysporum f. sp. cubense*. *Plant Disease* 101(4), 534–543.

Liu, J., Zhang, J., Hu, W., Miao, H., Zhang, J., Jia, C., Wang, Z., Xu, B. and Jin, Z. (2015). Banana ovate family protein MaOFP1 and MADS-Box protein MuMADS1 antagonistically regulated banana fruit ripening. *PLoS ONE* 10(4), e0123870.

Liu, J., Zhang, J., Miao, H., Wang, J., Gao, P., Hu, W., Jia, C., Wang, Z., Xu, B. and, Jin, Z. (2017a). Genome-wide analysis of banana MADS-box family closely related to fruit development and ripening. *Science Reports* 7, 3467.

Liu, X., Wu, S., Xu, J., Sui, C. and Wei, J. (2017b). Application of CRISPR/Cas9 in plant biology. *Acta Pharmaceutica Sinica B* 7(3), 292–302.

Martin, G., Baurens, F. C., Droc, G., Rouard, M., Cenci, A., Kilian, A., Hastie, A., Dolezel, J., Aury, J. M., Alberti, A., Carreel, F. and D'Hont, A. (2016). Improvement of the banana *Musa acuminata* reference sequence using NGS data and semi-automated bioinformatics methods. *BMC Genomics* 17, 243.

Mason, H. S. and Arntzen, C. J. (1995). Transgenic plants as vaccine production systems. *Trends in Biotechnology* 13(9), 388–392.

Matsumoto, K., Morais, L. S., Vianna, G. R., Aragão, F. J. L. and Rech, E. L. (2002). Genetic transformation of banana embryogenic cells through particle bombardment using an herbicide resistance gene as selectable marker. *Acta Horticulturae* 575, 61–67.

May, G. D., Afza, R., Mason, H. S., Wiecko, A., Novak, F. J. and Arntzen, C. J. (1995). Generation of transgenic banana (*Musa acuminata*) plants via *Agrobacterium*-mediated transformation. *Nature Biotechnology* 13(5), 486–492.

Menon, R., Cherian, A., Suma, A., Maicykutty, Mathew, P., Nair, S. and Aipe, K. C. (2011). Developing resistant banana and plantain cultivars through conventional breeding techniques. *Acta Horticulturae* 897, 207–213.

Mieulet, D., Aubert, G., Bres, C., Klein, A., Droc, G., Vielle, E., Rond-Coissieux, C., Sanchez, M., Dalmais, M., Mauxion, J. P., Rotahn, C., Guiderdoni, E. and Mercier, R. (2018). Unleashing meiotic crossovers in crops. *Nature Plants* 4(12), 1010–1016.

Molina, A. B., Sinohin, V. O., Fabregar, E. G., Ramillete, E. B., Loayan, M. M. and Chao, C. P. (2016). Field resistance of Cavendish somaclonal variants and local banana cultivars to tropical race 4 of Fusarium wilt in the Philippines. *Acta Horticulturae* 1114, 227–230.

Muthusamy, M., Uma, S., Backiyarani, S., Saraswathi, M. S. and Chandrasekar, A. (2016). Transcriptomic changes of drought-tolerant and sensitive banana cultivars exposed to drought stress. *Frontiers in Plant Science* 7, 1609. Available at: http://www.ncbi.nlm.nih.gov/pmc/articles/PMC5095140/.

Negi, S., Tak, H. and Ganapathi, T. R. (2016). Expression analysis of MusaNAC68 transcription factor and its functional analysis by overexpression in transgenic banana plants. *Plant Cell, Tissue and Organ Culture* 125(1), 59–70.

Noar, R. D. and Daub, M. E. (2016). Transcriptome sequencing of *Mycosphaerella fijiensis* during association with *Musa acuminata* reveals candidate pathogenicity genes. *BMC Genomics* 17, 690.

Nomura, E. S., Moraes, W. S., Damatto Jr, E. R., Fuzitani, E. J., Saes, L. A., Amorim, E. P. and Silva, S. O. (2013). Evaluation of banana genotypes over two crop cycles under subtropical conditions in the Ribeira valley, São Paulo, Brazil. *Acta Horticulturae* 986, 61–70.

Novak, F. J., Afza, R., van Duren, M. and Omar, M. S. (1990). Mutation induction by gamma irradiation of *in vitro* cultured shoot-tips of banana and plantain (*Musa* cvs.). *Tropical Agriculture* 67, 21–28.

Nyine, M. and Pillay, M. (2011). The effect of banana breeding on the diversity of East African Highland Banana (*Musa*, AAA). *Acta Horticulturae* 897, 225–229.

Orjeda, G. (2000). *Evaluating Bananas: A Global Partnership. Results of IMTP Phase II.* INIBAP, Montpellier, France.

Ortiz, R. and Swennen, R. (2014). From crossbreeding to biotechnology-facilitated improvement of banana and plantain. *Biotechnology Advances* 32(1), 158–169.

Ortiz, R. and Vuylsteke, D. (1994). Plantain breeding at IITA. In: Jones, D. R. (Ed.). *The Improvement and Testing of Musa: A Global Partnership*. INIBAP, Montpellier, France. pp. 130–156.

Osakabe, Y. and Osakabe, K. (2015). Genome editing with engineered nucleases in plants. *Plant and Cell Physiology* 56(3), 389–400.

Paul, J. Y., Khanna, H., Kleidon, J., Hoang, P., Geijskes, J., Daniells, J., Zaplin, E., Rosenberg, Y., James, A., Mlalazi, B., Deo, P., Arinaitwe, G., Namanya, P., Becker, D., Tindamanyire, J., Tushemereirwe, W., Harding, R. and Dale, J. (2017). Golden bananas in the field: elevated fruit pro-vitamin A from the expression of a single banana transgene. *Plant Biotechnology Journal* 15(4), 520–532.

Paul, J. Y., Harding, R., Tushemereirwe, W. and Dale, J. (2018). Banana21: from gene discovery to deregulated golden bananas. *Frontiers in Plant Science* 9(558), 1–8.

Peraza-Echeverria, S., Dale, J. L., Harding, R. M. and Collet, C. (2009). Molecular cloning and in silico analysis of potential Fusarium resistance genes in banana. *Molecular Breeding* 23(3), 431–443.

Pillay, M. and Tenkouano, A. (2011). *Banana Breeding: Progress and Challenges*. CRC Press, Boca Raton, FL.

Predieri, S. (2001). Mutation induction and tissue culture in improving fruits. *Plant Cell, Tissue and Organ Culture* 64(2/3), 185–210.

Price, E. J., Drapal, M., Perez-Fons, L., Amah, D., Bhattacharjee, R., Heider, B., Rouard, M., Swennen, R., Becerra Lopez-Lavalle, L. A. and Fraser, P. D. (2019). Metabolite database for root, tuber, and banana crops to facilitate modern breeding in understudied crops. *The Plant Journal for Cell and Molecular Biology* 101(6), 1258–1268.

Roux, N. (2004). In: Jain, S. M. and Swennen, R. *Mutation Induction in Musa - Review*. Science Publishers, Inc., Enfield, NH. pp. 23–32.

Roux, N. S., Toloza, A., Strosse, H., Busogoro, J. P. and Dolezel, J. P. (2009). Induction and selection of potentially useful mutants in banana. *Acta Horticulturae* 828, 315–322.

Roux, N., Rouard, M., Huang, X. L. and Smith, M. (2011). Opportunities for bridging the gap between genomics and genetic improvement in *Musa* spp. *Acta Horticulturae* 897, 509–515.

Rowe, P. (1987). Banana breeding in Honduras. In: Persley, G. J. and De Langhe, E. A. (Eds), *Banana and Plantain Breeding Strategies*. Proceedings of the an International Workshop, Cairns, Australia. ACIAR Proceedings No. 21, pp. 74–77.

Rowe, P. and Rosales, F. (1994). *Musa* breeding at FHIA. In: Jones, D. R. (Ed.). *The Improvement and Testing of Musa: A Global Partnership*. INIBAP, Montpellier, France. pp. 117–129.

Ruas, M., Guignon, V., Sempere, G., Sardos, J., Hueber, Y., Duvergey, H., Andrieu, A., Chase, R., Jenny, C., Hazekamp, T., Irish, B., Jelali, K., Adeka, J., Ayala-Silva, T., Chao, C. P., Daniells, J., Dowiya, B., Effa Effa, B., Gueco, L., Herradura, L., Ibobondji, L.,

Kempenaers, E., Kilangi, J., Muhangi, S., Ngo Xuan, P., Paofa, J., Pavis, C., Thiemele, D., Tossou, C., Sandoval, J., Sutanto, A., Vangu Paka, G., Yi, G., Van den Houwe, I., Roux, N. and Rouard, M. (2017). MGIS: managing banana (*Musa* spp.) genetic resources information and high-throughput genotyping data. *Database* 2017(1), 1-12.

Sági, L., Remy, S., Verelst, B., Panis, B., Cammue, B. P. A., Volckaert, G. and Swennen, R. (1995a). Transient gene expression in transformed banana (*Musa* cv. Bluggoe) protoplasts and embryogenic cell suspensions. *Euphytica* 85(1-3), 89-95.

Sági, L., Panis, B., Remy, S., Schoofs, H., Desmet, K., Swennen, R. and Cammue, B. P. A. (1995b). Genetic transformation of banana and plantain (*Musa* spp.) via particle bombardment. *Biotechnology* 13(5), 481-485.

Sánchez Timm, E., Hidalgo Pardo, L., Pacheco Coello, R., Chávez Navarrete, T., Navarrete Villegas, O. and Santos Ordóñez, E. (2016). Identification of differentially-expressed genes in response to *Mycosphaerella fijiensis* in the resistant *Musa* accession 'Calcutta-4' using suppression subtractive hybridization. *PLoS ONE* 11(8), e0160083.

Sardos, J., Rouard, M., Hueber, Y., Cenci, A., Hyma, K. E., Van Den Houwe, I., Hribova, E., Courtois, B. and Roux, N. (2016). A genome-wide association study on the seedless phenotype in banana (*Musa* spp.) reveals the potential of a selected panel to detect candidate genes in a vegetatively propagated crop. *PLoS ONE* 11(5), e0154448.

Sathiamoorthy, S. (1994). *Musa* improvement in India. In: Jones, D. R. (Ed.). *The Improvement and Testing of Musa: A Global Partnership*. INIBAP, Montpellier, France. pp. 188-200.

Schenk, P. M., Remans, T., Sági, L., Elliott, A. R., Dietzgen, R. G., Swennen, R., Ebert, P. R., Grof, C. P. and Manners, J. M. (2001). Promoters for pregenomic RNA of banana streak Badnavirus are active for transgene expression in monocot and dicot plants. *Plant Molecular Biology* 47(3), 399-412.

Shan, W., Kuang, J. F., Lu, W. J. and Chen, J. Y. (2014). Banana fruit NAC transcription factor MaNAC1 is a direct target of MaICE1 and involved in cold stress through interacting with MaCBF1. *Plant, Cell and Environment* 37(9), 2116-2127.

Shepherd, K. (1974). Banana research at I.C.T.A. *Tropical Agriculture (Trinidad)* 51, 482-490.

Shepherd, K., Dantas, J. L. L. and Silva, O. S. (1994). Breeding Prata and Maçã cultivars for Brazil. In: Jones, D. R. (Ed.). *The Improvement and Testing of Musa: A Global Partnership*. INIBAP, Montpellier, France. pp. 157-168.

Simmonds, N. W. (1966). *Bananas*. Longman, London.

Smith, M. K. (1988). A review of factors influencing the genetic stability of micropropagated bananas. *Fruits* 43, 219-223.

Smith, M. K., Hamill, S. D., Langdon, P. W., Giles, J. E., Doogan, V. J. and Pegg, K. G. (2006). Towards the development of a Cavendish banana resistant to race 4 of Fusarium wilt: gamma irradiation of micropropagated 'Dwarf Parfitt' (*Musa* spp., AAA group, Cavendish subgroup). *Australian Journal of Experimental Agriculture* 46, 1-7.

Smith, M. K., Langdon, P. W., Pegg, K. G. and Daniells, J. W. (2014). Growth, yield and Fusarium wilt resistance of six FHIA tetraploid bananas (*Musa* spp.) grown in the Australian subtropics. *Scientia Horticulturae* 170, 176-181.

South Green Collaborators (2016). The South Green portal: a comprehensive resource for tropical and Mediterranean crop genomics. *Current Plant Biology*. doi: 10.1016/j.cpb.

Stover, R. H. and Buddenhagen, I. W. (1986). Banana breeding: polyploidy, disease resistance and productivity. *Fruits* 41, 175–191.

Stover, R. H. and Simmonds, N. W. (1987). *Bananas*. Longman Scientific & Technical, London.

Tak, H., Negi, S., Ganapathi, T. R. (2017). Banana NAC transcription factor MusaNAC042 is positively associated with drought and salinity tolerance. *Protoplasma* 254(2), 803–816.

Tenkouano, A., Pillay, M. and Coulibay, O. (2011a). Hybrid distribution to farmers: adoption and challenges. In: Pillay, M. and Tenkouano, A. (Eds). *Banana Breeding, Progress and Challenges*. CRC Press, Boca Raton, FL. pp. 305–319.

Tenkouano, A., Pillay, M. and Ortiz, R. (2011b). Breeding techniques. In: Pillay, M. and Tenkouano, A. (Eds). *Banana Breeding, Progress and Challenges*. CRC Press, Boca Raton, FL. pp. 181-202.

Tenkouano, A., Lamien, N., Agogbua, J., Amah, D., Swennen, R., Traoré, S., Thiemele, D., Aby, N., Kobenan, K., Gnonhouri, G., Yao, N., Astin, G., Sawadogo-Kabore, S., Tarpaga, V., Issa, W., Lokossou, B., Adjanohoun, A., Amadji, G. L., Adangnitode, S., Igue, K. A. D. and Ortiz, R. (2019). Promising high-yielding tetraploid plantain-bred hybrids in West Africa. *International Journal of Agronomy* 2019, Article ID 3873198, 8 pp.

Tezanas du Montcel, H. (1986). Banana breeding in France and Guadeloupe. In: Persley, G. J. and De Langhe, E. A. (Eds), *Banana and Plantain Breeding Strategies*. Proceedings of the an International Workshop, Cairns, Australia. ACIAR Proceedings No. 21, pp. 84–86.

Tripathi, L. (2011). Biotechnology in *Musa* improvement. In: Pillay, M. and Tenkouano, A. (Eds). *Banana Breeding, Progress and Challenges*. CRC Press, Boca Raton, FL. pp. 219–236.

Tripathi, L., Tripathi, J. N. and Hughes, Jd'A. (2005). *Agrobacterium*-mediated transformation of plantain cultivar Agbagba (*Musa* spp.). *African Journal of Biotechnology* 4, 1378–1383.

Tripathi, L., Tripathi, J. N., Tenkouano, A. and Bramel, P. (2008). Banana and plantain. In: Kole, C. and Hall, T. C. (Eds), *Compendium of Transgenic Crop Plants: Transgenic Tropical and Subtropical Fruits and Nuts*. Blackwell Publishing, West Sussex, UK. pp. 77–107.

Tripathi, J. N., Ntui, V. O., Ron, M., Muiruri, S. K., Britt, A.. and Tripathi, L. (2019). CRISPR/Cas9 editing of endogenous banana streak virus in the B genome of *Musa* spp. overcomes a major challenge in banana breeding. *Communications Biology* 2, 46.

Uazire, A. T., Ribeiro, C. M., Mussane, C. R. B., Pillay, M., Blomme, G., Fraser, C., Staver, C. and Karamura, E. (2008). Preliminary evaluation of improved banana varieties in Mozambique. *African Crop Science Journal* 16(1), 17–25.

Uma, S., Mustaffa, M. M., Saraswathi, M. S. and Durai, P. (2011). Exploitation of diploids in Indian banana breeding programmes. *Acta Horticulturae* 897, 215–223.

Umber, M., Pichaut, J. P., Farinas, B., Laboureau, N., Janzac, B., Plaisir-Pineau, K., Pressat, G., Baurens, F. C., Chabannes, M., Duroy, P. O., Guiougou, C., Delos, J. M., Jenny, C., Iskra-Caruana, M., Salmon, F. and Teycheney, P. (2016). Marker-assisted breeding of *Musa balbisiana* genitors devoid of infectious endogenous banana streak virus sequences. *Molecular Breeding* 36(6), 1–11.

van Wesemael, J., Hueber, Y., Kissel, E., Campos, N., Swennen, R. and Carpentier, S. (2018). Homeolog expression analysis in an allotriploid non-model crop via integration of transcriptomics and proteomics. *Scientific Reports* 8(1), 1353.

Wang, Y., Cheng, X., Shan, Q., Zhang, Y., Liu, J., Gao, C. and Qui, J. L. (2014). Simultaneous editing of three homoalleles in hexaploid bread wheat confers heritable resistance to powdery mildew. *Nature Biotechnology* 32(9), 947–951.

Wang, Z., Zhang, J. B., Jia, C. H., Liu, J. H., Li, Y. Q., Yin, X. M., Xu, B. Y. and Jin, Z. Q. (2012). *De novo* characterization of the banana root transcriptome and analysis of gene expression under *Fusarium oxysporum f. sp*. Cubense tropical race 4 infection. *BMC Genomics* 13(1), 650.

Zhang, H., Zhang, J., Wei, P., Zhang, B., Gou, F., Feng, Z., Mao, Y., Yang, L., Zhang, H., Xu, N. and Zhu, J. K. (2014). The CRISPR/Cas9 system produces specific and homozygous targeted gene editing in rice in one generation. *Plant Biotechnology Journal* 12(6), 797–807.

Zorrilla-Fontanesi, Y., Rouard, M., Cenci, A., Kissel, E., Do, H., Dubois, E., Nidelet, S., Roux, N., Swennen, R. and Carpentier, S. C. (2016). Differential root transcriptomics in a polyploid non-model crop: the importance of respiration during osmotic stress. *Scientific Reports* 6, 22583.

Part 1

Classification

Chapter 2

Cytogenetics of structural rearrangements in Musa hybrids and cultivars

Fajarudin Ahmad, Indonesian Institute of Sciences (LIPI), Indonesia and Wageningen University & Research, The Netherlands; Peter M. Bourke and Henk Schouten, Wageningen University & Research, The Netherlands; Hugo Volkaert, Center for Agricultural Biotechnology – Kasetsart University, Thailand; Gert H. J. Kema, Wageningen University & Research, The Netherlands; and Hans de Jong, Kasetsart University, Thailand and Wageningen University & Research, The Netherlands

1 Introduction

Edible bananas and plantains constitute a group of crops whose fruits show starchy, sweet and aromatic flesh with characteristic flavour and structures. Their seedless tasty pulp makes them popular sources of daily nutrition for people all over the world. The absence of seeds, however, comes at a price as they are sterile and hence difficult for cross-breeding. Almost all cultivated bananas have seedless fruits, either as parthenocarpic diploids or sterile triploids of *Musa acuminata*, or interspecific hybrids of *M. acuminata* (AA) and *M. balbisiana* (BB), the latter known as AA, AB, BB, AAA, AAB, ABB, AAAB and AABB genome group bananas. Gamete sterility results from the failure of balanced chromosome

http://dx.doi.org/10.19103/AS.2020.0070.01

segregation in meiosis due to lack of pairing between the homoeologues, structural hybridity or triploidy or a combination of these (Bakry et al. 1990). Such mismatched chromosomes can lead to bridges, laggards or uneven separation of chromosomes, giving rise to infertile ovules and pollen. An additional factor that plays a role in seedless bananas is parthenocarpy, that is, the induced production of the fruit without fertilization of the ovule. This trait was likely preferred by farmers during early cultivation of the crop.

The sterility problems in most of the cultivars and interspecific hybrids of banana have necessitated breeders and geneticists to analyse their material for pollen fertility, chromosome numbers and aberrant meiotic behaviour including chromosome pairing switches, lack of crossover recombination and unbalanced chromosome segregation. The classical school of cytogenetic and taxonomic analyses of *Musa* species have produced an impressive series of research articles by Dodds (1943), Dodds and Simmonds (1946), Wilson (1946), Simmonds and Shepherd (1955) and others, and the valuable monograph of Shepherd (1999). Their scrutinous work on mitotic chromosome counts and meiotic configurations revealed a wealth of information on taxonomic relationships, ploidy levels and fertility of *Musa* species, hybrids and cultivars. Most of these studies were based on acetocarmine or orcein chromosome squash protocols and camera lucida drawings of selected cell complements, but some of them were able to produce high-quality photomicrographs (Agarwal 1983, 1988). For increased contrast and details of the chromosomes, Pillay and Adeleke (2001) and Adeleke et al. (2002) introduced silver nitrate incubation of air-dried chromosome preparations, a method known for its specific detection of active nucleolar organizer regions (Lacadena et al. 1984) and detailed staining of synaptonemal complexes (Albini 1994). Moreover, Bakry and Shepherd (2008) published a protocol with detailed information on required laboratory materials explaining all steps in making preparation and microscopy with useful hints for troubleshooting. Osuji et al. (2006) compared the karyotypes of the AA ('Calcutta 4', *M. accuminata* Colla) and BB ('Butohan 2', *M. balbisiana* Colla) mitotic chromosome complements concluding that these 1.4–3.6 μm small chromosomes of the parental species showed great similarity in structure and size.

Microscopic interpretation of meiotic chromosome configurations is limited by the rather small size of banana chromosomes, making it difficult to interpret chiasma positions in chromosome associations visible at diakinesis – metaphase I. In addition, pollen mother cells display dense cytoplasm with granular structures that mask minute acentric fragments accompanying anaphase bridges of inversion heterozygotes. Translocation complexes are clearly recognizable in adjacent quadrivalent and trivalent + univalent configurations, but interpretation of the more compact quadrivalent configurations with alternate-oriented centromeres can be challenging. The importance of such multivalent observations is illustrated in the analyses of banana hybrids by Shepherd and others (Wilson 1946; Simmonds and Dodds 1949; Shepherd 1999). Shepherd's study on F_1 hybrids obtained from crosses with diploid wild accessions of *M. acuminata* demonstrated multivalents including 4–8 chromosomes in diakinesis - metaphase I complements. In general, such hybrids display different levels of male and female fertility that are related to the number of structural change complexes, known as structural hybridity (Darlington 1929). Fertility may be from moderate to high when only one exchange is involved, depending on the ratio of adjacent (leading to unbalanced gametes) and alternate orientations (leading to balanced gametes), but fertility is progressively reduced, if the plant contains two or three translocation complexes (Shepherd 1999).

Some hybrids and partial sterile diploids are known to generate small numbers of larger size pollen grains, which likely result from meiotic restitutions. These so-called 2*n* gametes have been described extensively (reviewed in Ortiz (1995)) and may have derived from first division restitution (FDR) or second division restitution (SDR). The sterile trisomic polyploids 'Cavendish' and 'Gros Michel' are supposed to have arisen from 2*n* restitution gametes and a normal haploid gamete donor (Raboin et al. 2005). Offspring from triploid × diploid crosses may give rise to predominantly tetraploid progeny produced from 2*n* eggs resulted from SDR in the triploid mother (Ortiz 1995). FDR 2*n* eggs are also reported but they are likely even rarer in *Musa*.

2 Pollen fertility

Pollen fertility tests are often used instead of or in support of meiotic studies of pollen mother cells. They are fast and straightforward and provide a direct indication about sterility and aberrations, if any, during male meiosis. Various staining protocols are used for pollen fertility studies, including Alexander staining (Alexander 1969), lactophenol acid fuchsin (Sass 1964) and 2,3,5-Triphenyltetrazolium chloride (TTC) (Damaiyani and Hapsari 2017). Unbalanced chromosome segregation, especially in few female-fertile triploid cultivars (AAA, AAB and ABB) will in general give very few stainable, fertile pollen. Translocation heterozygotes, which occur regularly in banana, contain two translocation (e.g. 1T2 and 2T1) and two non-translocation chromosomes (1 and 2). Segregation of the chromosomes of the translocation complex will be balanced (1T2 + 2T1/1 + 2) giving rise to fertile gametes, or will be unbalanced (1T2 +1 / 2T1 + 2 or 1T2 +2 / 2T1 + 1) giving rise to sterile gametes. The second class of structural variants include heterozygoty for a paracentric inversion, which show variable levels of sterility depending on the frequency of chiasmata in the inversion loop. Unreduced and polyploid spores indicate restitutional divisions, either at anaphase I or II, or both. Damaiyani and Hapsari

(2017) conducted a 1% solution of the TTC staining for testing pollen grain viability of wild *M. acuminata* ssp. *rutilifes* and *M. balbisiana* and cultivars. Their study demonstrates considerable pollen viability among the tested banana accessions making it a useful tool for banana taxonomy.

3 Flow-cytometry

Flow-cytometry is a high-throughput alternative in determining ploidy levels and aneuploidy in banana species and hybrids. Lysak et al. (1999) used this technology for nuclear genome size variation in *M. acuminata*, *M. balbisiana* and various triploid clones with different genomic constitutions. The *M. balbisiana* genome was estimated at 537 Mbp, with no genome size variation and about 12% smaller than the *M. acuminata* accessions that showed genome sizes ranging from 591 Mbp to 615 Mbp. The even larger variation observed in the group of triploid *Musa* accessions (ranging from 559 Mbp to 613 Mbp) reflects both genomic constitutions as well as differences in the size between the A and the B genomes. A further refinement of the technology was published by Roux et al. (2003), who claim a rapid and accurate flow-cytometric detection of $2n$=31–32 aneuploids obtained from gamma-irradiated triploid shoot tips of the *M. acuminata* clone 'Grande Naine' ($2n$=$3x$=33). Oselebe et al. (2009) determined ploidy variation in progenies of 2x–2x, 2x–4x and 4x–2x hybrids. Ploidy was established using flow-cytometry analysis of nuclear DNA content and chloroplast characteristics, showing that progenies of $2x$–$2x$ crosses were predominantly diploid (99.7%), whereas those of $2x$–$4x$ crosses were mainly diploid (96.2%), and the $4x$–$2x$ crosses produced predominantly triploid progenies (94.1%). A comparable flow-cytometric analysis together with chromosome counts was carried out on tissue culture-induced off-type bananas by Msogoya et al. (2008) who found substantial loss of chromosomes or chromosome segments. The few examples of flow cytometry in banana show that the technology is fast and convenient, and with modern devices might be accurate enough to show aneuploidy even in triploid banana populations, segregating families or somaclonal samples, and hence will replace the much more time-consuming chromosome counting protocols. However, for the flow-cytometric estimates of unknown *M. acuminata* accessions or a segregating population, one should keep in mind that genome size differences can be up to 12% (Lysak et al. 1999) and so may unjustly suggest the occurrence of aneuploid individuals. Thus, chromosome counting of those plants still remains indispensable.

4 Fluorescent in situ hybridization (FISH)

Fluorescent in situ hybridization (FISH) is a molecular cytogenetic technique using single-strand fluorescently labelled DNA probes that bind to denatured chromosomal target DNA with a high degree of sequence complementarity. It was developed in the early 1980s mostly for human cytogenetics and found its way to plant research as an outstanding and versatile tool to map repeats and single-copy sequences on chromosomes and nuclei, bridging the disciplines of genetics and genomics. Applications of FISH to banana cytogenetics have elevated the research to a higher level, although some of its potentials are yet to be exploited. The small size of the chromosomes and the dense rigid cytoplasm in spread pollen mother cells are major challenges in successful hybridizations of probe DNAs to the chromosomal targets. The composition of parental chromosomes in *Musa* hybrids has been disclosed by a special type of FISH technology, known as genomic in situ hybridization (GISH) or genome painting, using total genomic DNAs of the parental species as probes in a FISH on cell spread slides of interspecific or intergeneric hybrids. In a double-labelling regime, Osuji et al. (1997) identified the A genome and B genome chromosomes in several amphiploid (AAB and ABB) cultivars, with most discriminating signals coming from satellite repeats and the pericentromeric and nucleolar organizer repeats in the heterochromatin regions. In a comparable GISH study of hybrids involving three wild species of section *Eumusa* (*M. acuminata*, *M. balbisiana*, *M. schizocarpa*) and one wild species of section *Australimusa* (*M. angustigemma*), D'Hont et al. (2000) distinguished the parental chromosomes in their hybrids, although their discriminating signals were confined to the heterochromatin regions. The strength of the fluorescent hybridization signals corresponds with the assumed phylogenetic relationship between these species. In a later study, D'Hont (2005) included the genome structure of more interspecific banana cultivars revealing that the chromosome constitutions in these hybrids in most cases correspond well with their presumed genome constituents. This method has been applied to analyse chromosome pairing in metaphase from triploid interspecific cultivars, and it clearly demonstrates that interspecific recombinations between *M. acuminata* and *M. balbisiana* chromosomes do occur and may be frequent in triploid hybrids. These results shed new light on *Musa* cultivar evolution and provide important implications for banana and plantain breeding. The potential of genome painting has further been explored in a study on homoeologous recombination in spread pollen mother cells at metaphase I. Bivalents and trivalents with A and B genome chromosomes could be distinguished by FISH fluorescence intensities, and so showed that A- and B-genome chromosomes display homoeologous recombination (Jeridi et al. 2011). In a follow-up analysis, Jeridi et al. (2012) studied the nature of chromosome configurations in a sterile AB Indian cultivar and in its fertile colchicine-induced allotetraploid (AABB). Genome painting demonstrated that the genomes of the AABB reveal the so-called segmental allopolyploidy (Sybenga 1975) with three chromosome sets showing a tetraploid-like pattern (i.e. showing quadrivalent pairing), whereas the remaining sets are diploid-like (bivalents only) or display undetermined patterns. Balanced and unbalanced diploid gametes were detected in progenies, with the chromosome constitution appearing to be more homogenous in pollen than in ovules. This segmental inheritance pattern exhibited by the AABB allotetraploid genotype confirms earlier findings on chromosome rearrangements between *M. acuminata* and *M. balbisiana*, which may provide the possibility of introgressing valuable alleles in banana breeding programmes.

A recent genome sequence analysis of *M. balbisiana* (Wang et al. 2019) enabled the comparison with the *M. acuminata* reference genome showing details of their synteny. Both genomes display a surprising collinearity and sequence similarity. Only two translocations and two inversions between the A- and B-genomes are observed, that were also supported by the re-arrangements based on genetic mapping data (Wang et al. 2019). The large-scale collinearity between the A- and B- genomes does explain very well why the A- and B-homoeologues can undergo homologous exchange and recombination in diploid and amphiploid hybrids.

5 Use of repetitive sequences in banana research

The first class of repetitive sequences that were localized on chromosomes by FISH were the ribosomal DNA genes using conserved parts of their sequences as probes. Doleželová et al. (1998) performed such a hybridization using parts of the 5.8S – 18S – 25S rDNAs of *Vicia faba* and a 5S rDNA on cell spreads of two banana cultivars and six wild relatives of *Musa*, all of them diploid with 2n=22 chromosomes. Next to only one locus for the 18S/25S rDNA tandem array on the NOR chromosome pair, a variable number of two, three or five loci of the 5S rDNA tandem array were detected, the latter possible resulting from heterozygosity of a structural re-arrangement (Doleželová et al. 1998). Extension of these analyses on 19 diploid accessions representing the four taxonomic sections *Eumusa*, *Rhodochlamys*, *Callimusa* and *Australimusa*, and the outgroup *Ensete gilletii* was performed by Bartoš et al. (2005) with 2–8 fluorescent spots for the 45S rDNA and 4–8 spots for the 5S rDNA. Hierarchical cluster analysis using genome size, chromosome number and 45S rDNA sites revealed a clear relationship between the accessions under study.

The second group of characteristic repetitive sequences in plant genomes are the retrotransposable elements, of which the *gypsy* type with its species-specific long terminal repeats that occupies most of the heterochromatin of the chromosomes is the most abundant. In *M. acuminata* cv. 'Grand Nain', Balint-Kurti et al. (2000) described the *gypsy*-like *monkey* element, of which the reverse transcriptase, RNase H and integrase genes resemble retroelements from plants, fungi and yeast. Southern analysis of genomic DNAs from nine different banana cultivars, digested with *Hind*III and probed with the different fragments of the major open reading frames demonstrated hybridizations with the A genome or with both A and B genomes. FISH demonstrates that this non-autonomous element co-localizes with the 18S-5.8S- 25S rRNA genes in the nucleolar organizer region and also occurs in small regions distributed over all chromosomes. In a more comprehensive study, Valárik et al. (2002) studied the chromosomal localization of various repetitive elements on cell complements of *M. acuminata* and *M. balbisiana*, and analysed sequence similarity to known DNA sequences. The characterized clones included 26S and 5S rDNA genes, and sequences that displayed fluorescent foci in centromeric regions. All repetitive sequences were more abundant in *M. acuminata* than in *M. balbisiana*, corresponding to the 12% larger genome size of *M. acuminata* compared to *M. balbisiana* suggesting that repetitive sequences to a greater part account for the difference in genome size between both species. In a follow-up on previous repeat research, Hřibová et al. (2007) performed DNA re-association kinetics to isolate the highly repeated fraction of the banana genome (*M. acuminata* 'Calcutta 4') and constructed two Cot libraries (Peterson et al. 2002) of different repeat complexity, and a third set of DNA clones with random sequences for comparison. 'BLAST' homology searches demonstrated that the majority of the repeats represent different types of retrotransposons, of which Ty3/*gypsy*-type *monkey* retrotransposon was the most prevalent. Tandemly organized sequences are mapped to the nucleolar organizers and proximal pericentromere regions of the *M. acuminata* chromosomes. In a low-depth 454 sequencing study of the repeat part of banana (*M. acuminata* 'Calcutta 4') Hřibová et al. (2010) produce partial sequence reconstruction and characterization of repetitive DNA. Most of the repeats are identified as Ty1/*copia* and Ty3/*gypsy* retroelements, occupying 16% and 7% of the genome, respectively. In addition, some LINEs, satellite repeats including rDNAs and unclassified retroelements are observed, while DNA transposons are relatively rare. The small genomic part of satellite repeats has further been explored by Čížková et al. (2013), who compared two main DNA satellites in a set of 19 *Musa* accessions, including representatives of A, B and S (*M. schizocarpa*) genomes and their interspecific hybrids. The two DNA satellites showed a high level of sequence similarity within, and a high homology between *Musa* species. Microscopic observations of these satellites in a FISH along with rRNA genes, LINE-like element and a single-copy BAC clone revealed characteristic fluorescence patterns in *M. acuminata* and *M. balbisiana*, in which up to five chromosomes could be identified. Next-generation sequencing and comparative analysis of the major repetitive sequences of representative species of the *Musaceae* family (*M. acuminata*, *M. ornata*, *M. textilis*, *M. beccarii*, *M. balbisiana*, and *Ensete gilletii*) demonstrated their overall intraspeific and interspecific similarities using graph-based clustering (Novák et al. 2014). The most abundant classes of highly repetitive DNA included the *Maximus*/SIRE and *Angela* lineages of Ty1/*copia* LTR retrotransposons and the chromovirus lineage of Ty3/*gypsy* elements. The differentiation of the repeats followed the taxonomic distances of the species in the *Musaceae* family, while the closely related species *M. acuminata* versus *M. ornata*, and *M. beccarii* versus *M. textilis* share more related populations of repetitive elements. This study did not reveal any repeat class that can help as a chromosome marker in karyotype analyses. It is still unknown if simple sequence repeats/microsatellites that are currently used in marker technology of *Musa* (Creste et al. 2003; Christelová et al. 2011) occur as large tandem arrays in the banana genome such that they could be used as microsatellite FISH (Bouilly et al. 2008; Cuadrado and Jouve 2010) for chromosome identification.

6 Chromosomal detection of single-copy sequences

In the next step of characterization of the genome and chromosomes of *Musa*, researchers have developed bacterial artificial chromosome (BAC) libraries. Vilarinhos et al. (2003) published the first publicly available BAC library obtained from genomic DNA of the wild diploid *M. acuminata* 'Calcutta 4' clone. The library consisted of 55 152 clones with an average insert size of 100 kb and has a coverage of 9× the haploid genome. The BAC clones were anchored on a still incomplete genetic map with eight linkage groups using a set of 13 RFLP probes. The BAC inserts were created for use as chromosome markers in FISH, but could be helpful as seed BACs during the assembly of the physical map of banana. Sequence information is now available from the BAC-end sequencing project of Cheung and Town (2007), and Arango et al. (2011), who generated 6252 and 46 080 reads, respectively. The obtained BAC-end sequences were screened against several databases and significant homology was found compared to mitochondria and chloroplasts (2.6%), transposons and repetitive sequences (36%) and proteins (11%). In addition, the sequencing revealed 352 potential simple sequence repeat (SSR) markers of which the most were AT-rich. After filtering mitochondria and chloroplast matches, thousands of BAC end-sequences had a significant BLASTN match to the *Oryza sativa* and *Arabidopsis* genome sequences. The syntenic relationships between *Musa* and rice have further been elaborated by Lescot et al. (2008).

Soon after Vilarinhos' study of the BAC library of *M. acuminata,* Šafář et al. (2004) released the BAC library of the second ancestral species of banana hybrid cultivars, *M. balbisiana* 'Pisang Klutuk Wulung'. Here the authors introduced an improved DNA isolation protocol and a novel flow cytometric-based nuclei isolation method as a strategy to circumvent interference of secondary metabolites (polyphenols and polysaccharides) and plastid DNA in leaf tissues for the isolation of high-molecular-weight nuclear DNA. In addition, the application of the inducible pCC1BAC vector improved the amount of BAC DNA. The two sub-libraries, for each of the two methods, were screened with twelve RFLP probes and were anchored to eight linkage groups of *M. acuminata*. Ten BAC clones were selected for FISH mapping on mitotic cell complements of *M. balbisiana*. Three BAC clones showed multiple fluorescent foci on the chromosomes, while a second group of three BAC clones, that were supposed to be rich in *gypsy* LTR elements, produced a uniform distribution mostly in the (peri)centromeric regions, and a third group likely rich in *copia* and/or *gypsy* elements produced almost complete labelling of the chromosome complement.

The major drawback of FISH in *Musa* species and hybrids is the small size of the mitotic chromosomes and hence the low resolution of adjacent or partly overlapping probe signal in multiple colour FISH essays. In de Jong et al. (1999), a comparison has been made between mitotic metaphase chromosomes, pachytene bivalents, interphase nuclei and extended DNA fibres in plant species. The choice of pachytene chromosome complements for BAC FISH is undeniably one of the best choices in species with small genomes including *Musa*. In Capdeville et al. (2008) the first attempts of such a BAC clone positioning on pachytene chromosomes of 'Calcutta 4' and the wild *M. velutina* is demonstrated. The preparation of spread pollen mother cells after digestion with pectolytic enzymes and maceration with acetic acid poses considerable technical challenges for reducing the thick cytoplasm with the polyphenolics and polysaccharides that hamper probe DNA to hybridize to the chromosomal targets. In spite of these hurdles, BAC FISH is powerful enough to support meiotic studies on pairing disturbances in hybrid cultivars with chromosomal rearrangements. In a more recent approach, single copy chromosomal markers can now be obtained directly from the highly improved genome map of *M. acuminata* (Martin et al. 2016) where specific unique sequences can be selected for making synthetic probes, as used recently in developing oligo painting FISH strategies (Han et al. 2015; Li et al. 2016; Braz et al. 2018; Hou et al. 2018).

7 Molecular markers in linkage studies

Geneticists and breeders have employed various DNA markers for the characterization and evaluation of genetic diversity in *Musa* species and map-based cloning in banana breeding programmes (Fauré et al. 1993; Pillay and Tripathi 2006). For a long period, the use of SSRs has been favoured (Jarret et al. 1994; Kaemmer et al. 1997; Creste et al. 2003; Buhariwalla et al. 2005). Furthermore, sequenced tagged microsatellite site (STMS) marker technology (Kaemmer et al. 1997), inter-simple sequence repeats or ISSR (Godwin et al. 1997; Padmesh et al. 2012; Silva et al. 2017) and inter-retrotransposon amplified polymorphic (IRAP) markers (Nair et al. 2005) have also been used. Arora et al. (2018) developed BanSatDB, a whole-genome-based database of putative and experimentally validated microsatellite markers of three *Musa* species. Hippolyte et al. (2010) applied the diversity array technology (DArT) in combination with SSRs for a linkage map of diploid *M. acuminata*, while Sardos et al. (2016) use DArT strategy for whole genome profiling to obtain deeper understanding of the evolution, taxonomy and domestication of edible banana.

The first linkage map of *M. acuminata* was obtained on the basis of a segregating F_2 population from a diploid AA clone (SF265) and a diploid AA Banksii (*M. acuminata* ssp. *banksii*) line (Fauré et al. 1993). Ninety-two individuals were scored for 58 RFLPs, four isozyme and 28 random amplified polymorphic DNA (RAPD) markers. Of the 90 loci that were identified, 77 could be grouped in 15 linkage groups, while the remaining 13 segregated independently. Segregation distortion was demonstrated for 36 loci, mostly favouring the male *ssp. banksii* parent, and is assumed the result of heterozygosity for a structural rearrangement of the parents. Lowering the LOD score between 3 and 4 allowed the merging of four out of the 15 linkage groups resulting in the expected 11 (chromosomal) linkage groups. To account for large segregation distortion, pseudo linkages and difficulties in ordering markers around putative translocation breakpoints, Hippolyte et al. (2010) designed novel genetic linkages strategies for a segregating F_1 population of 180 individuals from a cross between *M. acuminata*, 'Borneo' and 'Pisang Lilin', that were known to be genetically distant accessions. Microscopic analysis of chromosome configurations in pollen mother cells at diakinesis – metaphase I showed multivalent associations in a small number of cells, including trivalent, quadrivalent, quinquevalent and a hexavalent in the 'Borneo' accessions suggesting heterozygosity for two structural re-arrangements. The 'Pisang Lilin', however, displayed only an adjacent quadrivalent assuming one structural polymorphism and an anaphase I bridge which can be explained by a crossover in the loop of a paracentric inversion heterozygote. The linkage analysis was based on 489 segregating markers (167 SSRs and 322 DArTs) for the construction of gamete recombination of each parent, leading to the two parental maps with the expected eleven linkage groups and a total genetic map length of 1197 cm. Segregation distortion, demonstrated for 21.7% of the markers ($P < 0.05$) occurred in different linkage groups for each parent. The putative re-arrangements on the chromosomes 1 and 4, and on chromosome 10 were visualized by neighbour-joining trees that are designed from Kosambi's distance calculation. An extension of this genetic analysis was done for the study of the banana genome (D'Hont et al. 2012) using a set of 589 SSR markers, as much as possible equally dispersed along the genome, along with 63 DArT markers and genotyped over 180 individuals. Also, in this map the linkage groups 1 and 4 remain grouped and concentrated 72% of the distorted markers. Mbanjo et al. (2012) published a comparable linkage map of two half-sib diploid banana populations on the basis of allele-specific-polymerase chain reactions (AS-PCRs), diversity array technology (DArT) and SSR markers. The pseudo-testcross mapping strategy produced a maternal map of 15 linkage groups (LGs) covering 670 cm and a maternal map of 16 LGs with a total length of 698 cm and a combined map with 15 LGs and a total map length of 1004 cm. While sequence quality of the markers is generally good, incongruity in some cases suggests chromosomal rearrangements. Additional analysis of allelic ratios and patterns together with neighbour-joining trees (Hippolyte et al. 2010) enable the heterozygous status for the structural rearrangement to be resolved. The existence of this translocation is further substantiated in Martin et al. (2017), who used mate-pair sequencing, BAC-FISH, targeted PCR and marker (DArTseq) segregation in the progeny of a heterozygous *M. acuminata*. The heterozygous reciprocal translocation features two distal 3 Mb and 10 Mb segments from chromosomes 01 and 04 (chromosomes 1, 1T4, 4T1 and 4), respectively, and show high segregation distortion, reduced recombination and linkage between chromosomes 01 and 04 in its progeny. The rearranged chromosome structure can also be observed in triploid cultivars but is present only in wild *malaccensis* ssp. accessions, implying that this rearrangement occurred in *M. acuminata* ssp. *malaccensis*. The observed high transmission of the translocation chromosomes suggests a mechanism in which the non-translocated chromosome 1 is lost, and so favours spreading of this translocation chromosomes to other populations and accessions.

8 Case study

Various aspects of chromosome research and genetics are now illustrated on the basis of a study of a diploid heterozygous *Fusarium* wilt resistant accession of wild banana *Musa acuminata* Colla ssp. *malaccensis* (LIPI-010) from LIPI (Lembaga Ilmu Pengetahuan Indonesia, Indonesian Institute of Sciences), Bogor, Indonesia. A segregating population of 255 descendants obtained from a selfed heterozygote had been created for the mapping of QTLs (quantitative trait loci) conferring *Fusarium* resistance (Ahmad et al. 2020a). As structural hybridity was presumed, a cytogenetic analysis of male meiosis was carried out. To this end, we selected flower buds of the parent plant with anthers of 20 mm or less containing pollen mother cells at all stages of meiosis or pollen grains, and fixed the material in a fresh mixture of methanol:glacial acetic acid (3:1) for about 1 h before transferring the anthers to 70% ethanol and kept at 4°C for longer storage. Cell walls of the pollen mother cells were softened by incubating anthers in a mix of pectolytic enzymes followed by maceration in 45% acetic acid following Kantama et al. (2017). Chromosome slides were prepared by a standard squash or cell spreading technique. After selecting the slides with pollen mother cells at the appropriate stage under a phase contrast microscope we left the slides to dry overnight before mounting in Vectashield® containing 300 ng/µL DAPI. Slides were then studied under a fluorescence microscope with N.A. 1.4 Plan Apochromatic objectives and epifluorescence illumination containing a narrow band filter set for DAPI. Captured images were optimized for contrast, brightness and sharpness in Adobe Photoshop (Kantama et al. 2017). Pollen morphology and viability was studied in slides of pollen samples stained in a drop of lacto-phenol acid fuchsin (Sass 1964). Images of 100–150 pollen grains were captured using a bright-field microscope. Pollen grains were measured with a home-made pollen size ImageJ plugin and the obtained data for pollen size and morphology were processed in Microsoft Excel. Further technical details of the cytogenetic procedures are in Ahmad et al. (2020b).

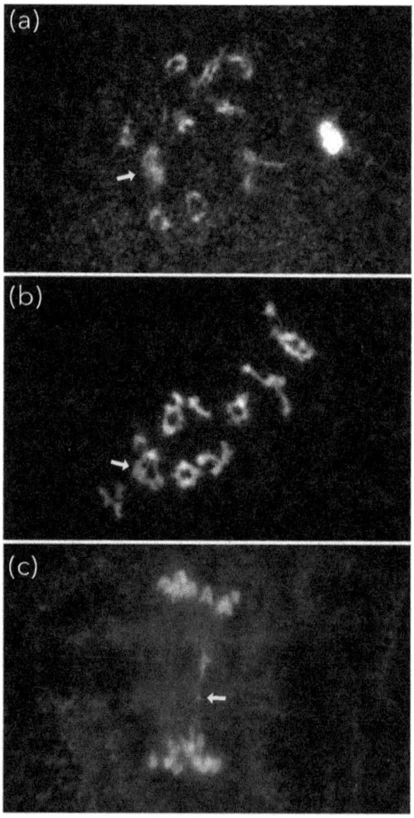

Figure 1 DAPI staining of spread pollen mother cells at meiosis of *Musa acuminata* ssp. *malaccensis*. (a) Cell complement at diakinesis. The arrow indicates two overlapping bivalents that resembles a ring quadrivalent. (b) Metaphase I complement. The arrow indicates a second example of overlapping bivalents resembling a quadrivalent. The arrowhead points at two overlapping bivalents. (c) Anaphase I cell with a dicentric chromosome bridge. The expected accompanying a-centric fragment could not be observed.

Banana anthers contain relatively few pollen mother cells at the same stage, while thick cloudy cytoplasm, with lots of polysaccharides and/or polyphenolics, masks details of the chromosomes. The opacity of the cytoplasm could partly be improved by using methanol instead of ethanol in the fixative, long pectolytic digestion of the cell walls and prolonged careful maceration in 45% acetic acid (Kantama et al. 2017; Ahmad et al. 2020b). We focussed on pollen mother cells at diakinesis, metaphase I and anaphase I, but only few of them displayed sufficiently clear details. Cells that we were able to interpret show only bivalents or overlapping chromosomes that resemble alternating ring quadrivalents (Fig. 1A and B). We did not detect complements with clear adjacent (open) ring quadrivalents, chain quadrivalents and trivalent + univalent combinations. Some of the configurations could be interpreted as overlapping bivalents or as compact alternate ring quadrivalents, and were further disregarded. In pollen mother cells at anaphase I, we found balanced chromosome segregations and very incidentally a dicentric chromosome bridge (lacking a clear acentric fragment) and some lagging chromosomes (Fig. 1C). The anaphase bridge was explained as the result of a crossover in the loop of a chromosome pair heterozygous for a paracentric inversion. Acentric fragments, if any, remained undetectable by the thick granular cytoplasm. Pollen was in general well-stained while pollen size values displayed a relatively narrow distribution (Fig. 2). The results were interpreted in a way that the parent was homozygous for the translocation, thus explaining the high fertility of the male gametes (Sybenga 1975, 1992).

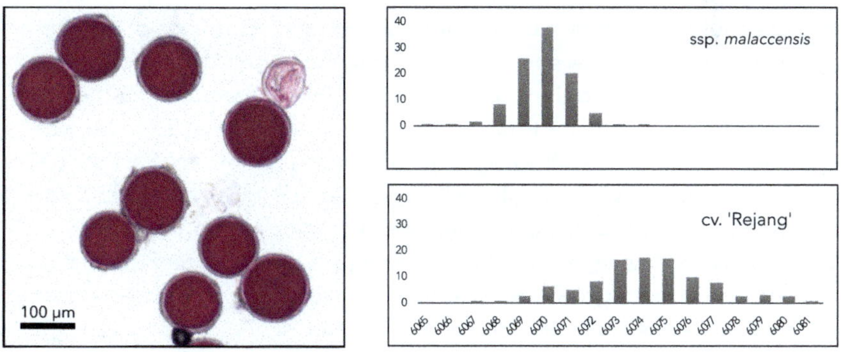

Figure 2 Pollen staining with lacto-phenol acid fuchsin (left). Most spores of the *M. acuminata* ssp. *malaccensis* (LIPI-010) are uniform and only very few are shrunken and colourless, and so representing dead pollen. Measurements of their size show little variation (right), in contrast to that of the *M. accuminata* cv. 'Rejang', which may include aneuploid and unreduced spores as well. The X-axis shows Image J pixel units of the pollen size; the Y-axis the number of measured pollen grains.

8.1 Genotyping and comparison of the genetic and physical maps

We used the DArTseq platform (https://www.diversityarrays.com) (Kilian et al. 2012) for genotyping the heterozygous parent and 217 descendants. Markers were putatively positioned on the genome assembly of the *M. acuminata* ssp. *malaccensis* DH 'Pahang', version 2 (http://banana-genome-hub.southgreen.fr/organism/Musa/acuminata) by aligning the DArTseq sequences to this reference genome.

We collected leaf samples from the segregating population and the parent from both leaf materials from plantlets in tissue culture or from cigar leaves in the greenhouse. We freeze-dried these samples and isolated DNA using the Wizard® Magnetic DNA Purification System from Food Promega. The DNA concentration was quantified using Quant-iT PicoGreen dsDNA Assay Kit from Invitrogen and the quality was checked using 1% gel electrophoresis. The DNA was sent to Diversity Arrays Technology Pty Ltd, Australia, for scoring SNP-markers, using the DArTseq platform (http://www.diversityarrays.com/). Based on the DNA sequences flanking the SNPs, the markers were putatively positioned on the genome assembly of the *Musa acuminata* ssp. *malaccensis* DH 'Pahang', version 2 (http://banana-genome-hub.southgreen.fr/organism/Musa/acuminata).

Eventually, 217 offspring and the parent were genotyped and 32 362 SNP markers were retrieved from the DArTseq (Ahmad et al. 2020a). The very high number of SNP-markers allowed us to apply a very stringent filtering, using the following criteria: (1) replicate value = 1; (2) sequences should hit only one position on the reference genome; (3) The polymorphism information content >0.3; (4) DNA markers should be based on more than 10 calls per allele; (5) We only considered markers that were heterozygous in the parent, and therefore could segregate in the progeny. The physical positions of the filtered, segregating markers were checked on the reference genome for detection of possible homozygous regions in the self-pollinated parent. Apparently, the whole genome of the parent must have been heterozygous as there was no region devoid of markers, allowing construction of a genetic map for the whole genome. For the construction of the linkage map we initially used JoinMap® 5 software (https://www.kyazma.nl/index.php/JoinMap/) using the setting of 'F_2' population type, thus regarding the heterozygous parent as an F_1 from a cross between two homozygous grandparents. Duplicate markers were identified and merged where possible (conflicting marker scores made missing). As well as six marker pairs tagging the same single nucleotide polymorphism (SNP) position, 166 marker pairs with identical scores in the population were identified and merged, leaving 2630 unique markers. A single individual with a large proportion of missing values (28%) was also removed, leaving a mapping population size of 216 individuals. Marker linkage analysis was performed using the POLYMAPR package (Bourke et al. 2018), with parental dosages specified as '1' at all markers for the selfed parent. Linkage groups were identified after clustering at a LOD threshold of 4, and marker ordering was performed using the MDSMap package with default settings (Preedy and Hackett 2016). Linkage groups were renumbered according to the *M. acuminata* ssp. *malaccensis* reference assembly (D'Hont et al. 2012), with the genetic positions reversed if the orientation conflicted with the order on the physical map. A comparison of the genetic and physical maps was visualized using R version 3.5.0 (R Core Team 2018). As there was evidence of map inflation (due to possible genotyping errors), the genetic map was further refined with an in-house implementation of the SMOOTH algorithm (Van Os et al. 2005), and marker information of 11 individuals with an exceptionally high number of re-combinations removed. Map smoothing was repeated using a decreasing delta threshold from 0.99 to 0.71 in steps of 0.02 (where delta is the threshold for declaring singletons), with marker clustering repeated only after the first round (clustering was unchanged) and marker order re-estimated at each round using the POLYMAPR pipeline and the MDSMAP ordering algorithm.

8.2 Linkage analysis

Eleven linkage groups were successfully built from the segregating population of *M. acuminata* ssp. *malaccensis* (LIPI-010) with values ranging from 103 cm to 184 cm, totalling 1640 cm (Table 1). This result is in the range of previously published genetic map lengths in the ssp. *malaccensis* DH 'Pahang' (D'Hont

Table 1 Overview of the genetic analysis with smoothing of the diploid heterozygous *Musa acuminata* Colla ssp. *malaccensis* using the polymapR package

	N	Original	Smoothed
LG1	274	180.16	163.47
LG2	215	113.39	103.31
LG3	295	192.66	178.76
LG4	170	168.07	138.04
LG5	245	163.71	147.08
LG6	300	202.79	184.31
LG7	212	166.89	143.43
LG8	282	182.51	162.46
LG9	247	178.17	152.05
LG10	200	162.11	142.98
LG11	176	140.66	123.67
Total	2616	1851.12	1639.56

et al. 2012; Martin et al. 2017). The linkage maps compared to the physical map clearly showed that interchanges involving chromosome 1 and 4 at position 118–128 cm (between markers chr01-4504873 and chr04-28778311) on chromosome 1 and at position 174–184 cm (between markers chr04-26534556 and chr01-763137) on chromosome 4. Furthermore, the comparison of the genetic maps and the physical maps of references showed markers on parts of chromosomes 1 and 4 did not map to their expected positions in linkage groups 1 and 4, also indicating the presence of a translocation between these chromosomes (Fig. 3). However, the genetic map did not indicate heterozygosity for the translocation. Although microscopic observations of diakinesis complements of the parent revealed few pollen mother cells with

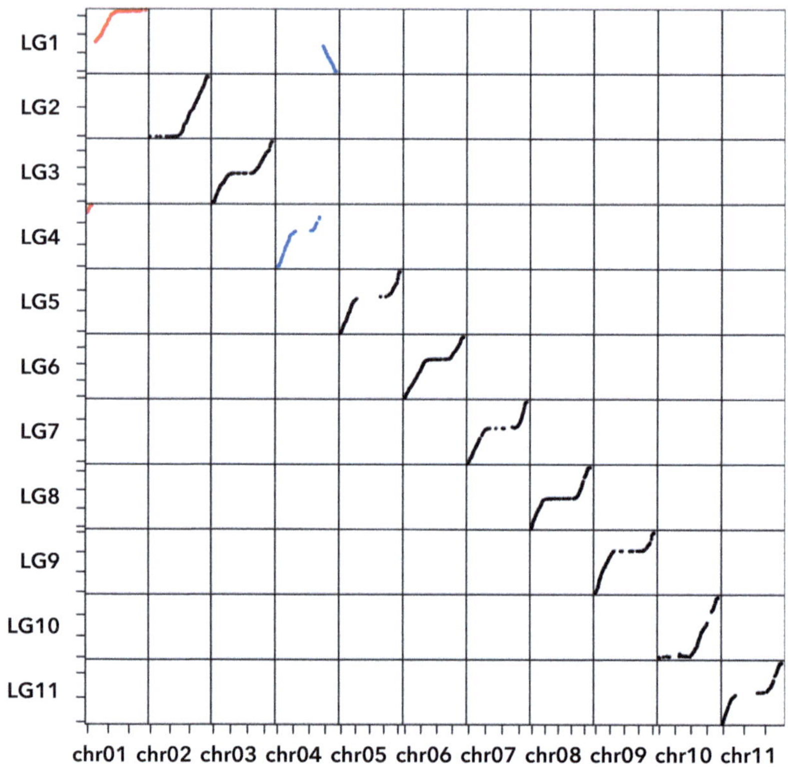

Figure 3 Comparison of the genetic maps with linkage groups LG1-11 (Y-axis) and physical (X-axis) positions according to DH 'Pahang' banana in Martin et al. (2017). In total 2,616 SNP markers are mapped in the segregating population of the heterozygous *M. acuminata* ssp. *malaccensis* (LIPI-010). The Y-axis tick marks correspond to 50 cm intervals, while X-axis tick marks correspond to 10 Mb intervals. Markers on linkage groups carrying the translocation between the chromosomes 1 and 4 have been highlighted in red and blue, respectively.

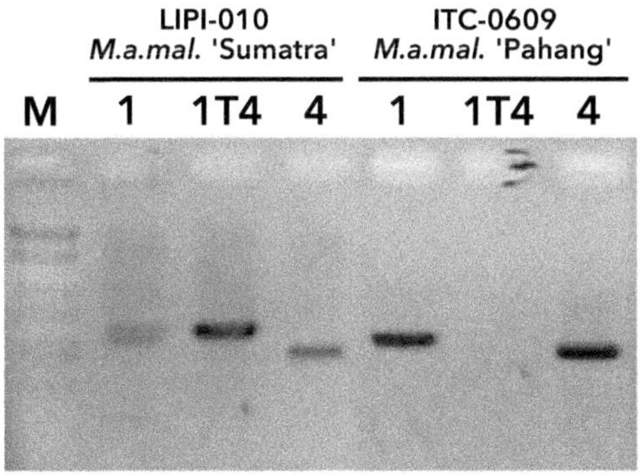

Figure 4 PCR of genomic DNA of the *M. acuminata* ssp. *malaccensis* (LIPI-010) and the reference genome *M. acuminata* ssp. *malaccensis* 'Pahang' (ITC-0609). The forward and reverse primers around the translocation breakpoint were described in Martin et al. (2017). The LIPI-010 plant has both chromosome 1 and 4, and the translocation chromosomes 1T4 and 4T1, and so is heterozygous for the structural rearrangement. The reference 'Pahang' does not have the translocation as previously described in D'Hont et al. (2012) and Martin et al. (2017).

four overlapping chromosomes that resemble an alternative ring quadrivalent, we were not sufficiently convinced about heterozygosity for the translocation of the ssp. *malaccensis* (LIPI-010) mother plant as adjacent quadrivalents were lacking and pollen fertility was high.

To further investigate the nature of the 1T4/4T1 translocation we performed an additional PCR with primers around the translocation breakpoints that confirmed the presence of both chromosomes 1 and 4, and the translocation chromosomes in the parent (Fig. 4). A simplified genetic map with the 11 linkage groups is shown in Fig. 5, along with the translocated segments and the markers flanking the translocation breakpoints. We conclude that the translocation in this study is identical to the 1T4/4T1 translocation in the PT-BA-00267 accession of *M. acuminata* ssp. *malaccensis* as described in Martin et al. (2017). Both share this translocation with two distinct genetic groups corresponding to ssp. *malaccensis* and the *Sucrier* cultivar subgroup, whereas heterozygous accessions were related to both *M. acuminata* ssp. *banksii*, *M. acuminata* ssp. *malaccensis*, and the *Sucrier* cultivar subgroup. The genetical data do not support heterozygosity for the translocation; the microscopic observations do not support a translocation complex either, whereas our PCR data are in favour of heterozygosity of the translocation in the malaccensis parent. Strong segregating distortion for chromosome 10 in the genetic analysis might explain heterozygosity for a paracentric inversion in which crossovers in the inverted regions lead to dicentric bridges and so loss of chromosomes at anaphase I (data not shown).

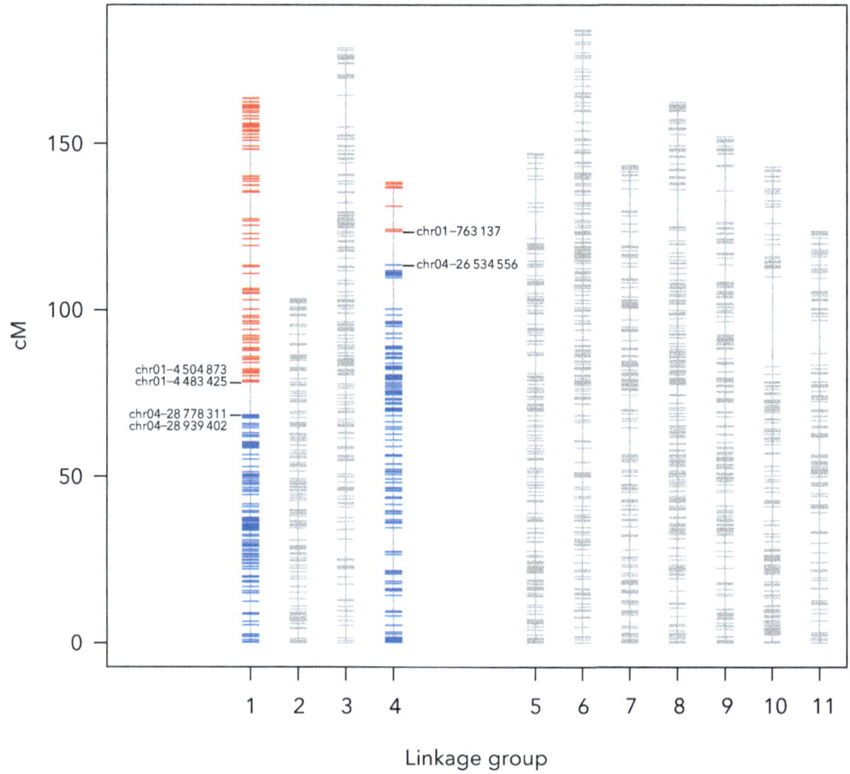

Figure 5 Genetic map of the *M. acuminata* ssp. *malaccensis* (LIPI-010) based on 2616 SNP markers mapped in the segregating population of the heterozygote. Markers proximal and distal of the translocation breakpoints are indicated and the corresponding translocation segments are displayed in red and blue.

9 Conclusion and future perspectives

Our study of the *M. acuminata* ssp. *malaccensis* has clearly proven the existence of the 1T4/4T1 translocation as documented in Martin et al. (2017). The observation is in line with the reports on structural hybridity as a general phenomenon in banana, which along with parthenocarpy, polyploidy and interspecific hybridizations form a serious limitation in breeding of this crop. It is currently accepted that transposons are the engines that create most structural rearrangements. Activation of these elements requires demethylation or other epigenetic changes of its DNA that enables transposition and so leads to breakage and ligation of chromosome parts (Bennetzen 2000; Lönnig and Saedler 2003; Zhang et al. 2011; Lysak and Schubert 2013). In this respect it is obvious to find the footprints of the transposon(s) that are responsible for the rearrangements in *M. acuminata* and subsequently, why these transposons are relatively active in this species. In the Columbia accession of *Arabidopsis thaliana,* a paracentric inversion in the short arm of chromosome 4 has breakpoints in F-box protein-coding genes of the RNI-like superfamily and *Vandal*5, a Mutator-like (Mule) transposon, suggesting that the inversion was created by the activity of the transposon (Fransz et al. 2016). As genomic sequences of the reference DH 'Pahang' and that of the 1–4 translocation are known, bioinformatic analyses provide excellent tools to identify such translocations at the DNA level, such as split-read mapping or paired-end-read mapping assuming that there are no long stretches of repetitive sequences around the breakpoints (Lin et al. 2015). Synteny-based comparisons of genomic sequences like the MUMmer plot tool (http://mummer.sourceforge.net/) can directly reveal the chromosomal-rearranged regions. Zooming into the breakpoints one can reveal the footprints of transposons that may have transposed in the regions of interest. If such breakpoints are caused by one and the same transposable element, it will be straightforward to identify breakpoints of yet unknown translocations and inversions. Alternatively, new, unknown translocations and inversions can be identified by FISH with chromosome-specific BACs or oligo painting pools (Han et al. 2015), microarrays (Bhat et al. 2007), whole-genome sequencing (Dong et al. 2018), long-range sequencing (Hu et al. 2018; de Coster et al. 2019). Genetic mapping is another useful tool for highlighting large structural re-arrangements. Now that most markers are sequence-based, it is possible to compare map orders and check whether there are cross-specific re-arrangements, or structural variation in comparison to the reference genome. The trend towards pangenomics would provide a convenient way to record these re-arrangements.

As structural hybridity is one of the major sources of sterility in banana, it is imperative to detect translocations and inversions in common diploid subspecies and cultivars of *M. acuminata* with favourable traits like disease resistances, using a combination of the methods described above. Accessions sharing the same rearrangements can be used for introgressive hybridization or traditional breeding work. In a final step, accessions differing in two or more translocations can be crossed to produce sterile diploids, or select triploid descendants from incidentally formed 2*n* gametes. Alternatively, recent studies have shown that CRISPR Cas9 technology is a useful alternative to correct translocations and inversions to their original wild-type arrangement (Lekomtsev et al. 2016; Jiang et al. 2016; Lynagh et al. 2018). However, the practical application of such genomic corrections still awaits acceptance of the public and policymakers for GMO plant materials for breeding purposes.

We are in an exciting time where microscopy techniques are probing chromosomes in ever finer detail while at the same time molecular techniques are upscaling to long-range sequences and sub-chromosomal information. However, there are still many caveats. The current reference genomes of banana (both 'Pahang' V2 and *M. balbisiana* 'Pisang Klutuk Wulung') are still incomplete and not guaranteed reliable along all of their length. There is still too little data on the genomic sequence and structural diversity in banana cultivars and wild relatives aside from some data on one single translocation between chromosomes 1 and 4. The details of the structural variants reported by Martin et al. (2017) may not be congruent with the genetic map from the wild Indonesian *Musa acuminata* ssp. *malaccensis* probably because of these gaps in our knowledge.

10 Where to look for further information

Cytogenetics of *Musa* species, hybrids and cultivars has long been used for determining ploidy, aneuploidy and structural variations. Besides, the study of chromosomes together with genetic analyses plays a vital role in supporting banana breeding, especially in introgressive hybridization programmes, providing essential information on meiotic pairing, recombination and transmission, and formation of unreduced gametes in interspecific hybrids and their backcross derivatives. Recently advanced fluorescent in situ hybridization technologies includes molecular protocols and genomics information enabling chromosomal positioning of single-copy and repetitive sequences on mitotic and meiotic cells and nuclei.

For understanding meiotic chromosome processes and aberrations in plant species and hybrids and especially in bananas we recommend:

- Shepherd, K. (1999). *Cytogenetics of the Genus Musa*. INIBAP.
  This monograph clearly describes the complexity of structural rearrangements between *Musa* (sub)species and cultivars.
- Sybenga, J. (1975). *Meiotic Configurations*. Springer-Verlag.
- Sybenga, J., (1992). *Cytogenetics in Plant Breeding*. Springer Verlag.
  Both monographs nicely clarify the complex meiotic behaviour of plants heterozygous for structural heterozygosity, polyploids and aneuploids.

Outstanding examples of recent advanced molecular cytogenetics of *Musa* are:

- Martin, G., Carreel, F., Coriton, O., Hervouet, C., Cardi, C., Derouault, P., Roques, D., et al. (2017). Evolution of the banana genome (*Musa acuminata*) is impacted by large chromosomal translocations. *Molecular Biology and Evolution* 34(9): 2140–2152. https://doi.org/10.1093/molbev/msx164.
  This study combines DNA sequencing, chromosome painting and genetic analysis to characterize a reciprocal translocation $1^4/4^1$ in an *M. acuminata* ssp. *malaccensis* banana accession.

- Šimoníková, D., Němečková, A., Čížková, J., Brown, A., Swennen, R., Doležel, J. and Hřibová, E. (2020). Chromosome painting in cultivated banana and their wild relatives (*Musa* spp.) reveals differences in chromosome structures, *bioRxiv*, preprint. doi: https://doi.org/10.1101/2020.08.01.232207.

  The authors used the recently developed powerful oligo-FISH painting technology to elucidate translocations in *M. acuminata*, *M. balbisiana* and their hybrids.

Important general websites about *Musa* research, breeding and agriculture:

- http://www.promusa.org/.

  This website is the general online platform of ProMusa a network of people promoting scientific discussions on bananas. The site also includes information on the global Musa genomics consortium.
- https://www.crop-diversity.org/mgis/organisations.

  The *Musa* Germplasm information system contains critical information on *Musa* germplasm diversity, including passport data, botanical classification, morpho-taxonomic descriptors, molecular studies, plant photographs and GIS information on 6548 accessions managed in 29 collections around the world.
- https://en.wikipedia.org/wiki/MusaNet.

  MusaNet is a global network of scientists and other stakeholders working on banana (*Musa* spp.) genetic resources.
- https://fusariumwilt.org/index.php/en/about-fusarium-wilt/.

  An informative website about banana and one of its most devastating threads in banana culture: fusarium wilt.

11 References

Adeleke, M. T. V., Pillay, M. and Okoli, B. E. (2002). An improved method for examining meiotic chromosomes in *Musa* L. *HortScience* 37(6): 959–961.

Agarwal, P. K. (1983). Cytogenetical investigations in Musaceae I. Meiotic studies in South Indian bananas. *Cytologia* 48(4): 847–852.

Agarwal, P. K. (1988). Cytogenetical investigations in Musaceae III. Meiotic studies in South Indian bananas. *Cytologia* 53(2): 359–363.

Ahmad, F., Martiwi, N. M., Poerba, Y. S., de Jong, H., Schouten, H. and Kema, G. H. J. (2020a). Genetic mapping of Fusarium Wilt resistance in a wild banana *Musa acuminata* ssp. *malaccensis* accession. In preparation.

Ahmad, F., Poerba, Y. S., Kema, G. H. J. and de Jong, H. (2020b). Meiotic studies of diploid Indonesian banana varieties and cultivars. In: F. Ahmad, PhD thesis. Wageningen University.

Albini, S. M. (1994). A karyotype of the *Arabidopsis thaliana* genome derived From synaptonemal complex analysis at prophase I of meiosis. *The Plant Journal* 5(5): 665-672.

Alexander, M. P. (1969). Differential staining of aborted and non-aborted pollen. *Stain Technology* 44(3): 117-122.

Arango, R. E., Togawa, R. C., Carpentier, S. C., Roux, N., Hekkert, B. L., Kema, G. H. J. and Souza, M. T. (2011). Genome-wide BAC-end sequencing of *Musa acuminata* DH Pahang reveals further insights Into the genome organization of banana. *Tree Genetics and Genomes* 7(5): 933-940. doi:10.1007/s11295-011-0385-3.

Arora, V., Kapoor, N., Fatma, S., Jaiswal, S., Iquebal, M. A., Rai, A. and Kumar, D. (2018). BanSatDB, a whole-genome-based database of putative and experimentally validated microsatellite markers of three *Musa* species. *The Crop Journal* 6(6): 642-650. doi:10.1016/j.cj.2018.01.005.

Bakry, F., Horry, J. P., Teisson, C., Tezenas Du Montcel, H. and Ganry, J. (1990). L'amélioration génétique des bananiers à l'IRFA/CIRAD, [Genetic improvement of banana plants at IRFA/CIRAD]. *Fruits* (numero speciale): 25-40.

Bakry, F. and Shepherd, K. (2008). Chromosome count on banana root tip squashes. *Fruits* 63(3): 179-181. doi:10.1051/fruits:2008008.

Balint-Kurti, P. J., Clendennen, S. K., Doleželová, M., Valárik, M., Doležel, J., Beetham, P. R. and May, G. D. (2000). Identification and chromosomal localization of the monkey retrotransposon in *Musa* sp. *Molecular and General Genetics: MGG* 263(6): 908-915.

Bartoš, J., Alkhimova, O., Doleželova, M., De Langhe, E. and Doležel, J. (2005). Nuclear genome size and genomic distribution of ribosomal DNA in *Musa* and *Ensete* (Musaceae): taxonomic implications. *Cytogenetic and Genome Research* 109(1-3): 50-57. doi:10.1159/000082381.

Bennetzen, J. L. (2000). The many hues of plant heterochromatin. *Genome Biology* 1(1): 107.1-4. doi:10.1186/gb-2000-1-1-reviews107.

Bhat, P. R., Lukaszewski, A., Cui, X., Xu, J., Svensson, J. T., Wanamaker, S., Waines, J. G. and Close, T. J. (2007). Mapping translocation breakpoints using a wheat microarray. *Nucleic Acids Research* 35(9): 2936-2943. doi:10.1093/nar/gkm148.

Bouilly, K., Chaves, R., Leitão, A., Benabdelmouna, A. and Guedes-Pinto, H. (2008). Chromosomal organization of simple sequence repeats in the pacific oyster (*Crassostrea gigas*): (GGAT)(4), (GT)(7) and (TA)(10) chromosome patterns. *Journal of Genetics* 87(2): 119-125.

Bourke, P. M., van Geest, G., Voorrips, R. E., Jansen, J., Kranenburg, T., Shahin, A., Visser, R. G. F., Arens, P., Smulders, M. J. M. and Maliepaard, C. (2018). polymapR - linkage analysis and genetic map construction from F1 populations of outcrossing polyploids. *Bioinformatics* 34(20): 3496-3502.

Braz, G. T., He, L., Zhao, H., Zhang, T., Semrau, K., Rouillard, J. M., Torres, G. A. and Jiang, J. (2018). Comparative oligo-FISH mapping: an efficient and powerful methodology to reveal karyotypic and chromosomal evolution. *Genetics* 208(2): 513-523. doi:10.1534/genetics.117.300344.

Buhariwalla, H. K., Jarret, R. L., Jayashree, B., Crouch, J. H. and Ortiz, R. (2005). Isolation and characterization of microsatellite markers From *Musa balbisiana*. *Molecular Ecology Notes* 5(2): 327-330. doi:10.1111/j.1471-8286.2005.00916.x.

Capdeville, G. de, Teixeira Souza Júnior, M., Szinay, D., Eugênio Cardamone Diniz, L., Wijnker, E., Swennen, R., Kema, G. H. J. and Jong, H. de (2008). The potential of high-resolution BAC-FISH in banana breeding. *Euphytica* 166(3): 431-443. doi:10.1007/s10681-008-9830-2.

Cheung, F. and Town, C. D. (2007). A BAC end view of the *Musa acuminata* Genome. *BMC Plant Biology* 7(1): 29. doi:10.1186/1471-2229-7-29.

Christelová, P., Valarik, M., Hřibova, E., Van den Houwe, I., Channelière, S., Roux, N. and Doležel, J. (2011). A platform for efficient genotyping in *Musa* using microsatellite markers. *AoB Plants* 2011: plr024. doi:10.1093/aobpla/plr024.

Čížková, J., Hřibova, E., Humplíková, L., Christelová, P., Suchánková, P. and Doležel, J. (2013). Molecular analysis and genomic organization of major DNA satellites in banana (*Musa* spp.). *PLoS ONE* 8(1): e54808. doi:10.1371/journal.pone.0054808.

Creste, S., Neto, A. T., Oliveira Silva, S. de and Figueira, A. (2003). Genetic characterization of banana cultivars (*Musa* spp.) From brazil using microsatellite markers. *Euphytica* 132(3): 259–268.

Cuadrado, Á. and Jouve, N. (2010). Chromosomal detection of simple sequence repeats (SSRs) using nondenaturing FISH (ND-FISH). *Chromosoma* 119(5): 495-503. doi:10.1007/s00412-010-0273-x.

Damaiyani, J. and Hapsari, L. (2017). Pollen viability of 19 Indonesian bananas (*Musa* L.) collection of Purwodadi Botanic Garden: preliminary study for breeding. *Proceeding of the International Conference on Tropical Plant Conservation and Utilization*, 1–10.

Darlington, C. D. (1929). Chromosome behaviour and structural hybridity in the Tradescantiae. *Journal of Genetics* 21(2): 207–286.

de Coster, W., De Rijk, P., De Roeck, A., De Pooter, T., D'Hert, S., Strazisar, M., Sleegers, K. and Van Broeckhoven, C. (2019). Structural variants identified by Oxford nanopore PromethION sequencing of the human genome. *Genome Research* 29(7): 1178-1187. doi:10.1101/gr.244939.118.

de Jong, H., Fransz, P. and Zabel, P. (1999). High resolution FISH in plants - techniques and applications. *Trends in Plant Science* 4(7): 258–263.

D'Hont, A., Paget-Goy, A., Escoute, J. and Carreel, F. (2000). The interspecific genome structure of cultivated banana, *Musa* spp. Revealed by genomic DNA in situ hybridization. *Theoretical and Applied Genetics* 100(2): 177–183.

D'Hont, A. (2005). Unraveling the genome structure of polyploids using FISH and GISH; examples of sugarcane and banana. *Cytogenetic and Genome Research* 109(1-3): 27–33. doi:10.1159/000082378.

D'Hont, A., Denoeud, F., Aury, J. M., Baurens, F. C., Carreel, F., Garsmeur, O., Noel, B., Bocs, S., Droc, G., Rouard, M., Da Silva, C., Jabbari, K., Cardi, C., Poulain, J., Souquet, M., Labadie, K., Jourda, C., Lengellé, J., Rodier-Goud, M., Alberti, A., Bernard, M., Correa, M., Ayyampalayam, S., Mckain, M. R., Leebens-Mack, J., Burgess, D., Freeling, M., Mbéguié-A-Mbéguié, D., Chabannes, M., Wicker, T., Panaud, O., Barbosa, J., Hribova, E., Heslop-Harrison, P., Habas, R., Rivallan, R., Francois, P., Poiron, C., Kilian, A., Burthia, D., Jenny, C., Bakry, F., Brown, S., Guignon, V., Kema, G., Dita, M., Waalwijk, C., Joseph, S., Dievart, A., Jaillon, O., Leclercq, J., Argout, X., Lyons, E., Almeida, A., Jeridi, M., Dolezel, J., Roux, N., Risterucci, A. M., Weissenbach, J., Ruiz, M., Glaszmann, J. C., Quétier, F., Yahiaoui, N. and Wincker, P. (2012). The banana (*Musa acuminata*) genome and the evolution of monocotyledonous plants. *Nature* 488(7410): 213–217.

Dodds, K. S. (1943). Genetical and cytological studies of *Musa*, V. Certain edible diploids. *Journal of Genetics* 45(2): 113–138.

Dodds, K. S. and Simmonds, N. W. (1946). Genetical and cytological studies of *Musa*, VIII. the formation of polyploid spores. *Journal of Genetics* 47: 223–241.

Doleželová, M., Valárik, M., Swennen, R., Horry, J. P. and Doležel, J. (1998). Physical mapping of the 18S-25S and 5S ribosomal RNA genes in diploid bananas. *Biologia Plantarum* 41(4): 497–505.

Dong, Z., Wang, H., Chen, H., Jiang, H., Yuan, J., Yang, Z., Wang, W. J., Xu, F., Guo, X., Cao, Y., Zhu, Z., Geng, C., Cheung, W. C., Kwok, Y. K., Yang, H., Leung, T. Y., Morton, C. C., Cheung, S. W. and Choy, K. W. (2018). Identification of balanced chromosomal rearrangements previously unknown Among participants in the 1000 genomes project: implications for interpretation of structural variation in genomes and the future of clinical cytogenetics. *Genetics in Medicine: Official Journal of the American College of Medical Genetics* 20(7): 697–707.

Fauré, S., Noyer, J. L., Horry, J. P., Bakry, F., Lanaud, C. and Gońzalez de León, D. (1993). A molecular marker-based linkage map of diploid bananas (*Musa acuminata*). *TAG. Theoretical and Applied Genetics. Theoretische und Angewandte Genetik* 87(4): 517–526. doi:10.1007/BF00215098.

Fransz, P., Linc, G., Lee, C. R., Alves Aflitos, S. A., Lasky, J. R., Toomajian, C., Ali, H., Peters, J., van Dam, P., Ji, X., Kuzak, M., Gerats, T., Schubert, I., Schneeberger, K., Colot, V., Martienssen, R., Koornneef, M., Nordborg, M., Juenger, T. E., de Jong, H. and Schranz, M. E. (2016). Molecular, genetic and evolutionary analysis of a paracentric inversion in *Arabidopsis thaliana*. *The Plant Journal: For Cell and Molecular Biology* 88(2): 159–178. doi:10.1111/tpj.13262.

Godwin, I. D., Aitken, E. A. and Smith, L. W. (1997). Application of inter simple sequence repeat (ISSR) markers to plant genetics. *Electrophoresis* 18(9): 1524–1528. doi:10.1002/elps.1150180906.

Han, Y., Zhang, T., Thammapichai, P., Weng, Y. and Jiang, J. (2015). Chromosome-specific painting in Cucumis species using bulked oligonucleotides. *Genetics* 200(3): 771–779. doi:10.1534/genetics.115.177642.

Hippolyte, I., Bakry, F., Seguin, M., Gardes, L., Rivallan, R., Risterucci, A. M., Jenny, C., Perrier, X., Carreel, F., Argout, X., Piffanelli, P., Khan, I. A., Miller, R. N., Pappas, G. J., Mbéguié-A-Mbéguié, D., Matsumoto, T., De Bernardinis, V., Huttner, E., Kilian, A., Baurens, F. C., D'Hont, A., Cote, F., Courtois, B. and Glaszmann, J. C. (2010). A saturated SSR/DArT linkage map of *Musa acuminata* addressing genome rearrangements Among bananas. *BMC Plant Biology* 10(1): 65. doi:10.1186/1471-2229-10-65.

Hou, L., Xu, M., Zhang, T., Xu, Z., Wang, W., Zhang, J., Yu, M., Ji, W., Zhu, C., Gong, Z., Gu, M., Jiang, J. and Yu, H. (2018). Chromosome painting and its applications in cultivated and wild rice. *BMC Plant Biology* 18(1): 110. doi:10.1186/s12870-018-1325-2.

Hřibová, E., Doleželova, M., Town, C. D., Macas, J. and Doležel, J. (2007). Isolation and characterization of the highly repeated fraction of the banana genome. *Cytogenetic and Genome Research* 119(3–4): 268–274. doi:10.1159/000112073.

Hřibová, E., Neumann, P., Matsumoto, T., Roux, N., Macas, J. and Doležel, J. (2010). Repetitive part of the banana (*Muhusa acuminata*) genome investigated by low-depth 454 sequencing. *BMC Plant Biology* 10: 204. doi:10.1186/1471-2229-10-204.

Hu, L., Liang, F., Cheng, D., Zhang, Z., Yu, G., Zha, J. and Wang, Y. (2018). Localization of balanced chromosome translocation breakpoints by long-read sequencing on the Oxford nanopore platform. *Biorxiv.org*. doi:10.1101/419531.

Jarret, R. L., Bhat, K. V., Cregan, P., Ortiz, R. and Vuylsteke, D. (1994). Isolation of microsatellite DNA markers in *Musa*. *InfoMusa* 3: 3–4.

Jeridi, M., Bakry, F., Escoute, J., Fondi, E., Carreel, F., Ferchichi, A., D'Hont, A. and Rodier-Goud, M. (2011). Homoeologous chromosome pairing Between the a and B genomes of *Musa* spp. Revealed by genomic in situ hybridization. *Annals of Botany* 108(5): 975–981. doi:10.1093/aob/mcr207.

Jeridi, M., Perrier, X., Rodier-Goud, M., Ferchichi, A., D'Hont, A. and Bakry, F. (2012). Cytogenetic evidence of mixed disomic and polysomic inheritance in an allotetraploid (AABB) *Musa* Genotype. *Annals of Botany* 110(8): 1593–1606. doi:10.1093/aob/mcs220.

Jiang, J., Zhang, L., Zhou, X., Chen, X., Huang, G., Li, F., Wang, R., Wu, N., Yan, Y., Tong, C., Srivastava, S., Wang, Y., Liu, H. and Ying, Q. L. (2016). Induction of site-specific chromosomal translocations in embryonic stem cells by CRISPR/Cas9. *Scientific Reports* 6: 21918. doi:10.1038/srep21918.

Kaemmer, D., Fischer, D., Jarret, R. L., Baurens, F.-C., Grapin, A., Dambier, D., Noyer, J.-L., Lanaud, C., Kahl, G. and Lagoda, P. J. L. (1997). Molecular breeding in the genus *Musa*: a strong case for STMS marker technology. *Euphytica* 96(1): 49- 63.

Kantama, L., Wijnker, E. and de Jong, H. (2017). Optimization of cell spreading and image quality for the study of chromosomes in plant tissues. In: Schmidt, A. (Ed.), *Plant Germline Development: Methods and Protocols, Methods in Molecular Biology* (Vol. 1669). Springer, 141–158. doi:10.1007/978-1-4939-7286-9_12.

Kilian, A., Wenzl, P., Huttner, E., Carling, J., Xia, L., Blois, H., Caig, V., Heller-Uszynska, K., Jaccoud, D., Hopper, C., Aschenbrenner-Kilian, M., Evers, M., Peng, K., Cayla, C., Hok, P. and Uszynski, G. (2012). Diversity arrays technology: a generic genome profiling technology on open platforms. *Methods in Molecular Biology* 888: 67–89.

Lacadena, J. R., Cermeno, M. C., Orellana, J. and Santos, J. L. (1984). Evidence for wheat-rye nucleolar competition (Amphiplasty) in Triticale by silver-staining procedure. *TAG. Theoretical and Applied Genetics. Theoretische und Angewandte Genetik* 67(2–3): 207–213.

Lekomtsev, S., Aligianni, S., Lapao, A. and Bürckstümmer, T. (2016). Efficient generation and reversion of chromosomal translocations using CRISPR/ Cas technology. *BMC Genomics* 17(1): 739. doi:10.1186/s12864-016-3084-5.

Lescot, M., Piffanelli, P., Ciampi, A. Y., Ruiz, M., Blanc, G., Leebens-Mack, J., da Silva, F. R., Santos, C. M., D'Hont, A., Garsmeur, O., Vilarinhos, A. D., Kanamori, H., Matsumoto, T., Ronning, C. M., Cheung, F., Haas, B. J., Althoff, R., Arbogast, T., Hine, E., Pappas, G. J., Sasaki, T., Souza, M. T., Miller, R. N., Glaszmann, J. C. and Town, C. D. (2008). Insights Into the *Musa* Genome: syntenic relationships to rice and Between *Musa* species. *BMC Genomics* 9(1): 58. doi:10.1186/1471-2164-9-58.

Li, K., Wang, H., Wang, J., Sun, J., Li, Z. and Han, Y. (2016). Divergence Between *C. melo* and African *Cucumis* species identified by chromosome painting and rDNA distribution pattern. *Cytogenetic and Genome Research* 150(2): 150–155. doi:10.1159/000453520.

Lin, K., Smit, S., Bonnema, G., Sanchez-Perez, G. and de Ridder, D. (2015). Making the difference: integrating structural variation detection tools. *Briefings in Bioinformatics* 16(5): 852–864. doi:10.1093/bib/bbu047.

Lönnig, W. E. and Saedler, H. (2003). Chromosome rearrangements and transposable elements. *Annual Review of Genetics* 36(4139): 389–410. doi:10.1146/annurev.genet.36.040202.092802.

Lynagh, P. G., Inagaki, S., Amundson, K. R., Marimithu, M. P. A., Pike, B. R., Henry, I. M., Tan, E. H. and Comai, L. (2018). Translocation and duplication From CRISPR-Cas9 editing in *Arabidopsis thaliana*. *bioRxiv*. doi:10.1101/400507.

Lysak, M. A., Doleželová, M., Horry, J. P., Swennen, R. and Dolez˘el, J. (1999). Flow cytometric analysis of nuclear DNA content in *Musa*. *Theoretical and Applied Genetics* 98(8): 1344–1350.

Lysak, M. and Schubert, I. (2013). Mechanisms of chromosome rearrangements. In: Leitch, I. J., Greilhuber, J., Doležel, J. and Wendel, J. F. (Eds), *Plant Genome Diversity* (Vol. 2), *Physical Structure, Behaviour and Evolution of Plant Genomes*. Springer-Verlag, Wien, 137–147. doi:10.1007/978-3-7091-1160-4_9.

Martin, G., Baurens, F. C., Droc, G., Rouard, M., Cenci, A., Kilian, A., Hastie, A., Doležel, J., Aury, J. M., Alberti, A., Carreel, F. and D'Hont, A. (2016). Improvement of the banana '*Musa acuminata*' reference sequence using NGS data and semi-automated bioinformatics methods. *BMC Genomics* 17(243): 243. doi:10.1186/s12864-016-2579-4.

Martin, G., Carreel, F., Coriton, O., Hervouet, C., Cardi, C., Derouault, P., Roques, D., Salmon, F., Rouard, M., Sardos, J., Labadie, K., Baurens, F. C. and D'Hont, A. (2017). Evolution of the banana genome (*Musa acuminata*) is impacted by large chromosomal translocations. *Molecular Biology and Evolution* 34(9): 2140–2152. doi:10.1093/molbev/msx164.

Mbanjo, E. G. N., Tchoumbougnang, F., Mouelle, A. S., Oben, J. E., Nyine, M., Dochez, C., Ferguson, M. E. and Lorenzen, J. (2012). Molecular marker-based genetic linkage map of a diploid banana population (*Musa acuminata* Colla). *Euphytica* 188(3): 369–386. doi:10.1007/s10681-012-0693-1.

Msogoya, T., Grout, B. and Roberts, R. (2008). Karyotypic and 2C nuclear DNA in vitro induced off-types of East African Highland banana (*Musa* AAA East Africa). *Biotechnology* 7(3): 578–581.

Nair, A. S., Teo, C. H., Schwarzacher, T. and Heslop-Harrison, P. H. (2005). Genome classification of banana cultivars from South India using IRAP markers. *Euphytica* 144(3): 285–290. doi:10.1007/s10681-005-7321-2.

Novák, P., Hřibova, E., Neumann, P., Koblížková, A., Doležel, J. and Macas, J. (2014). Genome-wide analysis of repeat diversity across the family Musaceae. *PLoS ONE* 9(6): e98918. doi:10.1371/journal.pone.0098918.

Ortiz, R. (1995). *Musa* genetics. In: Gowen, S. (Ed.), *Bananas and Plantains*. Chapman & Hall, London, 84–109.

Oselebe, H. O., Tenkouano, A. and Pillay, M. (2009). Ploidy variation of *Musa* hybrids From crosses. *African Journal of Biotechnology* 5: 1048–1053.

Osuji, J. O., Harrison, G., Crouch, J. and Heslop-Harrison, J. S. (1997). Identification of the genomic constitution of *Musa* L. Lines (bananas, plantains and hybrids) using molecular cytogenetics. *Annals of Botany* 80(6): 787–793.

Osuji, J. O., Okoli, B. E. and Edeoga, H. O. (2006). Karyotypes of the a and B genomes of *Musa* L. *Cytologia* 71(1): 21–24. doi:10.1508/cytologia.71.21.

Padmesh, P., Mukunthakumar, S., Vineesh, P. S., Skaria, R., Kumar, K. H. and Krishnan, P. N. (2012). Exploring wild genetic resources of *Musa acuminata* Colla distributed in the humid forests of Southern Western ghats of peninsular India using ISSR markers. *Plant Cell Reports* 31(9): 1591–1601. doi:10.1007/s00299-012-1273-5.

Peterson, D. G., Schulze, S. R., Sciara, E. B., Lee, S. A., Bowers, J. E., Nagel, A., Jiang, N., Tibbitts, D. C., Wessler, S. R. and Paterson, A. H. (2002). Integration of cot analysis,

DNA cloning, and high-throughput sequencing facilitates genome characterization and gene discovery. *Genome Research* 12(5): 795–807. doi:10.1101/gr.226102.
Pillay, M. and Adeleke, M. T. V. (2001). Silver staining of *Musa* L. chromosomes. *Cytologia* 66(1): 33–37.
Pillay, M. and Tripathi, L. (2006). Banana: an overview or breeding and genomics research in *Musa*. In: Kole, C. (Ed.), *Genome Mapping and Molecular Breeding in Plants* (vol. 4), *Fruits and Nuts*. Springer-Verlag, Heidelberg, Germany, 282–301. doi:10.1007/978-3-540-34533-6_15.
Preedy, K. F. and Hackett, C. A. (2016). A rapid marker ordering approach for high-density genetic linkage maps in experimental autotetraploid populations using multidimensional scaling. *TAG. Theoretical and Applied Genetics. Theoretische und Angewandte Genetik* 129(11): 2117–2132. doi:10.1007/s00122-016-2761-8.
R Core Team (2018). R: A language and environment for statistical computing. *R Foundation for Statistical Computing*, Vienna, Austria.
Raboin, L., Carreel, F., Noyer, J., Baurens, F., Horry, J., Bakry, F., Montcel, H. T. D., Ganry, J., Lanaud, C. and Lagoda, P. J. L. (2005). Diploid Ancestors of Triploid Export Banana Cultivars: molecular Identification of 2n Restitution Gamete Donors and N Gamete Donors. *Molecular Breeding* 16(4): 333–341. doi:10.1007/s11032-005-2452-7.
Roux, N., Toloza, A., Radecki, Z., Zapata-Arias, F. J. and Doležel, J. (2003). Rapid detection of aneuploidy in *Musa* using flow cytometry. *Plant Cell Reports* 21(5): 483–490. doi:10.1007/s00299-002-0512-6.
Šafář, J., Noa-Carrazana, J. C., Vrána, J., Bartoš, J., Alkhimova, O., Sabau, X., Simková, H., Lheureux, F., Caruana, M. L., Dolezel, J. and Piffanelli, P. (2004). Creation of a BAC resource to study the structure and evolution of the banana (*Musa balbisiana*) genome. *Genome* 47(6): 1182–1191. doi:10.1139/g04-062.
Sardos, J., Perrier, X., Doležel, J., Hřibova, E., Christelová, P., Van den Houwe, I., Kilian, A. and Roux, N. (2016). DArT whole genome profiling provides insights on the evolution and taxonomy of edible banana (*Musa* spp.). *Annals of Botany* 118(7): 1269–1278. doi:10.1093/aob/mcw170.
Sass, J. E. (1964). *Botanical Microtechnique*. Iowa State University Press, Ames, IA.
Shepherd, K. (1999). *Cytogenetics of the Genus Musa*. INIBAP, Montpellier, France.
Silva, A. V. C., Nascimento, A. L. S., Vitória, M. F., Rabbani, A. R. C., Soares, A. N. R. and Lédo, A. S. (2017). Diversity and genetic stability in banana genotypes in a breeding program using inter simple sequence repeats (ISSR) markers. *Genetics and Molecular Research: GMR* 16(1). doi:10.4238/gmr16019402.
Simmonds, N. W. and Dodds, K. S. (1949). Meiosis in seeded diploids of Musa. *Journal of Genetics* 49(3): 221–225.
Simmonds, N. W. and Shepherd, K. (1955). The taxonomy and origins of the cultivated bananas. *Journal of the Linnean Society of London, Botany* 55(359): 302–312.
Sybenga, J. (1975). *Meiotic Configurations*. Springer-Verlag. ISBN 978-3-642-80960-6.
Sybenga, J. (1992). *Cytogenetics in Plant Breeding*. Springer Verlag. ISBN 978-3-642-84083-8.
Valárik, M., Simkova, H., Hřibova, E., Šafář, J., Doleželová, M. and Doležel, J. (2002). Isolation, characterization and chromosome localization of repetitive DNA sequences in bananas (*Musa* spp.). *Chromosome Research: An International Journal on the Molecular, Supramolecular and Evolutionary Aspects of Chromosome Biology* 10(2): 89–100.
Van Os, H., Stam, P., Visser, R. G. and van Eck, H. J. (2005). SMOOTH: a statistical method for successful removal of genotyping errors from high-density genetic linkage data.

TAG. Theoretical and Applied Genetics. Theoretische und Angewandte Genetik 112(1): 187–194.

Vilarinhos, A. D., Piffanelli, P., Lagoda, P., Thibivilliers, S., Sabau, X., Carreel, F. and D'Hont, A. (2003). Construction and characterization of a bacterial artificial chromosome Library of Banana (*Musa acuminata* Colla). *TAG. Theoretical and Applied Genetics. Theoretische und Angewandte Genetik* 106(6): 1102–1106. doi:10.1007/s00122-002-1155-2.

Wang, Z., Miao, H., Liu, J., Xu, B., Yao, X., Xu, C., Zhao, S., Fang, X., Jia, C., Wang, J., Zhang, J., Li, J., Xu, Y., Wang, J., Ma, W., Wu, Z., Yu, L., Yang, Y., Liu, C., Guo, Y., Sun, S., Baurens, F. C., Martin, G., Salmon, F., Garsmeur, O., Yahiaoui, N., Hervouet, C., Rouard, M., Laboureau, N., Habas, R., Ricci, S., Peng, M., Guo, A., Xie, J., Li, Y., Ding, Z., Yan, Y., Tie, W., D'Hont, A., Hu, W. and Jin, Z. (2019). *Musa balbisiana* genome reveals subgenome evolution and functional divergence. *Nature Plants* 5(8): 810–821. doi:10.1038/s41477-019-0452-6.

Wilson, G. B. (1946). Cytological studies in the musae. II. Meiosis in some diploid clones. *Genetics* 31(5): 475–482.

Zhang, J., Yu, C., Krishnaswamy, L. and Peterson, T. (2011). Transposable elements as catalysts for chromosome rearrangements. In: Birchler, J. A. (Ed.), *Plant Chromosome Engineering: Molecular and Protocols, Methods in Molecular Biology* (Vol. 701). Humana Press, Totowa, NJ, 315–326. doi:10.1007/978-1-61737-957-4_18.

Chapter 3

Identifying and classifying banana cultivars

Jeff Daniells, Queensland Department of Agriculture and Fisheries, Australia; and Steven B. Janssens, Botanic Garden Meise, Belgium

1 Introduction

> *The first step in wisdom is to know the things themselves; this notion consists in having a true idea of the objects; objects are distinguished and known by classifying them methodically and giving them appropriate names. Therefore, classification and name-giving will be the foundation of our science.*
>
> –Carolus Linnaeus (Linnaeus, 1735).

Taxonomy, otherwise known as systematics, has been described as 'the science of naming, describing and classifying organisms ... using morphological, behavioural, genetic and biochemical observations' (Convention on Biological Diversity, 2019). Taxonomy is the foundation for discussion about the identity and diversity of organisms in the biological world. It provides a means of efficiently retrieving, recording and communicating information about organisms. If a researcher does not have a reliable name for a study object, it is impossible to analyse it accurately and share the results with others. If the taxonomy of a group is poor, this can seriously undermine effective research. If studies are conducted on different taxa (i.e. taxonomic group of any rank such

http://dx.doi.org/10.19103/AS.2020.0070.02

as a species, family or class), but are reported by the same name, this causes confusion and misunderstanding within and between studies.

The practice of classifying and naming organisms enables scientists both to organize large amounts of information about organisms and make deductions about their behaviour based on knowledge of similar organisms. A species' name and position within a larger classification system provides key information about its heritage and helps to explain its characteristics and behaviour. In the case of crops, taxonomy can facilitate the identification, breeding and introduction of new cultivars as well as enforce breeders' intellectual property rights over a new cultivar.

2 Banana cultivars: an overview

Bananas belong to the Musaceae family of flowering plants and the term banana can be used when referring to any members of the family. However, some terms are commonly used to describe additional crop subsets of the family – namely enset, abaca and plantains. Enset and abaca are dealt with a little later. The fruits of plantains are starchy at ripeness and are mostly cooked before consumption. In the narrow sense the term refers to a defined taxonomic grouping (AAB, Plantain subgroup). In the broad sense it contains the latter subgroup as well as several other bananas that are also starchy at ripeness and used for cooking purposes. Where plantain is used in this chapter, it is according to the narrow-sense definition.

Bananas are cultivated around the globe in tropical and subtropical locations and also occur naturally as wild populations of seeded species mainly in Southeast Asia, South Asia and Melanesia. In this chapter our attention is mostly on cultivated bananas. Although the number is somehow variable depending on the source, it is estimated that to date over 1000 banana cultivars occur worldwide, characterized by their own specific morphology (Fig. 1) (Heslop-Harrison and Schwarzacher, 2007). The International Code of Nomenclature for Cultivated Plants (ICNCP) states that cultivars can be characterized as 'assemblages of plants that (a) have been selected for a particular character or combination of characters, (b) are distinct, uniform and stable in those characters, and (c) when propagated by appropriate means, retain those characters' (ICNCP; Brickell et al., 2016). According to the ICNCP, a cultivar name can only be given to cultivated plants whose origin or selection is primarily the result of human interference. People often refer to cultivars as varieties, though this is taxonomically incorrect. Following the ICNCP, a variety is defined as 'a lower rank subdivision of a (wild) species, characterized by a restricted habitat'. However, the usage of the term 'variety' for a cultivar is widespread and unlikely to change. This chapter uses these terms according to their technical ICNCP definitions.

Figure 1 The diversity of fruit characteristics of banana cultivars. (Photo © A. Devouard).

We can easily think of banana cultivars as being just those with edible fruits either cooked or used as a dessert but they can have a broader context. As an example, there are cultivars of enset or Abyssinian bananas (*Ensete ventricosum*) used both as a food and fibre crop, and there are many abaca cultivars (*Musa textilis* and *Musa textilis/balbisiana* hybrids) producing fibre of commercial significance in countries such as the Philippines (Figs. 2 and 3). Some selections of wild species propagated as ornamental bananas (from sections Rhodochlamys and Callimusa in particular) might also be considered as cultivars. EMBRAPA's banana breeding programme in Brazil has an ornamental-breeding component (Fig. 4).

3 Classification at the genera level

Cultivars have been classified (i.e. arranged in categories) based on primary shared characteristics. These categories reflect known or assumed historical relationships. Most of the cultivars (except for Fe'i bananas) are derived from one or two species of the section *Eumusa* (genus *Musa*) of the banana family (Musaceae); *Musa balbisiana* and *Musa acuminata s.l.* (Fig. 5). In understanding taxonomic classification, it is worth remembering Linnaeus' view that 'It is the genus that gives the characters, and not the characters that make the genus' (Linnaeus, 1735). This chapter uses a top-down approach to explore the complexities of banana taxonomy.

Figure 2 Enset, also known as Abysinnian banana (*Ensete ventricosum*), is a staple food for nearly 20% of the Ethiopian population - seen here growing alongside bananas (*Musa* spp.) (E indicates enset, B indicates banana. (Photo courtesy: Guy Blomme, Bioversity International).

All wild and cultivated bananas are herbaceous monocotyledons that belong to the family of Musaceae, a member of the order Zingiberales. Within the Zingiberales, seven additional families are situated: Cannaceae, Heliconiaceae, Costaceae, Marantaceae, Zingiberaceae, Lowiaceae and Strelitziaceae (Kress, 1990). Important features that help to distinguish Musaceae species from the other families are that the stamens of their flowers taken collectively do not

Figure 3 Many cultivars of abaca are of commercial significance in the Philippines. Small bunch of *Musa textilis* showing characteristic male bud with predominant bract imbrication.

have the form or appearance of a petal. In addition, their leaves and bracts are spirally arranged with leaves emerging from the upper part of the pseudostem (Simmonds, 1962).

Musaceae are assumed to be the earliest diversified family within the order Zingiberales (Sass et al., 2016; Janssens et al., 2016; Kress, 1990). They are characterized by many morphological similarities with some of the other early diversified families such as Heliconiaceae and Strelitziaceae. Within the order Zingiberales, a trend can be observed from a large-sized habit and simplified inflorescence (e.g. Heliconiaceae, Musaceae and Strelitziaceae) to a more complex floral organization and a reduced habit for the more recently diversified families in Zingiberales (e.g. Zingiberaceae, Marantaceae, Costaceae and Lowiaceae) (Kirchoff et al., 2009). These differences also help to eliminate confusion with plants from other families in the Zingerberales, notably some *Heliconia* species.

Figure 4 Male buds of hybrids from EMBRAPA's ornamental banana breeding programme demonstrate some of the attractive diversity available of potential cultivars. (Photos courtesy Janay Santos-Serejo).

The Musaceae family contains in total three genera: *Musa*, *Ensete* and *Musella*. However, the taxonomic status of the latter remains to date a matter of dispute, yet is probably congeneric with *Ensete* (Li et al., 2010). From a morphological point of view, several characteristics are available to delineate the different *Musa* genera (Simmonds, 1962). *Musa* are perennial herbs rather

Figure 5 Fe'i bananas are very distinctive - usually with an upright bunch (Karat Kole pictured in Pohnpei, FSM) and pink/purple sap (Toraka Bonubonu pictured in Makira, Solomon Islands).

than monocarpic; their leaf sheaths are tightly clasping instead of rather lax or loosely clasping. Their bracts and flowers are inserted independently on the axis and mostly deciduous by abscission, rather than bracts and flowers being integral with each other and the axis as well as being persistent (deciduous only by rotting, not by abscission).

4 Classification at the section level

Section and/or series (a taxonomic rank lower than section) are typically used to help in the systematic organization of complex genera. Botanists wanting to distinguish groups of species sometimes prefer to create a taxon (a taxonomic group) at the rank of section or series to avoid making new scientific-name combinations which may lead to unnecessary confusion, particularly in cases where taxonomic uncertainties prevail, as is often the case within the genus *Musa*. As one of the leading taxonomists in this field, Simmonds seemed to use the terms sections and series interchangeably over time (e.g. Simmonds, 1954; Stover and Simmonds, 1987).

Within the genus *Musa*, historically four sections have been recognized based on overall morphology, biogeography and cytological data:

- Australimusa;
- Callimusa;
- Eumusa; and
- Rhodochlamys.

More recently, molecular data (Wong et al., 2002; Nwakanma et al., 2003) was used to suggest a reduction from four to two sections in the genus *Musa*. Häkkinen (2013) proposed the two following sections:

- *Musa* (combining Eumusa and Rhodochlamys); and
- Callimusa (combining former Callimusa, Australimusa and Ingentimusa).

However, more recent molecular phylogenetic research suggests the proposed new arrangement has oversimplified the taxonomic reality. Molecular phylogenetic results (Janssens et al., 2016; Li et al., 2010) have demonstrated that representatives of section Rhodochlamys are scattered among the Eumusa lineages and, as such, should be regarded as polyphyletic. Therefore, from a systematic point of view, it is better to merge the Rhodochlamys representatives in one single natural grouping together with Eumusa species in order to create a more natural classification. Moreover, the polyphyletic nature of Rhodochlamys underlines the huge complexity that is present within the genus as specific morphological characters that were used to discriminate between the two former sections are now proven to contain a high degree of homoplasy.

The same molecular phylogenetic studies also showed that the situation of former sections Callimusa and Australimusa is completely different. Results indicate that former section Australimusa can be regarded as a monophyletic clade, positioned among the representatives of former section Callimusa (incl. *M. ingens* as member of Ingentimusa), making the latter a paraphyletic grouping. Following these results, we can assume that the most recent common ancestor of the former section Australimusa is characterized by a unique set of morphological characters, chromosome number and biogeography and that, likewise, a significant degree of homology has been assessed for this group. Moreover, the same is true for members of the former Callimusa section - not taking the position of *M. ingens* into account - with the main difference that Callimusa does not form a monophyletic but a paraphyletic group.

Even though the currently simplified sectional delineation of the genus Musa has decreased the degree of artificial groupings within the genus, it also diminished somehow the discriminative power that was initially present in the original system - especially within the current Callimusa section (Jones and Daniells, 2019). In addition, from a taxonomical perspective, using Musa as a new section name unfortunately promotes unnecessary confusion as now both section and genus share the same name. It is clear that a more detailed study needs to be undertaken before the four-section approach is completely abandoned (Ploetz et al., 2007).

5 Classification at the species level

The polyphyletic nature of groups and lineages within the genus *Musa* does not stop at the sectional level. It is also present at species level. It is generally known that the large diversity in *Musa acuminata* probably led to the immense morphological diversity that is present within cultivated bananas. A number of studies have shown that several of the *M. acuminata* subspecies were involved in the domestication process of bananas (Carreel et al., 2002; Boonruangrod et al., 2009; Perrier et al., 2009).

Based on differences mainly in floral characteristics, multiple subspecies have been recognized in *M. acuminata* of which the following are considered to be taxonomically well-supported: *M. acuminata* ssp. *banksii*, *M. acuminata* ssp. *errans*, *M. acuminata* ssp. *malaccensis*, *M. acuminata* ssp. *microcarpa*, *M. acuminata* ssp. *truncata*, *M. acuminata* ssp. *zebrina*, *M. acuminata* ssp. *burmannica* and *M. acuminata* ssp. *siamea* (Simmonds, 1962; Boonruangrod et al., 2008). To date this classification is still a matter of debate as some studies have pointed out that the latter two should probably be merged into one subspecies *burmannica* (Perrier et al., 2009). Moreover, also *M. acuminata* ssp. *microcarpa* represents a special case within the *M. acuminata* group as this subspecies is characterized by a large degree of heterozygosity compared to the other *M. acuminata* subspecies. In fact, its status as subspecies is still unclear as it is often suggested that *M. acuminata* ssp. *microcarpa* populations are possibly the result of hybridization between different gene pools (e.g. subspecies *banksii* and subspecies *zebrina*) and not that the high degree of heterozygosity is caused by the presence of unfixed alleles (Perrier et al., 2009; Martin et al., 2020). Furthermore, based on the results of Janssens et al. (2016) it is clear that subspecies delineation of *Musa acuminata* is rather complicated as there is a clear indication of complex evolutionary patterns and relationships within the species.

The most striking taxonomic discrepancy found within *Musa acuminata* lies in the position of former Rhodochlamys species *M. rosea*, *M. laterita* and *M. rubra*, which fall as a monophyletic group close to *M. burmannica* s.l. (including former subspecies *burmannicoides* and *siamea*). The same situation occurs for the former Rhodochlamys species *M. siamensis* and *M. ornata* that appear to be closely related to *M. acuminata* ssp. *malaccensis*. This indicates that *M. acuminata* s.l. is also a polyphyletic group, which should eventually be taxonomically reassessed after a thorough morphological revision.

Despite its taxonomic complexity, *M. acuminata* s.l. should still be considered as the most important taxonomic unit that has led to the vast number of cultivated bananas currently known to exist. *Musa acuminata* provided the alleged A genome in banana cultivars (discussed below). The species can be discriminated from the other important *Musa* species for banana cultivation (*M. balbisiana*) by having features which include petiole canals with erect or spreading margins, a hairy peduncle and revolute bracts on the male bud (compared to enclosed petiole margins, glabrous peduncle and bracts on the male bud which lift but don't roll). Of the currently recognized subspecies within the *Musa acuminata*, genetic traces of the following four have been found in commercially grown edible banana: *M. acuminata* ssp. *malaccensis, M. acuminata* ssp. *banksii, M. acuminata* ssp. *zebrina* and *M. acuminata* ssp. *burmannica*.

Although the *M. acuminata* subspecies are morphologically rather similar, some clear morphological traits are present to distinguish between them. Especially the position and orientation of the bunch and male rachis as well as bract behaviour can be used to discriminate between the different subspecies, yet even then one has to be well-trained not to make mistakes in the identification of the different subspecies.

6 Current banana cultivar groups

Unaware of the huge diversity of cultivated bananas, Linnaeus applied his binomial system – useful for wild species – to denominate the edible bananas known at that time:

- *M. paradisiaca* L. (syn. *M. cliffortiana;* pre-Linnaean name) for plantains; and
- *M. sapientum* L. for dessert and/or cooking bananas.

However, over the years many more cultivars were discovered and it quickly became clear that the use of the original binomial system for the denomination of *Musa* cultivars had become impracticable as different cultivars were assigned under the same name. The discovery of the hybrid origin of cultivated bananas also made clear that the binomial denomination of cultivated bananas was an oversimplification of the complexity present within and among *Musa* cultivars.

The current system of assigning banana cultivars into cultivar groups is based on a system worked out by Simmonds and Shepherd (1955) who understood that the original Latin denomination had become unworkable to further document newly discovered cultivars. Their system was based on two elements

1 number of chromosomes (diploid, triploid and tetraploid); and
2 the origin of the chromosomes (*M. balbisiana* and/or *M. acuminata* species), where A stands for chromosomes derived from an *M. acuminata* ancestor and B for an *M. balbisiana* ancestor.

However, the distinction into A or B genomes and its associated morphological characterization is now considered as an oversimplification of banana cultivar taxonomy. A full segregation between the different genomes actually does not exist (Baurens et al., 2019), and so it is not fully correct to talk about the different cultivar groups based on chromosome ancestry (e.g. AAB, ABB), which is explained further below. Interestingly in an article by Ken Shepherd (Shepherd, 1990) he says (referring to his paper with Simmonds in 1955) 'The intention then ... was only to elucidate the genomic structure of the cultivars and not to create an international taxonomic tool as the system ... has tended to become.' Nevertheless, despite novel insights in the origin and evolution of banana cultivars in recent years, the system that Simmonds and Shepherd created more than 65 years ago remains in use today to discriminate between *Musa* cultivars (e.g. Atom et al., 2015) and therefore we will briefly explain their system. When developing their banana cultivar assignment method, Simmonds and Shepherd (1955) selected 15 morphological characters that could be visually assessed and from where a cultivar could be scored into different genome combinations (see Table 1 and Fig. 6).

Their scoring system combined with determination of ploidy gave rise to the different groups of cultivated bananas that we know today:

Table 1 Characters used in taxonomic scoring of banana cultivars

Character	*M. acuminata*	*M. balbisiana*
Pseudostem color	More or less heavily marked with brown or black blotches	Blotches slight or absent
Petiole canal	Margin erect or spreading, with scarious wings below, not clasping pseudostem	Margin enclosed, not winged below, clasping pseudostem
Peduncle	Usually downy or hairy	Glabrous
Pedicels	Short	Long
Ovules	Two regular rows in each loculus (Fig. 6)	Four irregular rows in each loculus (Fig. 6)
Bract shoulder	Usually high (ratio < 0.28 (Fig. 6))	Usually low (ratio > 0.30 (Fig. 6))
Bract curling*	Bracts reflex and roll back after opening (Fig. 6)	Bracts lift but do not roll (Fig. 6)
Bract shape	Lanceolate or narrowly ovate, tapering sharply from the shoulder (Fig. 6)	Broadly ovate, not tapering sharply (Fig. 6)
Bract apex	Acute (Fig. 6)	Obtuse (Fig. 6)
Bract color	Red, dull purple or yellow outside; pink, dull purple or yellow inside	Distinctive brownish- purple outside; bright crimson inside
Color fading	Inside bract color fades to yellow towards the base	Inside bract color continuous to base
Bract scars	Prominent (Fig. 6)	Scarcely prominent (Fig. 6)
Free tepal of male flower	Variably corrugated below tip (Fig. 6)	Rarely corrugated (Fig. 6)
Male flower color	Creamy white	Variably flushed with pink
Stigma color	Orange or rich yellow	Cream, pale yellow or pale pink

* In varieties with persistent male bracts, curling is weak or absent, regardless of genotype.
Source: From Simmonds and Shepherd (1955).

- AA (diploid cultivars, containing genetic material only from *M. acuminata*);
- AB (diploid cultivars containing genetic material of both *M. acuminata* and *M. balbisiana*);
- AAA (triploid cultivars containing genetic material only from *M. acuminata*);
- AAB (triploid cultivars, containing genetic material of both *M. acuminata* and *M. balbisiana*, including the Plantain subgroup);
- ABB (triploid cultivars, containing genetic material of both *M. acuminata* and *M. balbisiana*);
- AAAA (tetraploid cultivars containing genetic material only from *M. acuminata*); and
- AAAB, AABB and ABBB (tetraploid cultivars containing genetic material of both *M. acuminata* and *M. balbisiana*).

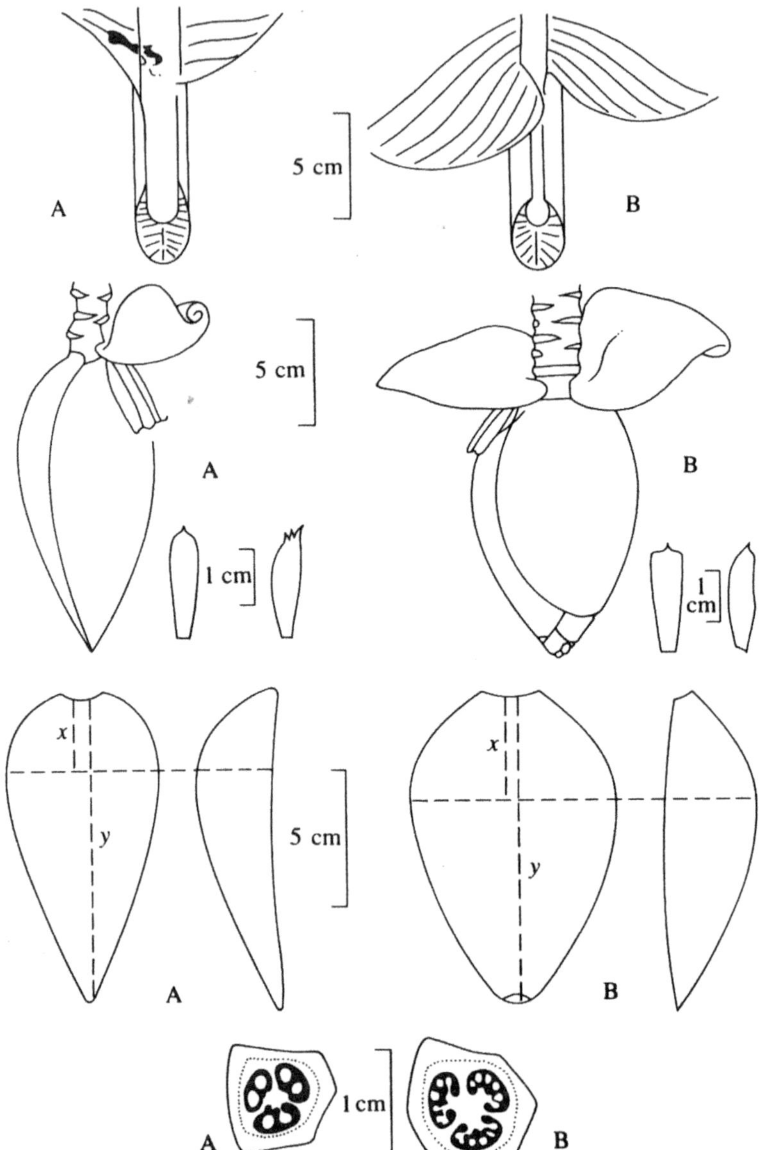

Figure 6 Characteristics of (A) *Musa acuminata* and (B) *Musa balbisiana*: Row 1: petioles; Row 2: male buds (showing shape, bract curling and insertion) and free tepals of male flowers; Row 3: bracts of male buds (ratio = x/y); Row 4: ovule arrangement. (Source: Redrawn from Simmonds and Shepherd, 1955).

If ploidy has been determined by chromosome counts or flow cytometry and the specimen is triploid (either AAA, AAB or ABB) then one can focus on just a few of the descriptors - petiole canal, pseudostem blotching (macules) and internal bract colour to discriminate between the three triploid groups (Shepherd, 1990; Jean-Pierre Horry and Christophe Jenny unpub. pers. comm.).

Within these distinct groups, subgroups are delineated based on morphological characters (e.g. AAA Cavendish subgroup). A subgroup has been identified as: 'A subdivision of a group for a set of cultivars that are assumed to have been generated from a common ancestor by somatic mutation. The ancestor is assumed to be a hybrid between two distinct parents' (MusaNet, 2016). It is important to note that epigenetic variation also plays an important role.

The different cultivars within those subgroups are then given a specific cultivar name (e.g. Williams, Grande Naine, Dwarf Cavendish), which refers to the different clones under which they are traded and grown. So, the formal reference to one of the most cultivated dessert bananas worldwide would then be: *Musa* (AAA Group, Cavendish Subgroup) 'Grande Naine'.

When a new hybrid 'genotype' is created at some point in time (either naturally or via conventional cross-breeding) this hybrid can be denoted as a cultivar when selected for particular characteristics. It can eventually lead to recognition as a subgroup once genetic variants derived from it by mutation or epigenetically are likewise selected as cultivars. Let us suppose, for example, that the first member of what has become the Cavendish subgroup was what is known as Dwarf Cavendish. This has given rise through one or more steps to numerous variants, some of which are named cultivars such as Williams, Formosana and Jaffa. It is important to note that not all the variants would be referred to as cultivars since many (with undesirable characteristics) would be mostly selected against rather than for in breeding or selection programmes.

As alluded to above the time for change has arrived in the way we classify banana cultivars. It is the recent studies on AAB and ABB cultivars using high-throughput sequencing methods which generate whole genome data that showed there is, in fact, no clear segregation between A and B chromosomes (Baurens et al., 2019). Frequent interspecific recombination between both genomes has led to a mosaic genome structure in which elements of the B genome have been introgressed through several generations in the A genome. As such, it is not fully correct to talk about the different cultivar groups based on chromosome ancestry (e.g. AAB, ABB), as a full segregation between the different genomes actually does not exist. A ProMusa-led initiative is currently underway, which in the near future, will set out a suggested way forward in such classifications that both deals with the artificially higher taxonomic ranking conferred upon the current genome groupings and better complies with the requirements of the ICNCP (Vezina, 2019). Meanwhile there is clearly a need for more genetic and genomic research to resolve the ancestry and genetic composition of our banana cultivars.

7 Identifying cultivars

The following section provides practical guidance on identifying cultivars based on the author's experience. How accurate an identification is depends on the quantity and the quality of the plant material and/or information available for the specimen in question and the context in which it is encountered. As with all identification, accuracy depends on experience. Therefore, wherever possible, one should seek advice from experts, either on-site or on-line.

Before one starts the process of identifying a cultivar, one needs to know the exact purpose of the identification required. Is it related to a biosecurity issue, an intellectual property matter or simply to ensure we are talking about the same thing when communicating with others? Other issues include 'How quickly do I need an answer?' 'What resources are available to get the answer?' A shortage of time or money will influence just how much effort can realistically be expended to achieve an answer. As an example, using the 100 or so descriptors in 'Descriptors for Banana' (IPGRI-INIBAP/CIRAD, 1996) requires a massive input of time to get a complete description of a cultivar.

The process to follow also depends on the operator's level of experience. For those with less experience, an important early step is to compare with specimens in a labelled collection. For bananas one ideally needs to grow the specimen alongside reference cultivars. This process takes into account environmental effects on phenotype (the set of observable characteristics of an individual resulting from the interaction of its genotype with the environment). The taxonomic reference collection (http://www.promusa.org/Taxonomic+Reference+Collection) came about partly to provide such reference cultivars for comparative purposes. Typically herbarium specimens are of limited value for identifying banana cultivars because the vast majority of specimens held are wild species and most herbaria do not give cultivar detail but deal with species-level descriptions. Furthermore, because of the immense size of living banana plants and the high moisture content of most plant parts, dried herbarium specimens are relatively of limited use. Often one is required to compare with pictures or descriptions. There are various resources with photo sets available. They include the Musa Germplasm Information System (MGIS https://www.crop-diversity.org/mgis/), which contains key information on Musa germplasm diversity, including passport data, botanical classification, morpho-taxonomic descriptors, molecular studies, plant photographs and GIS information on over 6 500 accessions managed in 29 collections around the world. There are also numerous catalogues available on-line (http://www.musanet.org/).

Identification keys should be consulted where they are available. However, the identification obtained from a key is by no means proof of identity. Keys have their limitations. They are only as good as the information used to create them. They are not necessarily accurate for identification and classification of cultivars not included in the construction of the key. The other methods discussed can assist in verification. Keys tend to have their best application when developed for regional purposes and where the total number of cultivars is limited.

For those with more experience, a key question is 'Does it look like something I already know?' One can compare what one sees with one's prior experiences. It is about recognizing the whole. This is something like what happens with image recognition technology which uses a range of features to make an identification (e.g. PlantNet https://identify.plantnet.org/). The latter does not cover bananas in detail but is one area worthy of future research/development for banana identification purposes.

Often the specimen is like something one already knows. One might then have a closer look at a particular characteristic (e.g. male bud shape, bunch and individual fruit characteristics, predominant underlying colour of the pseudostem and presence of particular diseases). Some characteristics are distinctive and can make it easier to distinguish between specimens. The context of the banana is important: 'What is known about the diversity in that location?' It could, for example, be 1 of only 20 cultivars known to occur in that location. Which ones of these are most common? Answering these questions can narrow down the process of identification.

Irrespective of one's level of experience, if the specimen does not appear to be like something one knows, and can't be determined by comparison with pictures, descriptions and keys, then one will need to describe the specimen. Photography is a good starting point for description - 'A picture is worth a thousand words'. If one can only take one photograph, then an in situ photo of a mature bunch with male bud attached would provide the best information for an identification (Fig. 7). If one can take photographs of different features, refer to the 'Guidelines for documenting the minimum set of photos' (https://drive.google.com/file/d/0B6WMCDtu_LjpblZZVktjSWpvdmM/view), which describes how to photograph 14 key characteristics. Figure 8 is an example of photo descriptors applied to the cultivar Gros Michel. The importance of having good photographic skills and equipment cannot be overstated.

Additional information can be gained if one goes beyond descriptive morphology and drills down to analysis at chromosome and molecular level. The ploidy of the specimen can be determined by either flow cytometry, for which a fresh sample of the cigar leaf is required (Fig. 9), or chromosome counting for which root tips are most commonly examined. The cigar leaf can also be used for the molecular characterization. The signature of a specimen at the molecular level then can be compared with an existing database of cultivars such as that of the Musa Genotyping Centre in the Czech Republic (Christelová et al., 2011). Currently microsatellite markers (SSR) can be used to identify to the subgroup level, though identification to the cultivar level is still complex and prohibitively expensive (https://sites.google.com/a/cgxchange.org/musanet/genotyping-centre).

Figure 7 A mature bunch complete with male bud photographed from the side is the single best image to facilitate cultivar identification. Lady Finger (AAB, Pome) pictured in north Queensland.

If time permits more detailed morphological descriptions can be made using the 'Minimum List of Descriptors for Musa' which provides a standardized procedure for morphological characterization of plants based on 34 descriptors with photographs to assist in descriptor scoring (http://www.musalit.org/seeMore.php?id=14474). Also available are minimum descriptors for East African Highland bananas and plantains (https://sites.google.com/a/cgxchange.org/musanet/documentation/technical-guidelines). These latter two sets of minimum descriptors were developed in response to the need for additional descriptors, relevant to a particular subgroup to aid in discriminating within the subgroup. MusaTab is an Android© application for recording such Musa characterization data in the field. It allows data to be scored in the field directly on a mobile device and then later uploaded to a computer (https://sites.google.com/a/cgxchange.org/musanet/projects/musatab). Collecting such descriptor

Figure 8 Photo descriptors applied to the cultivar Gros Michel.

Figure 9 Collecting fresh cigar-leaf (arrowed) samples for both SSR genotyping and determining ploidy can be invaluable for identifying and classifying cultivars (photo courtesy: Agus Sutanto).

information can be a major undertaking and researchers need to appreciate that it is just one of the means to an end and not an end in itself. The end you are mostly seeking is identification and classification of the specimen(s).

MusaID is a taxonomy decision-making application that helps users to identify an unknown taxon by comparison with a reference collection of known Musa taxa described by a set of qualitative descriptors (https://sites.google.com/a/cgxchange.org/musanet/projects/musaid). However, MusaID currently has limitations. These include not having enough accessions in the database and not enough diversity of accessions against which to compare a specimen. There are plans to upgrade the application by adding more characterization data from more diverse genotypes and diverse environments. This should help to validate the methodology behind the tool. The software also needs improvement in its reliability.

In some situations, there may be several cultivars within a subgroup with minor differences that are hard to distinguish. The diversity within the Plantain subgroup is quite immense as well illustrated by the CARBAP collection

Musalogue (http://www.musanet.org/). This catalogue contains photos of 145 Plantain accessions which make it possible to identify to the cultivar level in many cases, but in some cases it may only be possible to narrow down the identity of a specimen to a subset of the subgroup, with uncertainty remaining as to the exact cultivar.

8 Future trends and conclusions

There are those who lament the imperfect knowledge existing on banana taxonomy, griping about how confusing it all is to them. However, they fail to realize they can be part of the solution and make a positive contribution by becoming more aware of the information resources nowadays available, utilizing them, and so ensuring in their communication on banana cultivars that they are properly understood by their audiences. The level of global connectedness which the internet has provided can be used to seek assistance/collaboration to solve identification/classification problems together. The recently published CARBAP Plantain collection is a great example of the progress that can be made by a concerted group effort to describe cultivars in detail with good-quality photographs. This has created a platform for better understanding and we look forward to the development of more such cultivar guides. However, like many things it remains a work in progress which can be improved further by linking such information via an on-line key to then more quickly identify an unknown specimen, particularly given the large number (145) of described Plantains it contains.

Recent collecting missions in the Indonesian Triangle (Sutanto et al., 2016) and Bougainville (Sardos et al., 2018) are also good examples of several organizations working together and combining morphological and molecular characterization methods to better understand the wealth of diversity in remote far-off lands while contributing to its conservation for future generations.

Research findings in recent years in banana taxonomy have challenged some of our underlying assumptions regarding cultivar origins and so our knowledge quest is very much ongoing. Additionally, far greater latent variation may be present in certain cultivars than was ever imagined until manifested through agents such as mutagenesis. The broadening of phenotypic diversity by plant improvement programmes, while anticipated to deliver more robust cultivars, may potentially bring additional challenges to our identification efforts.

The 'Global strategy for the conservation and use of *Musa* (banana) genetic resources' (MusaNet, 2016) describes much of the work underway to better understand banana taxonomy and facilitate cultivar identification, as well as additional recommendations for further work. These activities need to continue to be amply resourced with appropriate funding and staffing, including a next-generation well-trained nurtured staff for the task ahead as they are foundational to our broader research efforts on bananas.

9 Where to look for further information

ProMusa is a network of people promoting scientific discussions on bananas. Of particular interest concerning cultivars would be Musapedia: an online, collaboratively built compendium of knowledge on bananas. The ProMusa website is also linked to three databases: Musalit, the largest repository of references on bananas; Musarama, the image bank that provides access to a range of photos on bananas; and Musacontacts, the place to find people working on bananas.

MusaNet is a global network aimed at ensuring the long-term conservation, and facilitating the increased utilization of Musa genetic resources. It provides useful links to Musa collections and diversity publications.

A selection of references illustrating the diversity of cultivars is:

- Arnaud, E. and Horry, J. P. (eds) 1997. *Musalogue: A Catalogue of Musa Germplasm. Papua New Guinea Collecting Missions, 1988–1989*. INIBAP, Montpellier, France.
- Daniells, J. W. 1995. Illustrated guide to the identification of banana varieties in the South Pacific. ACIAR Monograph No. 33.
- Daniells, J. W. and Bryde, N. J. 2001. Banana varieties – The ACIAR years 1987–1996. Queensland DPI Information Series QI01013.
- Irish, B. M., Rios, C., Daniells, J. W. and Goenaga, R. 2016. Catalog of banana (*Musa* spp.) accessions maintained at the USDA-ARS Tropical Agriculture Research Station.
- Karamura, D. A., Karamura, E. and Tinzaara, W. (editors) 2012. *Banana Cultivar Names, Synonyms and Their Usage in East Africa*. Bioversity International, Uganda.
- Kepler, A. K. and Rust, F. G. 2011. *The World of Bananas in Hawai'i: Then and Now – Traditional Pacific & Global Varieties, Cultures, Ornamentals, Health & Recipes*. Pali-O-Waipi'o Press, Haiku, Hawai'i.
- Omar, S., Jamaluddin, S. H., Tahir, M. and Bahari, U. M. 2012. *Collection of Bananas in Malaysia*. MARDI, Malaysia.
- Rosales, F. E., Arnaud, E. and Coto, J. 1999. *A Tribute to the Work of Paul H. Allen: A Catalogue of Wild and Cultivated Bananas*. INIBAP, Montpellier, France.
- Simmonds, N. W. 1966. *Bananas*. Longman, London.
- Sutanto, A. and Edison. H. S. 2005. *Diskripsi Pisang Indonesia*. Solok, Balai Penelitian Tanaman Buah.
- Uma, S. and Sathiamoorthy, S. 2002. *Names and Synonyms of Bananas and Plantains of India*. ICAR, Tiruchirapalli, India.
- Valmayor, R. V., Espino, R. R. C. and Pascua, O. C. 2002. *The Wild and Cultivated Bananas of the Philippines*. PARRFI and BAR, Los Banos, Philippines.

10 References

Atom, A. D., Lalrinfela, P. and Thangjam, R. 2015. Genome classification of banana genetic resources of Manipur using morphological characters. *Science Vision* 15: 189-195.

Baurens, F. C., Martin, G., Hervouet, C., Salmon, F., Yohoume, D., Ricci, S., Rouard, M., Habas, R., Lemainque, A., Yahiaoui, N. and D'Hont, A. 2019 Recombination and large structural variations shape interspecific edible bananas genomes. *Molecular Biology and Evolution* 36(1): 97-111.

Boonruangrod, R., Desai, D., Fluch, S., Berenyi, M. and Burg, K. 2008. Identification of cytoplasmic ancestor gene-pools of *Musa acuminata* Colla and *Musa balbisiana* Colla and their hybrids by chloroplast and mitochondrial haplotyping. *Theoretical and Applied Genetics* 118: 43-55.

Boonruangrod, R., Fluch, S. and Burg, K. 2009. Elucidation of origin of the present day hybrid banana cultivars using the 5′ ETS rDNA sequence information. *Molecular Breeding* 24(1): 77-91.

Brickell, C. D., Alexander, C., Cubey, J. J., David, J. C., Hoffman, M. H. A., Leslie, A. C., Malecot, V. and Jin, X. 2016. *International Code of Nomenclature for Cultivated Plants* (9th edn). International Society for Horticultural Science, Leuven, Belgium.

Carreel, F., Gonzalez de Leon, D., Lagoda, P., Lanaud, C., Jenny, C., Horry, J. P. and Tezenas du Montcel, H. 2002. Ascertaining maternal and paternal lineage within Musa by chloroplast and mitochondrial DNA RFLP analyses. *Genome* 45(4): 679-692.

Christelová, P., Valárik, M., Hřibová, E., Van den Houwe, I., Channelière, S., Roux, N. and Doležel, J. 2011 A platform for efficient genotyping in Musa using microsatellite markers. *AoB Plants* 2011: plr024.

Convention on Biological Diversity 2019. What is taxonomy? Available at: https://www.cbd.int/gti/taxonomy.shtml.

Häkkinen, M. 2013. Reappraisal of sectional taxonomy in Musa (Musaceae). *Taxon* 62(4): 809-813.

Heslop-Harrison, J. S. and Schwarzacher, T. 2007. Domestication, genomics and the future of the banana. *Annals of Botany* 100(5): 1073-1084.

IPGRI-INIBAP/CIRAD 1996. *Descriptors for Banana (*Musa *spp.)*. IPGRI, Rome, Italy; INIBAP, Montpellier, France; CIRAD, France.

Janssens, S. B., Vandelook, F., De Langhe, E., Verstraete, B., Smets, E., Vandenhouwe, I. and Swennen, R. 2016. Evolutionary dynamics and biogeography of Musaceae reveal a correlation between the diversification of the banana family and the geological and climatic history of Southeast Asia. *New Phytologist* 210(4): 1453-1465.

Jones, D. R. and Daniells, J. W. 2019. Introduction to banana, abaca and enset. In: Jones, D. R. (Ed.), *Handbook of Diseases of Banana, Abaca and Enset*. CABI Publishing, Boston, MA.

Kirchoff, B. K., Lagomarsino, L. P., Newman, W. H., Bartlett, M. E. and Specht, C. D. 2009. Early floral development of *Heliconia latispatha* (Heliconiaceae), a key taxon for understanding the evolution of flower development in the Zingiberales. *American Journal of Botany* 96(3): 580-593.

Kress, W. J. 1990. The phylogeny and classification of the Zingiberales. *Annals of the Missouri Botanical Garden* 77(4): 698-721.

Li, L. F., Häkkinen, M., Yuan, Y. M., Hao, G. and Ge, X. J. 2010. Molecular phylogeny and systematics of the banana family (Musaceae) inferred from multiple nuclear and chloroplast DNA fragments, with a special reference to the genus Musa. *Molecular Phylogenetics and Evolution* 57(1): 1-10.

Linnaeus, C. 1735. *Systema Naturae*, trans. Engel-Ledeboer, M. S. J. and Engel, H. (1964). Johann Willem Groot (Publishers), Leyden, The Netherlands.

Martin, G., Cardi, C., Sarah, G., Ricci, S., Jenny, C., Fondi, E., Perrier, X., Glaszmann, J. C., D'Hont, A.. and Yahiaoui, N. 2020. Genome ancestry mosaics reveal multiple and cryptic contributors to cultivated banana. *The Plant Journal*. doi: 10.1111/tpj.14683.

MusaNet 2016. *Global Strategy for the Conservation and Use of Musa (Banana) Genetic Resources*. Montpellier, France. 226p.

Nwakanma, D. C., Pillay, M., Okoli, B. E.and Tenkouano, A. 2003. Sectional relationships in the genus *Musa* L. inferred from the PCR-RFLP of organelle DNA sequences. *Theoretical and Applied Genetics* 107(5): 850-856.

Perrier, X., Bakry, F., Carreel, F., Jenny, C., Horry, J. P., Lebot, V.and Hippolyte, I. 2009. Combining biological approaches to shed light on the evolution of edible bananas. *Ethnobotany Research and Applications* 7: 199-216.

Ploetz, R. C., Kepler, A. K., Daniells, J. W. and Nelson, S. C. 2007. Banana and plantain: an overview with emphasis on Pacific island cultivars. In: Elevitch, C. R. (Ed.), Species Profiles for Pacific Island Agroforestry, Permanent Agriculture Resources, Holualoa. Hawaii. 26pp. Available at: http://agroforestry.org/images/pdfs/Banana-plantain-overview.pdf.

Sardos, J., Christelová, P., Cızková, J., Paofa, J., Sachter-Smith, G. L., Janssens, S. B., Rauka, G., Ruas, M., Daniells, J. W., Dolezel, J. and Roux, N. 2018. Collection of new diversity of wild and cultivated bananas (*Musa* spp.) in the Autonomous Region of Bougainville, Papua New Guinea. *Genetic Resources and Crop Evolution* 65(8): 2267-2286.

Sass, C., Iles, W. J. D., Barrett, C. F., Smith, S. Y. and Specht, C. D. 2016. Revisiting the Zingiberales: using multiplexed exon capture to resolve ancient and recent phylogenetic splits in a charismatic plant lineage. *PeerJ* 4: e1584.

Shepherd, K. 1990. Observations on the Simmonds-Shepherd system of genomic classification of banana cultivars. In: *Musa Conservation and Documentation: Proceedings of the a Workshop Held in Leuven*, Belgium, 11-14 December 1989. INIBAP/IBPGR.

Simmonds, N. W. 1954. Isolation in *Musa*, sections Eumusa and Rhodochlamys. *Evolution* 8(1): 65-74.

Simmonds, N. W. 1962. *The Evolution of the Bananas*. Longmans, London.

Simmonds, N. W. and Shepherd, K. 1955. The taxonomy and origins of the cultivated bananas. *Journal of the Linnean Society of London, Botany* 55(359): 302-312.

Stover, R. H. and Simmonds, N. W. 1987. Classification of banana cultivars. In: Stover, R. H. and Simmonds, N. W. (Eds) *Bananas*, 3rd edn. Wiley, New York, pp. 97-103.

Sutanto, A., Edison, H. S., Nasution, F., Hermanto, C., Cizkova, J., Hribova, E., Dolezel, J., Roux, N., Horry, J.-P., Daniells, J. W. and De Langhe, E. 2016. Collecting banana diversity in eastern Indonesia. *Acta Horticulturae* 1114: 19-26.

Vezina, A. 2019. Modernizing Simmonds and Shepherd's legacy. Available at: http://www.promusa.org/blogpost614-Modernizing-Simmonds-and-Shepherd-s-legacy. (accessed 23 Sept 2020).

Wong, C., Kiew, R., Argent, G., Set, O., Lee, S. K. and Gan, Y. Y. 2002. Assessment of the validity of the sections in *Musa* (Musaceae) using AFLP. *Annals of Botany* 90(2): 231-238.

Chapter 4

Exploiting current *Musa* collections

V. Guignon, Alliance of Bioversity International and CIAT, France

1 Introduction

Research on germplasm and genetic improvement relies on material that usually is held in collections. The primary roles of most collections are to maintain, characterize and distribute genetic material. While those roles seem quite straightforward, organizations responsible for these collections often face old and new challenges. In this chapter, we discuss the following aspects: conservation, characterization, diversity coverage, environmental and experimental data, in situ data integration and finally, efficient tools for data mining.

Bananas are vegetatively propagated and varieties used for human consumption are seedless, requiring germplasm for genetic improvement to be conserved in field collections. The plants are therefore exposed to external factors such as climatic hazards, pests and diseases. Several strategies have been adopted to reduce the risk of loss, including growing plants in greenhouses and employing in vitro conservation in tissue culture or cryopreservation. Another important risk management strategy is exchanging material between collections, such that lost plants can be reintroduced from another collection. This however requires adherence to regulatory and phytosanitary requirements as well as arrangements for sharing of germplasm.

After collecting and conserving germplasm, the next challenge is to document and characterize it. Different tools and guidelines are available to

http://dx.doi.org/10.19103/AS.2020.0070.04

help users to systematically characterize *Musa* germplasm. A minimum set of morphological descriptors for banana are required at first to characterize the germplasm (Taxonomic Advisory Group (TAG), 2010). In order to compare data between different accessions, standards based on ontologies like the Multi-crop Passport Descriptors (MCPD) (Alercia et al., 2012) (URL ref. MCPD) are employed. However, characterization of Musa accessions is a lengthy process and at present not all accessions in collections are well-documented. To simplify the work of collection curators, tools such as MusaTab (URL ref. MusaTab) have been recently deployed to facilitate data collection in a standardized manner. To encourage curators to collect data and increase accession documentation, metrics have been developed to rate the level of detail provided for an accession (van Hintum et al., 2011) (URL ref. PDCIm).

While curators of field collections can visually characterize their accessions, this is not possible for in vitro, seed collections or cryopreserved material. When the material is duplicated and conserved in laboratories, there is always a risk of somaclonal mutation and even sometimes, human error. Therefore, such material must be verified regularly in the field in order to make sure it corresponds to its recorded characterization data.

Collections are not the only places where characterization takes place. Accessions are also characterized in research programs and projects, providing valuable new data. These data need to be included in information systems, but several problems have to be addressed first: identifying researchers conducting studies on banana accessions, being able to make a link between a study accession and a collection accession and finally, extracting the results and adding it to the collection database to make them available to all users. Fortunately, several solutions already exist, such as Musalit (URL ref. Musalit) for ProMusa (URL ref. ProMusa), DOI (URL ref. DOI) and the *Musa* Germplasm Information System (MGIS) (Ruas et al., 2017) (URL ref. MGIS) including BrAPI (Selby et al., 2019) and the popular FAIR data (Wilkinson et al., 2016) approach that enforces the use of these solutions.

After focusing on characterization data, another challenge is the lack of agronomic data. In the face of climate change, agronomic data on performance under different environmental conditions is essential to achieve sustainable cultivation of bananas. To select the appropriate accessions, one should know which climatic conditions they tolerate and what other environmental and biotic constraints due to pest and disease affect plant productivity. Environmental data can be extracted from collection locations but may be of limited value. Data more precisely recorded during experiments is often more valuable.

Well-characterized accessions and their agronomic data are not enough to provide what is needed for crop improvement – diversity of the material is also important. With the knowledge accumulated and improved over years on collection accessions, it is possible to perform gap analyses to get an idea of the genetic material that is missing in collections. For instance, MGIS provides several tools that show available or missing material at each taxonomic level.

Banana collections seldom have more than 500 accessions due to high maintenance costs, human resource limitations and/or field-area constraints. In order to broaden the diversity of conserved banana plants, other approaches can complement field collections. Bananas can be conserved in situ, on farms or in household gardens. They can also be observed (but left) in the wild environment during collecting missions or by people who show an interest in bananas. But how to collect this type of data? Several solutions have been investigated such as the use of local surveys (Ocimati et al., 2013; Yemataw et al., 2016), through new technologies and crowdsourcing through iNaturalist (URL ref. iNaturalist), Flickr (URL ref. Flickr) and others. But even if tools, such as MusaID (URL ref. MusaID), are available to identify and characterize an observed banana plant, those data often need to be curated by specialists before being integrated into an information system that can link them to related collection accessions. Another difficulty is obtaining agronomic data from such accessions as this need to be collected under standardized conditions. The International Musa Testing Programme (IMTP), led by Bioversity International, aims to solve that issue.

A well-documented and diversified collection can only be optimally used if it has an efficient information system. While each collection could have its own information system, it is more convenient to have a centralized system. For instance, Genesys (URL ref. Genesys) is a CGIAR-supported program that includes among other crops, banana accessions from many collections. But even if Genesys provides a wide range of useful tools, it has been designed for multiple crops and lacks specific tools for the banana community. MGIS has been developed for that purpose; it offers many unique features and enables data crossing with other information systems including Genesys (through DOI), the Banana Genome Hub (BGH) (Droc et al., 2013) for genomic data, MusaBase (URL ref. MusaBase) and PHIS (Neveu et al., 2019) for breeding and experimental data, iNaturalist for in situ data and offers APIs (Application Programming Interfaces) for data exchanges such as BrAPI.

Once the collection material of interest has been identified, the last challenge is to make it available to researchers. The International *Musa* Germplasm Transit Center (ITC) (URL ref. ITC) managed by Bioversity International currently plays this role via the MGIS Online Ordering System (MOOS). ITC holds the world's largest collection of *Musa* accessions and provides virus-free material freely to research institutes under the Standard Material Transfer Agreement (SMTA) (URL ref. SMTA).

Exploiting the full diversity of *Musa* involves collections around the world in different environments that conserve a wide variety of well-documented banana accessions available to scientists. This base must be enriched with studies, environmental and agronomic data, and in situ data, all gathered in a main information system with efficient tools to enable data mining. With the new sequencing technologies and the upcoming mass phenotyping technologies, information systems and their user interfaces will play a major role in utilizing collections of *Musa* for germplasm and genetic improvement programs.

2 Collections and accession conservation

2.1 Overview of current Musa *collections*

Banana genebanks are the main source of genetic material for *Musa* genetic improvement. Most consist of field collections in tropical countries where bananas are grown. They are spread across Africa, Asia, the Pacific and the Americas. The world's largest banana genebank is the International *Musa* Germplasm Transit Center (ITC) held by Bioversity International and hosted at the Katholieke Universiteit Leuven (KU Leuven) in Belgium.

Region	Country	Collection	Field	Tissue culture or in vitro
Europe	Belgium	ITC		>1600

There are around 13 banana collections in West and Central Africa and around 17 collections in East and South Africa (URL ref. MusaNet-Af). Some of the most well-known collections in Africa are the Centre Africain de Recherche sur Bananiers et Plantains (CARBAP, Douala, Cameroon) (URL ref. CARBAP) with 617 accessions, the Agricultural Plant Genetic Resources Conservation and Research Center (APGRC, Kassala, Sudan) (URL ref. APGRC) with 358 accessions, the National Agricultural Research Organization (NARO, Kampala, Uganda) (URL ref. NARO) with 442 accessions and the International Institute of Tropical Agriculture (IITA, Ibadan, Nigeria) (URL ref. IITA) with 393 accessions (August 2019).

Region	Country	Collection	Field	Tissue culture or in vitro
West and Central Africa	Benin	INRAB	>10	
	Cameroon	CARBAP	>500	
	Congo (DRC)	UNIKIS	>100	
	Congo (DRC)	INERA	>50	
	Ivory Coast	CNRA	>50	
	Gabon	CENAREST	>10	

Region	Country	Collection	Field	Tissue culture or in vitro
	Ghana	CSIR/PGRRI	n/a	
	Guinea	IRAG	n/a	
	Nigeria	IITA/IBADAN	>200	
	Nigeria	IITA/ONNE	>200	
	Nigeria	NIHORT	n/a	
	Sierra Leone	MAFMR	n/a	
	Togo	ITRA	>10	
East and South Africa	Burundi	IRAZ	>200	
	Comoros	INRAPE	n/a	
	Ethiopia	Melkassa	>10	>50
	Ethiopia	Jimma	>100	
	Kenya	KISII	>100	
	Kenya	THIKA	>50	<10
	Madagascar	FOFIFA	>10	
	Malawi	Bvumbwe	>50	
	Mauritius	AREU	>50	>10
	Mozambique	-	n/a	
	Rwanda	RAB	>100	
	South Africa	ARC/ITSC	>50	
	Sudan	APGRC	>200	<10
	Tanzania	ARI	>100	
	Uganda	NARO	>200	
	Uganda	IITA/NAGURU	>200	
	Zanzibar	Kizimbani	n/a	

Asia and the Pacific have around 13 collections in South Asia, 24 collections in East and Southeast Asia and 13 collections in Oceania and the Pacific (URL ref. MusaNet-OP). Notably, there is the National Research Centre on Banana (NRCB, Tiruchirappalli, India) (URL ref. NRCB) with 196 accessions, the Bureau of Plant Industry - Davao National Crop Research, Development and Production Support Center (BPI-DNCRDPSC, Davao City, Philippines) (URL ref. BPI-DNCRDPSC) with 80 accessions, the Fruit and Vegetable Research Institute (FAVRI, Phu Tho, Vietnam) (URL ref. FAVRI) with 84 accessions, the Institute of Plant Breeding - University of the Philippines Los Baños (IPB-UPLB, Los Baños, Philippines) (URL ref. IPB-UPLB) with 183 accessions and the Department of Agriculture and Fisheries - South Johnstone (DAF-SJ, South Johnstone, Australia) (URL ref. DAF-SJ) with 42 accessions.

Region	Country	Collection	Field	Tissue culture or in vitro
South Asia	Bangladesh	BARI	> 10	
	India	AAU	> 50	
	India	APAU	> 100	
	India	BCKV	> 10	
	India	KAU	> 200	
	India	MPKV	> 50	
	India	NBPGR	> 200	
	India	NRCB	> 200	> 200
	India	RAU	> 100	
	India	TNAU	> 100	
	India	UAS	> 50	
	Sri Lanka	FCRDI	> 10	
	Sri Lanka	HORDI	> 10	
Oceania and the Pacific	Australia	DAF	> 50	> 200
	Cook Islands	MoA	> 10	
	Fiji	SPC		> 100
	Fiji	Sigatoka	> 10	< 10
	French Polynesia	-	> 100	> 10
	Hawaii	Waimea Valley	> 10	
	Marshall Islands	-	> 10	
	Papua New Guinea	NARI	> 200	
	Samoa	MAF	> 10	> 10
	Solomon Islands	KGA	n/a	
	Tonga	-	n/a	
	Vanuatu	VARTC	> 100	
East and Southeast Asia	Cambodia	CARDI	> 100	
	China	CATAS-ITCGR	> 50	
	China	IFTR/GDAAS	> 200	> 50
	China	SCAU	> 100	> 100
	China (Taiwan)	TBRI	> 200	> 200
	Indonesia	CIBINONG	> 200	> 10
	Indonesia	HERBARIUM	> 10	
	Indonesia	ICHORD	> 200	> 100
	Indonesia	ITFRI	> 100	> 10
	Indonesia	PURWODADI	> 100	
	Indonesia	RIF-SOLOK		> 200

Region	Country	Collection	Field	Tissue culture or in vitro
	Indonesia	SUBANG	> 10	
	Philippines	BPI	> 50	< 10
	Philippines	IPB/UPLB	> 50	
	Philippines	UPLB	> 100	
	Laos	-		< 10
	Malaysia	MARDI	> 100	
	Myanmar	DAR	> 10	
	Thailand	BangkokHerbarium	n/a	
	Thailand	IHR/DOA	> 100	
	Vietnam	FAVRI	> 50	
	Vietnam	SOFRI	n/a	
	Vietnam	VASI	n/a	

In the Americas, there are around 16 collections in Central America and the Caribbean and 6 collections in South America (URL ref. MusaNet-Am). Some well-known collections are Centre de coopération Internationale en Recherche Agronomique pour le Développement (CIRAD) - Centres de Ressources Biologiques (CRB) Plantes Tropicales (INRA-CIRAD, Guadeloupe, France) (URL ref. INRA-CIRAD) with 381 accessions, the United State Department of Agriculture - Tropical Agriculture Research Station (USDA-TARS, Mayaguez, USA/Puerto Rico) (URL ref. USDA-TARS) with 152 accessions and the Brazilian Agricultural Research Corporation (Embrapa, Bahia, Brazil) (URL ref. Embrapa) with around 300 accessions.

Region	Country	Collection	Field	Tissue culture or in vitro
North and Central America	Costa Rica	CATIE	n/a	
	Costa Rica	CORBANA	> 100	> 100
	Cuba	INIFAT	n/a	n/a
	Cuba	INIVIT	> 200	> 100
	Dominican Republic	CIAZA	n/a	
	Dominican Republic	ISA	n/a	
	Dominican Republic	LNB		n/a
	Dominican Republic	MIP	n/a	

Region	Country	Collection	Field	Tissue culture or in vitro
	Dominican Republic	IDIAF	n/a	
	Guadeloupe (France)	CRB at CIRAD	> 200	> 100
	Honduras	FHIA	> 200	
	Jamaica	Banana Board	> 100	< 10
	Jamaica	SRC		n/a
	Mexico	CICY	> 100	
	Mexico	INIFAP	> 10	
	USA/Puerto Rico	USDA	> 100	> 100
South America	Brazil	EMBRAPA	> 200	> 200
	Colombia	CORPOICA/Agrosavia	> 100	> 100
	Colombia	FEDEPLATANO	> 100	
	Venezuela	INIAP/CENIAP	> 100	> 10
	Venezuela	FA/UCV(MARACAY)	n/a	
	Venezuela	FA/UCV(MOCHO)	n/a	

Finally, the world's largest genebank of banana germplasm, the ITC, is in Belgium, where it is not threatened by banana pests and diseases. Nowadays, it holds 1604 accessions which are kept in vitro under slow-growth conditions at 16°C. To better secure the material, samples are also cryopreserved both at the ITC and at the 'Institut de recherche pour le développement' (IRD) in France as safety backup (around 70% of the collection).

2.2 Collection conservation strategies

Almost all *Musa* collections hold material in fields (around 95%, source: MusaNet) which is exposed to environmental factors and pests and diseases. In order to avoid loss of material, some collections have laboratories where they hold accessions in vitro (around 30% of the collections, source: MusaNet) or use cryopreservation techniques to preserve the material for long term. It is also possible to grow bananas in controlled greenhouses or even in a Growtainer™ (URL ref. Growtainer), but this is often too expensive.

Not all collections have the opportunity or budget resources to use such techniques. Fortunately, there are other low-cost approaches to limit material loss. Based on agreements, collections can exchange their material and rely on other collections as a backup in case of loss. The Standard Material Transfer

Agreement (SMTA) from the International Treaty on Plant Genetic Resources for Food and Agriculture (ITPGRFA) (URL Ref. ITPGRFA) is the reference document for germplasm exchange provided by the Food and Agriculture Organization (FAO).

The ITC plays a major role in both conserving banana germplasm and exchanging material between collections. As a 'transit center', it receives material from other collections and collecting missions to safeguard long term. The in vitro collection is protected from climatic hazards and is indexed for and cleaned of viruses if required. Furthermore, it is held under the Multilateral System of Access and Benefit Sharing of the ITPGRFA, ensuring that the germplasm is accessible to everyone.

ITC makes its material freely available for international distribution for scientific research. Distributed material is accompanied by a health statement, phytosanitary certificate and a copy of the SMTA. Not all *Musa* collections have agreed to share their material and associated passport information by signing a Data Sharing Agreement (DSA); nearly 45% of collections have signed a DSA or provide free material. The number of collections signing the DSA keeps increasing as it provides a better international visibility to collections, which also benefits from MGIS tools.

2.3 Accession characterization

Another important role of *Musa* collections is to characterize their material. Without a standard characterization, the germplasm held by a collection is more or less invisible and therefore seldomly used. Germplasm characterization can become challenging if the resources are insufficient or missing. Many reasons may hinder characterization: a lack of human resources, time, budget, knowledge or efficient tools.

A methodology for describing banana accessions has been jointly developed by several institutes and is today considered as the standard. It has been published under the title 'Descriptors for Banana (*Musa* spp.)' (IPGRI et al., 1996) (URL ref. Desc4Banana). The 123 descriptors can be divided into plant passport, management, environment and site, plant (anatomical traits) and evaluation descriptors (non-anatomical traits). An abridged version of the 34 most discriminating morphological descriptors has been published as the 'Minimum List of Descriptors for *Musa*' (Taxonomic Advisory Group (TAG), 2010) (URL ref. MLDM).

The banana descriptors can be mapped on the Multi-Crop Passport Descriptor (MCPD) ontology and the Banana Ontology (URL ref. CO_325), both from the Crop Ontology (CO) project (Shrestha et al., 2012) of the Generation

Challenge Programme (GCP). Conforming to standard ontologies guarantees the efficient use of collected data and enables scientists to find accessions of interest.

To facilitate the use of ontologies and accelerate the characterization process, tools like MusaTab have been developed. MusaTab is an Android© application for recording characterization data directly from the field using a tablet or mobile phone. Its user interface is intuitive and easy to use by collection curators with minimal training. MusaTab has been deployed during international banana characterization workshops organized by MusaNet, the Global *Musa* Genetic Resources Network (URL ref. MusaNet). The workshops provide collection curators the opportunity to learn how to capture and store information and to exchange tips on other subjects such as collection management. Tablets with MusaTab installed were offered to collections signing the data sharing agreement.

To promote well-documented accessions and encourage curators to provide more information on insufficiently documented accessions, a metric was developed to measure the passport data completeness index (PDCI). As this metric relies on the MCPD ontology and some information for banana are not taken into account, the variant metric PDCIm (PDCI *Musa* or RDIC for Relative Data Item Count) was developed to better describe *Musa*.

2.4 Accessions field verification

Accessions can be characterized in the field but are not always conserved there. Material stored in vitro as tissue culture plantlets, as seeds or even cryopreserved may be altered or mislabeled during manipulations. Altered or mislabeled material used in experiments could lead to miscellaneous results, a waste of time and/or money. It is important that collections ensure the in vitro accession still matches its original characterization through regular verification that the in vitro material is true-to-type. In the case of the ITC collection, there is a field verification process applied to accessions that have been in vitro for more than 10 years. The process involves many people, in laboratories for material duplication and transfer, in the field for plantation and characterization and at the office for expert validation (Fig. 1).

While it is also a good opportunity to gather more data on insufficiently characterized accessions, the field verification process raises many challenges. The cost in human resources and logistics is considerable. Furthermore, expert validation, based on field verification pictures, is done by taxonomists with little time for this task. The process from start to finish for an accession can take several years. Therefore, new approaches to make field verification more efficient are currently under investigation.

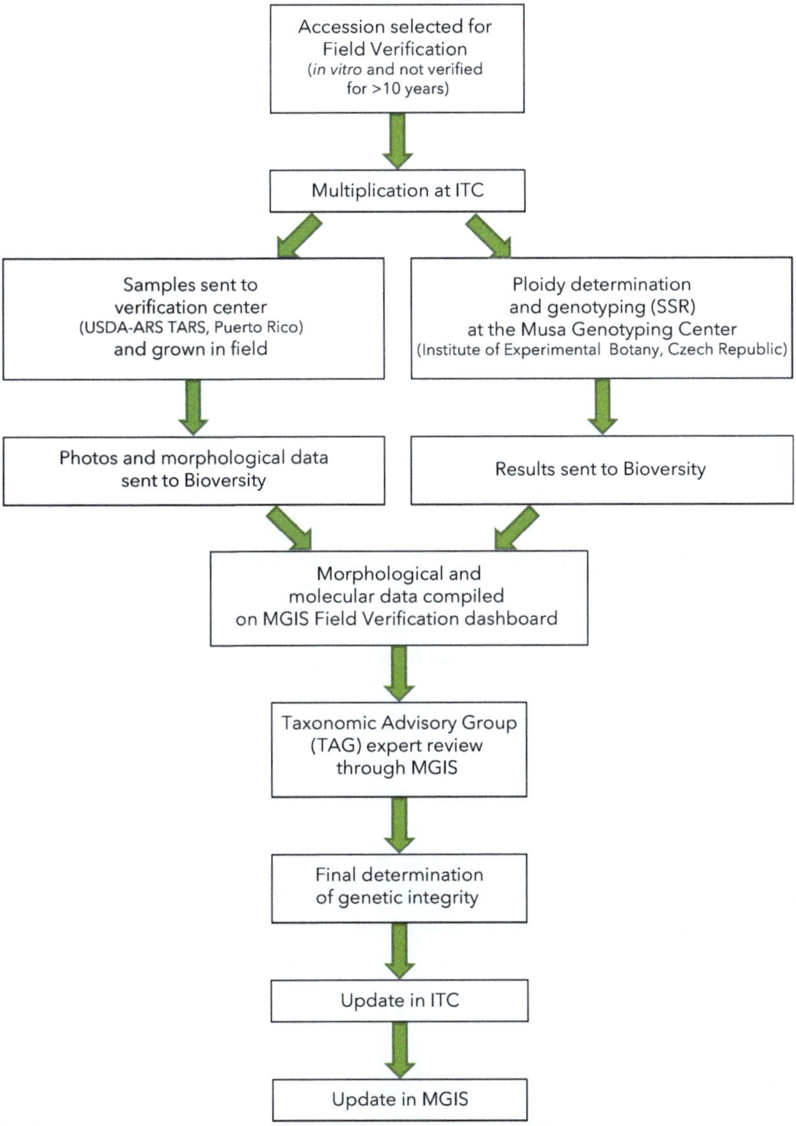

Figure 1 Overview of the field verification process to check the genetic integrity of the distributed material.

3 Adding value to collection data

3.1 Experimental data: linking studies to accessions

Characterization data from collection curators and the ITC field verification process can be further enriched with experimental data from research studies.

When researchers are selecting accessions for a study it would be convenient if they could easily find accessions used in similar studies. Breeders may also be interested in studies on their accessions of interest. Thus, there is a need to access more information linked to available accessions.

Authors of research papers are not required to feed information systems such as MGIS when they publish their results. Fortunately, most open-access studies are added to Musalit (www.musalit.org), an online repository of publications on bananas. However, finding particular accessions in a study is still challenging and it can be even more difficult to find studies on a given accession.

Looking for accession information is also limited by the fact that there is no standard way of designating accessions. For example, an accession number could be written in different ways in different studies (e.g. 'Tani' accession as 'ITC1120' in Risterucci et al. (2009), but as 'ITC 1120' or just '1120' in Novák et al. (2014), as 'ITC.1120' (Kamaté et al. (2001) or as 'ITC-1120' (Umali and Nakamura, 2003)). Data mining from such publications is even more complicated if a same given accession number corresponds to different accessions in different collections (e.g. 'APO' is 'Apem Omniaba' in CARBAP and CIRAD collections while it is 'Apoka Magbuke' in UNIKIS collection); in that case, the accession number needs to be associated with its collection in order to identify the corresponding accession.

Previously, one had to search manually within individual publications for information on given accessions. Publications in Musalit have been linked to their accessions in MGIS by manual curation. The MGIS website provides a user interface (URL ref. MGIS-accessions) to filter accessions by publications and studies and links each accession to its related studies. MGIS also provides a user interface for a given set of publications and studies (URL ref. MGIS-studies) that enables the user to browse linked accessions. This makes finding data on accessions easier although, the process is still not automated and requires some manual operations. Obtaining data on germplasm has been a problem for years in many domains; however, a solution has been found and is progressively adopted in plant research: the use of digital object identifiers or DOIs. Such identifiers can be used for accessions in order to identify them uniquely with no ambiguity among collections, reports and publications. While manual curation is still required for older publications, the use of DOIs in new publications will greatly simplify the management of links between accessions and publications. While DOIs solve the accession identification issue, information from studies is still spread in different places. Some studies are published in papers, others are held on websites, some are public, and others can have restricted access. At first glance, it may look like an unsolvable issue. However, to create interoperable systems, a new standard supported by a large community of institutes was developed: The Plant Breeding Application

Programming Interface (BrAPI), which will be described later in this chapter in the 'building data bridges' section.

In addition, beyond the DOI, a new approach is now being widely promoted to share data and simplify the work of people and systems using it: the FAIR data principle. FAIR stands for 'Findability, Accessibility, Interoperability, and Reusability'. It's a principle that researchers are encouraged to apply which will improve the use of DOIs and also improve the access to their related data.

3.2 Including environmental data

For genetic improvement, environmental data are useful for selecting accessions, especially when the data is related to climate change. Environmental data includes geolocation, habitat, climate, soils, nutrients, proximity of other species and more. As the data can change over time, collecting it is often challenging. Many climatic data are publicly available from recording stations all around the world as well as from satellites. In order to take advantage of these resources, it is only necessary to know when and where samples are located, which is relatively easy for collected accessions and field collections. Geolocated in situ observations may also shed light on climate conditions suitable for certain accessions. For other types of environmental data, there is still a lot of work to be done. Such information is easy to collect from controlled environments such as greenhouses or Growtainers™ designed for this purpose. With the rise of new technologies such as drones and automated sensors, the amount of environmental data will increase in the near future. The challenge is to link all this data to agronomic performance of accessions.

3.3 Measuring diversity in Musa collections

A well-characterized accession is just one ingredient for genetic improvement. To create a new banana variety, researchers also need a large number of accessions with different genetic backgrounds in order to identify useful parental combinations. Banana collections provide a certain amount of genetic diversity but there are always gaps, overrepresented and underrepresented *Musa* species and cultivars. To report the range of the genetic diversity, collection content must be evaluated. Then in order to reduce gaps and increase diversity some collecting mission must be planned to collect new wild and cultivated forms.

Several types of 'diversity' can be measured. In the context of this chapter, it could be morphological, taxonomic or genomic diversity. Morphological diversity can be evaluated through accession passport descriptors and associated pictures. Taxonomic and genomic diversity evaluation relies on other tools.

MGIS provides tools to evaluate collection content based on *Musaceae* taxonomy. The 'taxonomy browser' (see Fig. 2) (URL ref. MGIS-taxonomy) shows

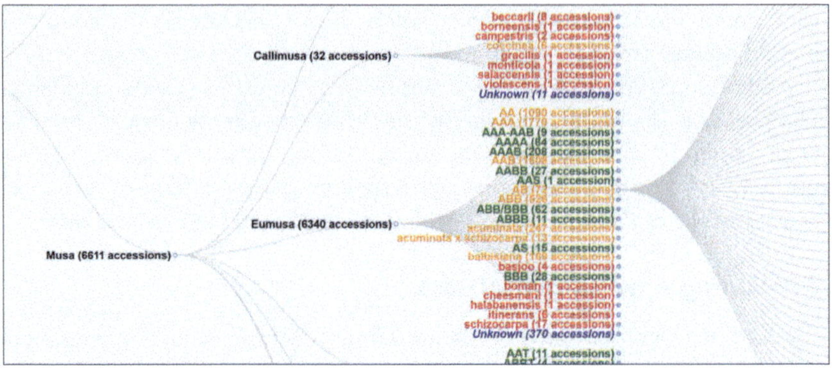

Figure 2 Taxonomy browser from the Musa Germplasm Information System (MGIS).

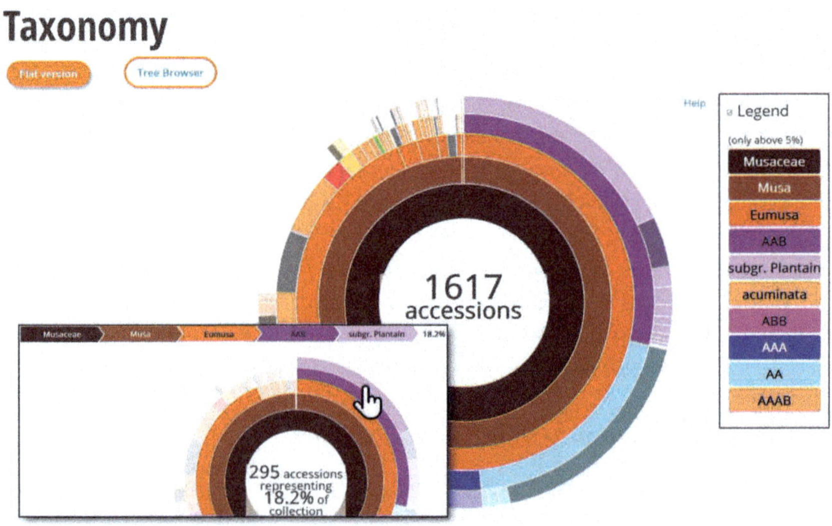

Figure 3 Taxonomy wheel for the International Musa Germplasm Transit Center (ITC) collection. Bottom-left part shows sector highlight using the mouse pointer.

how many accessions are available at each taxonomic level up to the accession name. It allows MGIS users to either browse the taxonomy for all the collections or for a specific one. Each taxonomy node displays how many accessions are in the MGIS database and nodes with no accessions can also be displayed in order to view gaps.

Another tool provided by MGIS is the taxonomy wheel, which displays each collection's taxonomic composition (see Fig. 3). For instance, one can see that a quarter of the CARBAP (Cameroon) collection is dedicated to plantain and another quarter to the AA group. More than half of the NARO (Uganda) collection is composed of the AAA group in which the subgroup

Mutika/Lujugira represents more than 40% of the whole collection. The ITC collection (Belgium), the CIRAD collection (Guadeloupe, France), the INIVIT collection (Cuba) and the ICHORD collection (Indonesia) are quite diverse with AAA, AAB, AA and ABB groups but also *acuminata, balbisiana* and other species. On the other hand, the APGRC collection (Sudan) is 98% subgroup Cavendish. In some cases, like the CARDI collection (Cambodia), most of the material is currently known only at a low classification level (Eumusa), but efforts are being made to improve the level of characterization of these collections.

The content of collections is also of interest for what is not yet conserved. For collections like the ITC, which has a safeguarding role, it is useful to know what material is missing in the collection but available in other collections. To visualize this, MGIS provides an internal taxonomy tree tool.

Another approach for measuring the diversity coverage is to display collecting sites on a world map to view regional gaps. Sites like Genesys or MGIS can provide such maps. Mixing these maps with maps showing geolocated in situ observations of wild banana plants could identify sites where new material should be collected. Such tools are presently under active development and should shortly be available.

Next-generation sequencing, providing new genetic markers, can also shed new light on banana diversity. With tools like Gigwa (Sempéré et al., 2019) it is possible to browse genetic-marker information and find rare alleles. Molecular studies go further by using sequencing information to analyze the diversity and link it to subspecies/subgroup and/or geographical regions (Christelová et al., 2017; Sardos et al., 2016). Such information can be used as a basis to measure diversity.

3.4 Beyond collections: integrating in situ data

Banana collections have their limits in terms of genetic and spatial diversity and conservation, often due to maintenance costs, field limitations and institute interests. To broaden the knowledge of banana diversity and go beyond collections, new approaches can bring in more information. One of these approaches is to integrate in situ observations into information systems managing collections.

There are several sources of in situ data that can be exploited, such as surveys in regions where banana plants are cultivated and wild. The surveys can collect a lot of information, but the local vocabulary used by the surveyed people is often subject to interpretation. Another source is crowdsourcing applications used by enthusiastic people interested in bananas. iNaturalist and Flickr are two examples of online tools where people can freely collaborate and provide banana plant observations anywhere in the world. However, expertise is required in order to enter this data into information systems and link them to existing species/subspecies and accessions.

Tools have been developed to enable non-experts with basic knowledge on bananas to better identify varieties. While catalogues like Musalogue (Daniells et al., 2001) or sites like MGIS provide a lot of photos and characterization data to identify a banana plant, more convenient tools can make the task a lot easier. MusaID, a mobile application designed for collection accessions, proposes taxonomic classifications via targeted questions about the observed plant.

4 Data mining in collections

4.1 Information systems

Musa collections hold material for research and plant breeding. Well-documented collection accessions require appropriate user interfaces for optimal use. Such interfaces are provided by information systems that can provide a wide variety of filters in order to search all sources of accession-related data. Several solutions are available to researchers, breeders and collection curators. While each collection could manage a dedicated information system, researchers and others would have to access every collection information system to find the material of interest. Using an application programming interface (API) to bridge collection information systems could solve this problem, but at the cost of low performance and complexity of API evolution. The best solution at present is to use a centralized information system that simplifies the research.

Genesys is a database which allows users to explore the world's crop diversity conserved in genebanks supported by the CGIAR Genebank platform. The system provides useful user interfaces to browse collections and available accessions. It offers features like list views, map views, histograms, 'diversity trees', many filters, APIs and more. However, such tools are designed for many crops and are not crop-specific. Furthermore, any improvement must be tested and validated for every crop which can slow down developments.

For the banana community, the *Musa* Germplasm Information System (MGIS) provides these generic tools but also banana-specific tools. To name a few specific features, there is the ordering system – MOOS – that will be described later in this chapter, publications and studies on bananas that are linked to their related accessions and integrated into filters, Gigwa on banana NGS data, an accession comparator that includes banana-specific characterization data, the ITC field verification management interface, phenotyping data filter using a schema displaying the banana anatomy, in situ data integration from specific sources and PDCIm integration.

4.2 Dedicated user interfaces

Banana-specific tools make the work of *Musa* researchers simpler. For instance, generic tools to search for scientific publications exist, such as Google Scholar (URL ref. GScholar), Bielefeld Academic Search Engine (BASE) (URL ref. BASE) or Baidu Scholar (URL ref. BScholar), but even with the appropriate terms, those engines can return non-banana related papers. Musalit has a banana-specific publications search engine to solve that problem. Another use of banana-specific tools concerns characterization and phenotyping data. Such data uses a specific ontology, the Banana Ontology, which is based on the Banana Anatomy Ontology (URL ref. CO_125). It simplifies a search by filtering accession traits using an anatomical diagram of banana plant (URL ref. MGIS-trait) in order to quickly find available traits of interest. Data provided by external sources not dedicated to banana can also be integrated. Integrating in situ observations from iNaturalist into a banana website like MGIS is feasible because iNaturalist hosts a project dedicated to banana, the banana natural diversity mapping project (URL ref. iNaturalist-BNDM).

Currently, it would be very hard to create efficient generic tools like Genesys that are able to deal with each crop's specificities and dedicated projects. Each crop has a specific history which includes its community, its research and projects and this forms a kind of culture. For this reason, banana has many dedicated user interfaces provided by sites like Musalit, MGIS and others to exploit collections and mine their data.

4.3 Building data bridges

Depending on the research subject and the information systems a researcher knows, he or she may start at one point and will need to follow other data sources. That is why bridges between systems are needed. Such bridges can also be used to enrich information systems with external data, cross data and help the systems provide valuable information to users.

MGIS holds collection data and provides tools to perform efficient data mining. Its mission is not to hold breeding or experimental data. In fact, other websites have been designed for this specific task such as MusaBase or PHIS. However, when using MGIS to select accessions for genetic improvement, it might be convenient to integrate data from other websites as well.

The BrAPI enables automated communication between websites dealing with breeding and genebank data, making it possible to dynamically exchange information. Servers providing the BrAPI service can make their data available, either public or private through client authentication, to consumer end points such as tablets or websites. This technology is used by MGIS to retrieve studies associated with accessions from servers holding information, such as MusaBase or PHIS.

When integrating in MGIS external data, such as in situ data from iNaturalist, it is important to give credit to its source. MGIS provides back links to iNaturalist observations, thus promoting and encouraging people to use such systems. In return, people provide new data to the system that can be then integrated by MGIS for the benefit of its end users.

Sometimes links to outside resources are needed to allow researchers to discover new information systems. For instance, DOIs can be used to bridge Genesys and MGIS which both hold genebank data, but one could be more focused on genome data. The Banana Genome Hub holds such data and links to and from MGIS are available in order to jump easily from one system to the other during the research process. Following the same idea, Musalit links ITC accessions found in publication to their MGIS passport data page. These links save a lot of time for researchers.

4.4 From collections to research: ordering accessions

After mining collections for accessions of interest, the next step is to obtain the genetic material. While there are ways to obtain it from national and regional collections, the most convenient option is to order it from the ITC. Indeed, one of the roles of ITC is to distribute banana genetic material to researchers. Easy ordering can be achieved through the MGIS Online Ordering System (MOOS), which automatically generates an SMTA containing the requested material list.

ITC can provide several types of material, such as proliferating tissue, rooted plantlets and lyophilized leaf tissue. Since it is a free service for researchers, there is a maximum amount of accessions that can be ordered per researcher per year. The material is distributed with a health statement, phytosanitary certificate and a copy of the SMTA to simplify the delivery in any country.

5 Case study

The primary objective of this fictitious case study is to demonstrate how to use some of the tools described in this chapter to select accessions required for a specific study. Here, we want to find relevant accessions linked to nematode resistance in Asia and more specifically in Laos. We follow a logical approach based on previous studies on the topic and available accession characterization data.

The first step consists of looking for existing publications related to nematode resistance in bananas and more specifically in Asia. A good place to start with is Musalit. Searching 'nematode' in the title provides 750 publications, 'Laos' only 4 and combining both gives no results but combining the words 'nematode' and 'Vietnam' (Laos's neighbor) results in 12 publications. One relevant publication entitled 'Screening of Vietnamese *Musa* germplasm for resistance to root knot and root lesion nematodes in the greenhouse' (Van den Bergh et al., 2002) provides a list of 26 accessions that have been screened.

A similar approach can be done on MGIS to see if this publication has been integrated. To do so, we can use the search box at the bottom of the page and type its full title or keywords such as 'nematodes' or 'Vietnamese' that are specific enough to reduce the number of results and help us find the publication. The publication is in MGIS and its page (URL ref. MGIS-nematode-paper) provides the list of the 26 accessions used that is available in MGIS.

From the list of accessions in the publication we can see that 18 are available for distribution from ITC, 16 have photos, while only 10 of them meet both criteria. All of them can be easily added to the basket (see Fig. 4) for further actions such as export, data comparison and ordering (when available).

The data export page allows users to export information from each selected accession like passport data fields or associated studies in an Excel or a CSV file format. We can download that file and later extend it with other notes from our search. We can also use the data export interface to get more information on the targeted accessions. For instance, if we want to know if some of those accessions have already been used for breeding studies, we can check the box 'Breeding material (MusaBase)' and apply the new data selection. It will update the list at the bottom of the page, dynamically query MusaBase website (URL ref. MusaBase-BrAPI) using the BrAPI and shows that two accessions have been used in studies on MusaBase: ITC1122 'Gros Michel' and ITC1358 'Ngu'.

Following the links to MusaBase shows us the related experiments done: resistance to *Fusarium* wilt (ITC1122) and a drought trial (ITC1358). It is possible to visit each accession page on MGIS, but it could also be more convenient to compare these two accessions using the comparator. To do so, we go back to the 'My list' page, select ITC1122 and ITC1358 only, and click on the 'Compare' button. The comparison page is split into two vertical parts displaying the

Figure 4 MGIS 'My list' page.

accessions (see Fig. 5). It should be noted that it is possible to invert the two accessions or compare more than two accessions using the tabs. However, only two accessions can be displayed at the same time.

The passport data comparison shows that ITC1122 and ITC1358 have different origins and taxonomic classifications. For both, characterization photos are also available, as requested earlier during our first selection, but only ITC1122 has morphological characterization data. However, if needed, many morphological descriptors can be guessed from the photos provided for ITC1358.

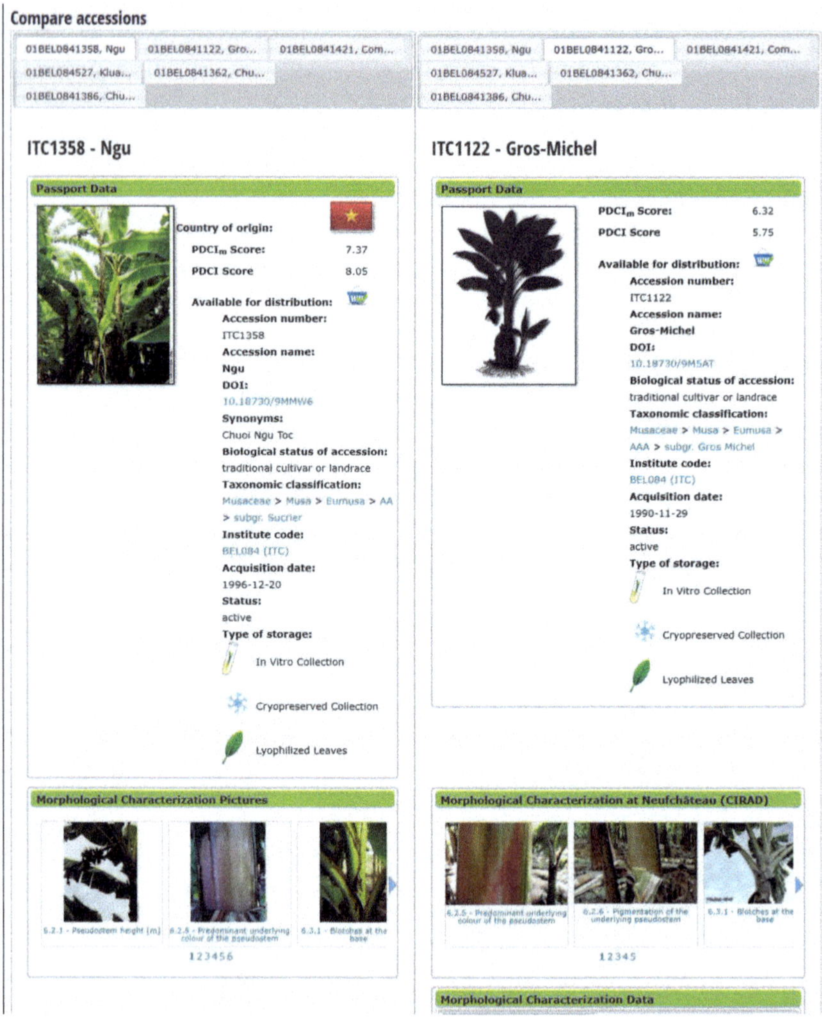

Figure 5 MGIS accession comparison tool.

On the comparator or on accession pages, displayed trees from published studies may also help in the selection process of accession as they show the genetic proximity with other accessions. The genetic markers that may reveal the genetic diversity among a set of accessions can also be extracted using the Gigwa tool available from the menu 'Genotyping, NGS markers' by providing access to genetic markers (i.e., single-nucleotype polymorphism) along the chromosomes.

Finally, and if the accessions selected using the described tools are conserved in the ITC, it is possible to order the material using the MOOS. For that purpose, we will use the 'Tools, Online ordering (MOOS)' menu and the 'Order Germplasm Material' button. The user needs to register and enter personal contact information before generating the SMTA that is mandatory to receive accessions from the ITC.

6 Summary and future trends

For banana collections, the trend is to share datasets (under Data Sharing Agreement - DSA) and benefit from a range of tools developed with MGIS to have a better worldwide visibility of existing diversity while keeping the ownership of their data and material. Once in MGIS, the collection update system and the PDCI and PDCIm scores encourage curators to better characterize their accessions and use the official taxonomy. This leads to a wider catalog of accessions visible by MGIS users and an improved collection content, thus facilitating accession mining for crop improvement.

Another trend is the use of the FAIR data principle in research results. To do this, the use of DOIs is largely promoted and recommended. In MGIS, for instance, an icon and a popup window play this role and DOI are automatically included in SMTAs. The use of DOIs will enable material-use traceability, simplify data mining from scientific publications and enable reliable data exchange between systems through tools like the BrAPI.

Finally, next-generation sequencing (NGS) technologies continue to provide *Musa* researchers with more genomic information. New technologies now allow users to have multifaceted high-throughput plant phenotyping data. User interfaces exist to mine genomic data, but new interfaces are needed to capture and mine phenotyping data. A bridge also must be built between genotyping and phenotyping data in order to better filter accessions of interest for germplasm and genetic improvement.

7 Where to look for further information

Websites:

- *Musa* Germplasm Information System (MGIS): www.crop-diversity.org/mgis.

Contains key information on *Musa* germplasm diversity, collections, accessions and tools.

- Genesys: www.genesys-pgr.org.
  Contains information on germplasm of many crops, including banana.
- MusaNet: musanet.org.
  Website of the global collaborative network for *Musa* genetic resources which includes information on banana collections around the world and holds international workshops on *Musa* characterization and documentation.
- ProMusa: www.promusa.org.
  Website of the network of people promoting scientific discussions on bananas which provides links to useful tools such as Musalit.
- Crop Ontology: www.cropontology.org.
  Provide several ontologies used for banana: the Multicrop Passport Descriptor (MCPD, CO_020), the Banana Ontology (CO_325) and the Banana Anatomy Ontology (CO_125).

Publications:

- Descriptors for Banana (*Musa* spp.) (IPGRI et al., 1996) (URL ref. Desc4Banana).
- Key access and utilization descriptors for banana genetic resources (2009) (URL ref. Desc4BananaKAU).
- The Impact of the *Musa* International Transit Center (2010) (URL ref. ITC-impact).
- MGIS: managing banana (*Musa* spp.) genetic resources information and high-throughput genotyping data (Ruas et al., 2017) (URL ref. MGIS).
- BrAPI - an application programming interface for plant breeding applications (Selby et al., 2019).

8 Acknowledgements

I would like to thank the following people for their great suggestions and reviews of this chapter: Catherine Breton, Julie Sardos, Mathieu Rouard, Max Ruas and Rachel Chase (Alliance of Bioversity International and CIAT staff).

9 References

9.1 Publications

Alercia, A., Diulgheroff, S. and Mackay, M. 2012. Source/contributor: FAO (Food and Agriculture Organization of the United Nations), Bioversity International. In: FAO/Bioversity Multi-Crop Passport Descriptors (MCPD V.2). Available at: http://www.bioversityinternational.org/nc/publications/publication/issue/faobioversity_mul ti_crop_passport_descriptors_v2_mcpd_v2.html. https://doi.org/10.13140/RG.2.1.5015.3689.

Christelová, P., De Langhe, E., Hřibová, E., Čížková, J., Sardos, J., Hušáková, M., Van Den Houwe, I., Sutanto, A., Kepler, A. K., Swennen, R., Roux, N. and Doležel, J. 2017. Molecular and cytological characterization of the global Musa germplasm collection provides insights into the treasure of banana diversity. *Biodiversity and Conservation* 26, 801–824. https://doi.org/10.1007/s10531-016-1273-9.

Daniells, J., Jenny, C., Karamura, D., Tomekpe, K., Arnaud, E. and Sharrock, S. 2001. *Musalogue*: a catalogue of *Musa* germplasm. Diversity in the genus *Musa*. International Network for the Improvement of Banana and Plantain, Montpellier, France.

Droc, G., Larivière, D., Guignon, V., Yahiaoui, N., This, D., Garsmeur, O., Dereeper, A., Hamelin, C., Argout, X., Dufayard, J. F., Lengelle, J., Baurens, F. C., Cenci, A., Pitollat, B., D'Hont, A., Ruiz, M., Rouard, M. and Bocs, S. 2013. The banana genome hub. *Database* (Oxford) 2013, bat035–bat035. https://doi.org/10.1093/database/bat035.

IPGRI, INIBAP and CIRAD 1996. *Descriptors for Banana (Musa spp.)*. IPGRI/INIBAP/CIRAD.

Kamaté, K., Brown, S., Durand, P., Bureau, J. M., Nay, D. D. and Trinh, T. H. 2001. Nuclear DNA content and base composition in 28 taxa of Musa. *Genome* 44(4), 622–627. https://doi.org/10.1139/g01-058.

Neveu, P., Tireau, A., Hilgert, N., Nègre, V., Mineau-Cesari, J., Brichet, N., Chapuis, R., Sanchez, I., Pommier, C., Charnomordic, B., Tardieu, F. and Cabrera-Bosquet, L. 2019. Dealing with multi-source and multi-scale information in plant phenomics: the ontology-driven Phenotyping Hybrid Information System. *New Phytologist* 221(1), 588–601. https://doi.org/10.1111/nph.15385.

Novák, P., Hřibová, E., Neumann, P., Koblížková, A., Doležel, J. and Macas, J. 2014. Genome-wide analysis of repeat diversity across the family Musaceae. *PLoS ONE* 9(6), e98918. https://doi.org/10.1371/journal.pone.0098918.

Ocimati, W., Blomme, G., Karamura, D., Ragama, P., Lepoint, P., Kanyaruguru, P. J., Ngezahayo, F., Ndungo, V. and Hakizimana, S. 2013. On-farm Musa germplasm diversity in different agro- ecologies of Burundi 5, 751–760. https://doi.org/10.5897/IJBC2013.0599.

Risterucci, A.M., Hippolyte, I., Perrier, X., Xia, L., Caig, V., Evers, M., Huttner, E., Kilian, A. and Glaszmann, J. C. 2009. Development and assessment of diversity arrays technology for high-throughput DNA analyses in Musa. *Theoretical and Applied Genetics* 119(6), 1093–1103. https://doi.org/10.1007/s00122-009-1111-5.

Ruas, M., Guignon, V., Sempere, G., Sardos, J., Hueber, Y., Duvergey, H., Andrieu, A., Chase, R., Jenny, C., Hazekamp, T., Irish, B., Jelali, K., Adeka, J., Ayala-Silva, T., Chao, C. P., Daniells, J., Dowiya, B., Effa Effa, B., Gueco, L., Herradura, L., Ibobondji, L., Kempenaers, E., Kilangi, J., Muhangi, S., Ngo Xuan, P., Paofa, J., Pavis, C., Thiemele, D., Tossou, C., Sandoval, J., Sutanto, A., Vangu Paka, G., Yi, G., Van den Houwe, I., Roux, N. and Rouard, M. 2017. MGIS: managing banana (Musa spp.) genetic resources information and high-throughput genotyping data. Database 2017. https://doi.org/10.1093/database/bax046.

Sardos, J., Rouard, M., Hueber, Y., Cenci, A., Hyma, K. E., van den Houwe, I., Hribova, E., Courtois, B. and Roux, N. 2016. A genome-wide association study on the seedless phenotype in banana (Musa spp.) reveals the potential of a selected panel to detect

candidate Genes in a vegetatively propagated crop. *PLoS ONE* 11(5), e0154448. https://doi.org/10.1371/journal.pone.0154448.
Selby, P., Abbeloos, R., Backlund, J. E., Basterrechea Salido, M., Bauchet, G., Benites-Alfaro, O. E., Birkett, C., Calaminos, V. C., Carceller, P., Cornut, G., Vasques Costa, B., Edwards, J. D., Finkers, R., Yanxin Gao, S., Ghaffar, M., Glaser, P., Guignon, V., Hok, P., Kilian, A., König, P., Lagare, J. E. B., Lange, M., Laporte, M.-A., Larmande, P., LeBauer, D. S., Lyon, D. A., Marshall, D. S., Matthews, D., Milne, I., Mistry, N., Morales, N., Mueller, L. A., Neveu, P., Papoutsoglou, E., Pearce, B., Perez-Masias, I., Pommier, C., Ramírez-González, R. H., Rathore, A., Raquel, A. M., Raubach, S., Rife, T., Robbins, K., Rouard, M., Sarma, C., Scholz, U., Sempéré, G., Shaw, P. D., Simon, R., Soldevilla, N., Stephen, G., Sun, Q., Tovar, C., Uszynski, G., Verouden, M., 2019. BrAPI–an application programming interface for plant breeding applications. *Bioinformatics*. https://doi.org/10.1093/bioinformatics/btz190.
Sempéré, G., Pétel, A., Rouard, M., Frouin, J., Hueber, Y., De Bellis, F. and Larmande, P. 2019. Gigwa v2–extended and improved genotype investigator. *GigaScience* 8(5). https://doi.org/10.1093/gigascience/giz051.
Shrestha, R., Matteis, L., Skofic, M., Portugal, A., McLaren, G., Hyman, G. and Arnaud, E. 2012. Bridging the phenotypic and genetic data useful for integrated breeding through a data annotation using the crop ontology developed by the crop communities of practice. *Frontiers in Physiology* 3, 326. https://doi.org/10.3389/fphys.2012.00326.
Taxonomic Advisory Group (TAG) 2010. *Minimum List of Descriptors for Musa*. Bioversity International, Montpellier, France.
Umali, R. P. and Nakamura, I. 2003. Identification of dCAPS markers that discriminate A and B cytoplasms in banana (Musa spp.). *Plant Biotechnology* 20(2), 159–164. https://doi.org/10.5511/plantbiotechnology.20.159.
Van den Bergh, I., Nguyet, D. T. M., Tuyet, N. T., Nhi, H. H. and De Waele, D. 2002. Screening of Vietnamese Musa germplasm for resistance to root knot and root lesion nematodes in the greenhouse. *Australasian Plant Pathology* 31(4), 363–371. https://doi.org/10.1071/AP02041.
van Hintum, T., Menting, F. and van Strien, E. 2011. Quality indicators for passport data in ex situ genebanks. *Plant Genetic Resources* 9(3), 478–485. https://doi.org/10.1017/S1479262111000682.
Wilkinson, M. D., Dumontier, M., Aalbersberg, I. J., Appleton, G., Axton, M., Baak, A., Blomberg, N., Boiten, J.-W., da Silva Santos, L. B., Bourne, P. E., Bouwman, J., Brookes, A. J., Clark, T., Crosas, M., Dillo, I., Dumon, O., Edmunds, S., Evelo, C. T., Finkers, R., Gonzalez-Beltran, A., Gray, A. J. G., Groth, P., Goble, C., Grethe, J. S., Heringa, J., 't Hoen, P. A. C., Hooft, R., Kuhn, T., Kok, R., Kok, J., Lusher, S. J., Martone, M. E., Mons, A., Packer, A. L., Persson, B., Rocca-Serra, P., Roos, M., van Schaik, R., Sansone, S.-A., Schultes, E., Sengstag, T., Slater, T., Strawn, G., Swertz, M. A., Thompson, M., van der Lei, J., van Mulligen, E., Velterop, J., Waagmeester, A., Wittenburg, P., Wolstencroft, K., Zhao, J. and Mons, B., 2016. The FAIR Guiding Principles for scientific data management and stewardship. *Scientific Data* 3, 160018.
Yemataw, Z., Tesfaye, K., Zeberga, A. and Blomme, G. 2016. Exploiting indigenous knowledge of subsistence farmers' for the management and conservation of Enset (Ensete ventricosum (Welw.) Cheesman) (Musaceae family) diversity on-farm. *Journal of Ethnobiology and Ethnomedicine* 12(1), 34. https://doi.org/10.1186/s13002-016-0109-8.

9.2 Web references

APGRC: http://arcsudan.sd/agricultural-plant-genetic-resources-conservation-and-research-center/
BASE: https://www.base-search.net/
BPI-DNCRDPSC: http://bpi.da.gov.ph/bpi/index.php/bpi-office-divisions/bpi-centers/bpi-dncrdpsc
BScholar: https://xueshu.baidu.com/
CARBAP: http://www.carbapafrica.org/
CO_125: https://www.cropontology.org/ontology/CO_125/Banana%20Anatomy
CO_325: https://www.cropontology.org/ontology/CO_325/Banana
DAF-SJ: https://www.daf.qld.gov.au/contact/offices/stations-facilities/south-johnstone
Desc4Banana: https://www.bioversityinternational.org/index.php?id=244&tx_news_pi1%5Bnews%5D=314&cHash=6cb076b75c0ca05e3a09d06f3e01aa49
Desc4BananaKAU: https://www.bioversityinternational.org/index.php?id=244&tx_news_pi1%5Bnews%5D=1158&cHash=aa6f3e2de74d91639d8863a9b6c32206
DOI: https://www.doi.org/
Embrapa: https://www.embrapa.br/
FAVRI: http://favri.org.vn/
Flickr: https://www.flickr.com/
Genesys: https://www.genesys-pgr.org/
Growtainer: http://www.growtainers.com/
GScholar: https://scholar.google.com/
IITA: http://my.iita.org/accession2/collection.jspx?id=2
iNaturalist: https://www.inaturalist.org
iNaturalist-BNDM: https://www.inaturalist.org/projects/banana-natural-biodiversity-mapping
INRA-CIRAD: http://intertrop.antilles.inra.fr/Portail/accessions/find/4
IPB-UPLB: http://cafs.uplb.edu.ph/institute-of-plant-breeding/
ITC: https://www.bioversityinternational.org/banana-genebank/
ITC-impact: https://www.bioversityinternational.org/index.php?id=244&tx_news_pi1%5Bnews%5D=1262&cHash=80af29da7504a47c613799cb51f36a44
ITPGRFA: http://www.fao.org/plant-treaty/en/
MCPD: http://www.cropontology.org/ontology/CO_020/FAO-Bioversity%20Multi-Crop%20Passport%20Descriptors
MGIS: https://www.crop-diversity.org/mgis/
MGIS-accessions: https://www.crop-diversity.org/mgis/accession-search
MGIS-nematode-paper: https://www.crop-diversity.org/mgis/content/screening-vietnamese-musa-germplasm-resistance-root-knot-and-root-lesion-nematodes
MGIS-studies: https://www.crop-diversity.org/mgis/genotyping-study-list
MGIS-taxonomy: https://www.crop-diversity.org/mgis/taxonomy
MGIS-trait: https://www.crop-diversity.org/mgis/trait-filter
MLDM: https://drive.google.com/file/d/1ieKJ7poxwvoOQ_XBMRkDfx3zP__IckmQ/view
MusaBase: https://www.musabase.org/
MusaBase-BrAPI: https://www.musabase.org/brapi/v1/calls
MusaID: https://www.crop-diversity.org/musaid/
Musalit: http://www.musalit.org/

MusaNet: http://www.musanet.org/

MusaNet-Am: https://sites.google.com/a/cgxchange.org/musanet/musa-collections/collections-in-the-americas

MusaNet-OP: https://sites.google.com/a/cgxchange.org/musanet/musa-collections/collections-in-asia-and-pacific

MusaNet-Af: https://sites.google.com/a/cgxchange.org/musanet/musa-collections/collections-in-africa

MusaTab: https://www.crop-diversity.org/mgis/content/musatab

NARO: http://www.banana.go.ug/

NRCB: http://nrcb.res.in/about-us.php

PDCIm: https://www.crop-diversity.org/mgis/content/musa-pdci-or-rdic-relative-data-item-count

ProMusa: http://www.promusa.org/

SMTA: http://www.fao.org/plant-treaty/areas-of-work/the-multilateral-system/the-smta/en/

USDA-TARS: https://www.ars.usda.gov/southeast-area/mayaguez-pr/tropical-crops-and-germplasm-research/

Part 2

Broadening the genetic base

Chapter 5

Scope of collecting wild *Musa* species germplasm

Julie Sardos, Alliance of Bioversity International and CIAT, Montpellier Office, France

1 Introduction

The value of Crop Wild Relatives as sources of new genes that can help secure crop production is increasingly recognized. In banana (*Musa* spp.), past collecting and conservation efforts focused mainly on the two direct ancestors of cultivated bananas, *M. acuminata* Colla and *M. balbisiana* Colla. However, more than 70 taxa of wild *Musa* species exist in the wild and hold through their diversity and the different habitats they occupy, high potential for adaptation. This chapter presents how much wild banana diversity has been underexplored in the past and how the clonal propagation of the banana crop has prevented the efficient collecting and conservation of wild *Musa* species. The change in focus from the collection of single clonal individuals to seed populations are presented and practical issues linked to collecting populations are discussed within the scope of wild *Musa* species reproductive biology.

http://dx.doi.org/10.19103/AS.2020.0070.06

2 Genetic vulnerability of banana plantations

Since the beginning of the banana industry in the late 19th century, the cultivation practices associated with commercial plantations, such as the use of large-scale monoclonal plantations, have favoured outbreaks of devastating diseases. That's what happened in the 1950s, when Fusarium Race 1 wiped out the cultivation of the first commercial bananas, the Gros Michel variety. Since the resistant cultivar Cavendish replaced its cousin Gros Michel in the western world's supermarkets and in tropical countries' plantations, the focus on the production of a single variety hasn't changed, and commercial banana production is still highly vulnerable because of their extremely narrow genetic base. This is the reason why the current global diffusion of Fusarium Tropical Race 4 (TR4), to which Cavendish is susceptible, is expected to severely impact the banana industry, with dramatic economic consequences both at the national and livelihood levels. However, this does not equate to the extinction of the banana, as many media outlets have suggested. In tropical countries, an unexpected number of different bananas are indeed cultivated as landraces in smallholders' fields, and represent a much wider amount of diversity than suggested by the clonal nature of the crop (Fig. 1). However, most of them do not meet the requirements for commercial exploitation. They also exhibit high levels of sterility (Fig. 2), which prevents their easy use in breeding. On

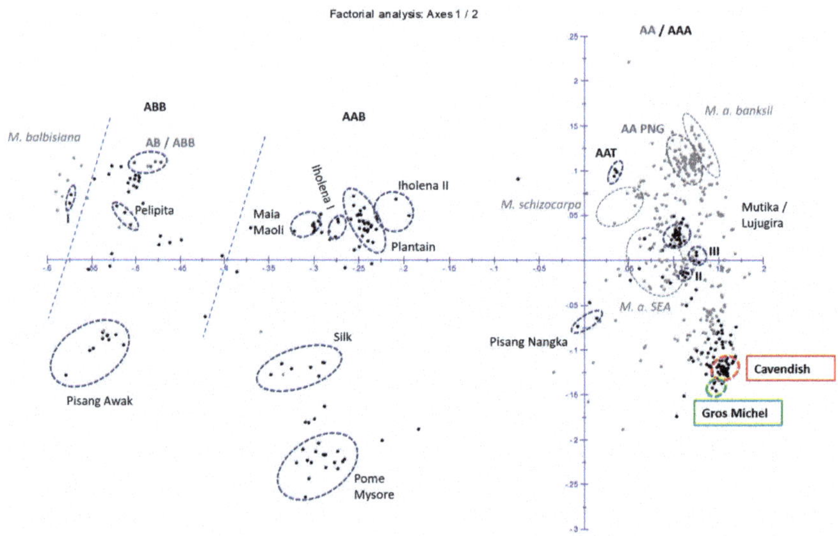

Figure 1 PCA built on dissimilarity matrix calculated on DArT markers, representing the diversity of cultivated bananas and their close ancestors *M. acuminata* and *M. balbisiana*, modified from Sardos et al. (2016b). Commercial bananas Cavendish and Gros Michel are highlighted in red and green respectively.

Figure 2 Fruits of one of the cultivated banana ancestors, *M. acuminata* ssp. *banksii* (left) and of a highly selected cooking diploid from Papua New Guinea, 'Pagal' (right) collected in West New Britain province (Sardos et al., 2019b).

the other hand, a totally underexplored amount of banana diversity lies in the many different species of wild *Musa* that can be found in the South-east Asia/ West Oceania region.

3 The importance of crop wild relatives

The value of Crop Wild Relatives (CWR) as sources of new genes for their cultivated counterparts is increasingly recognized, notably for crop improvement through breeding (Dempewolf et al., 2017). On site studies also have highlighted the importance of gene flow between crops and their CWR *in situ* to maintain the adaptability of the crops in the face of change (reviewed in Bellon et al., 2017). Due to high levels of sterility in cultivated bananas, a mandatory condition to obtain fruits that are free of seeds, banana breeding programmes started incorporating fully fertile ancestors of cultivated banana, namely *Musa acuminata* Colla and *M. balbisiana* Colla, in their crossing schemes a long time ago. This is notably the case of the Fundación Hondureña de Investigación Agrícola (FHIA), who started in the second part of the 20th century to use both wild and cultivated diploids to produce selected hybrids (SH series of improved bananas). The SH diploids were created for use as parents in crosses to produce improved triploids and tetraploids, and some of these hybrids, such as SH-3362, are currently still used in banana breeding programmes. However, the SH series were produced using mostly two wild accessions, belonging to *Musa acuminata* ssp. *burmanica* ('Calcutta 4') and ssp. *malaccensis* (Rowe and Richardson, 1975). In addition, 'Calcutta 4' which is highly resistant to Black Leaf Streak (BLS), to several races of Fusarium wilt and tolerant to certain nematodes, has been widely used in other breeding programmes (e.g. Ortiz and Vuylsteke, 1994; Batte et al., 2019). Fundamental research, notably on trait inheritance, was also conducted using the wild and fully fertile accessions of bananas. Here again, the accession 'Calcutta 4' was widely used (e.g. Bakry et al., 2009; Batte et al., 2019; Ortiz and Vuylsteke, 1992). Considering that more than 70 species exist in the genus *Musa*, these few concrete examples show how underexploited the diversity of wild banana available for breeding and research is.

Wild bananas, in addition to being fully fertile, hold potentially new and very interesting traits. These traits could be of high value to improve cultivated bananas, especially in the face of upcoming changes to climates and emerging diseases. For example, the recent evaluation of more distant wild banana species against TR4 has revealed different sources of resistance to this disease in six of the species tested, with *M. itinerans* and *M. basjoo* seeming fully resistant. This study emphasizes the value of underexplored wild banana germplasm (Li et al., 2015). However, wild *Musa* species represent only 13% of global banana genebanks' accessions (MusaNet, 2016), and not even half of the recognized species are conserved *ex-situ* and, when they are, it is often as a single accession/genotype (Ruas et al., 2017; https://www.crop-diversity.org/mgis). These huge gaps in the poor representation of wild *Musa* species in *ex-situ* facilities is worrying, not only because it hampers the use of these resources for research and breeding, but also because these taxa are increasingly threatened in the wild due to habitat loss in their native range. In India, for example, more than 50% of the wild banana taxa inventoried are considered threatened and two taxa are already extinct (Sabu et al., 2016). Another taxon gone extinct is *M. fitzalanii* F. Muell., one of the three *Musa* species native to Australia (WCSP, 2020). In such context, launching new initiatives for the collection and the conservation of wild *Musa* species is critical. However, it is a challenging task for different reasons. First, exact data on species distribution is sparse and scattered. Second, with the noticeable exception *M.*

acuminata and its subspecies, the magnitude and location of wild species intra-specific diversity is barely described. Third, the reproductive biology of these wild species and the dynamics of the wild banana populations are poorly studied.

4 Which wild *Musa* species to collect

4.1 Distribution, diversity and taxonomy of wild Musa species

A mandatory condition for the effective collection and conservation of wild banana taxa is to have a fair idea of the extent and distribution of the diversity existing in the wild. Wild bananas belong to the genus *Musa*, which is native to the South-East Asia/Western Oceania region (Fig. 3). The native range of the genus spans from India at the north-west and as far as the Solomon Islands at the south-east where a species resembling a new form of *M. bukensis* Argent was observed on Makira and Malaita islands, but not yet firmly identified (Sachter-Smith, 2011; Daniells et al., 2016). Some seeded bananas are also found outside of the native range of the genus, but their presence is due to past human introduction. For example, feral populations of *M. balbisiana* can be found in the southern islands of Japan (Hendrickx, 2007) and specimens of *M. acuminata* ssp. *banksii* were also observed in Samoa (Simmonds, 1956a; Sardos et al., 2019a). In East Africa, populations of *M. acuminata* were reported in the islands of Pemba and Funzi in Tanzania and are invasive in Madagascar (Perrier et al., 2019).

In his reappraisal of wild *Musa* taxonomy, Häkkinen (2013) listed 70 species in the genus *Musa*. However, not all species published in literature are included

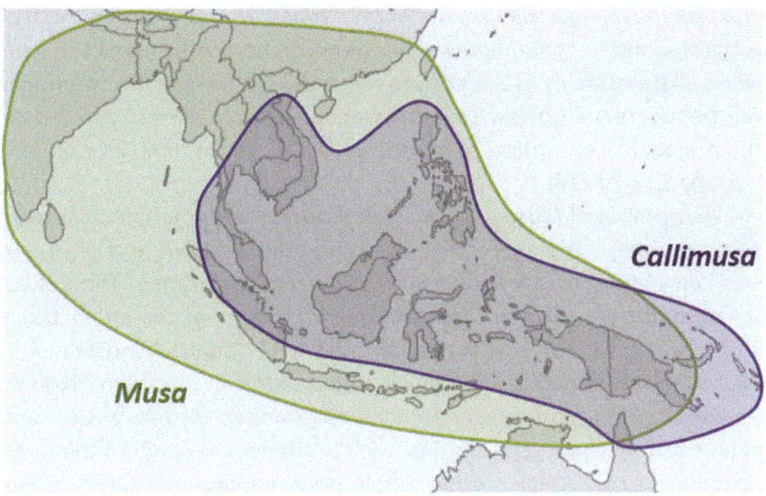

Figure 3 Distribution of the two sections of the genus *Musa*, adapted from Pollefeys et al. (2004), MusaNet (2016) and De Langhe et al. (2009).

in the list. Some were indeed described after the publication of this chapter, such as *M. argentii* Gogoi and Borah (Gogoi and Borah, 2014), *M. cylindrica* Joe, Sreejith and Sabu (Joe et al., 2014), *M. haekkinenii* N.S. Lý & Haev. (Lý et al., 2012) and the critically endangered *M. nanensis* Swangpol & Traiperm (Swangpol et al., 2015). In addition to unknown species, some synonymy may occur, that is, the same species may have been described by different teams and then published under different names. For example, Joe et al. (2016) recently published that *M. kattuvazhana* K.C. Jacob was a south Indian synonym of *M. acuminata* ssp. *burmanica* N.W. Simmonds and of the once described *M. banksii* var. *singampatti* T.G. Nayar. A monographic work was started by Markku Häkkinen to help clarify the number and occurrence of wild banana species, but Markku passed away in 2015 before completing it. The monographic work has since been taken over, and the project is still on-going (Sachter-Smith, *in prep.*).

Until recently, the more than 70 species of this genus were divided into four main botanical sections, *Eumusa* Cheesman, *Rhodochlamys* (Baker) Cheesman, *Callimusa* Cheesman and *Australimusa* Cheesman (Cheesman, 1947), and a minor one, *Ingentimusa* Argent, comprising only the giant species *Musa ingens* N. W. Simmonds (Argent, 1976). Based on a review including molecular analyses and more particularly on phylogenetic evidence published by Li et al. (2010) and Christelová et al. (2011a), Häkkinen (2013) reorganized the genus into two sections. Using 38 *Musa* taxa for a wide phylogeographic study, Janssens et al. (2016) further confirmed the occurrence of only two clades, corresponding to these two sections, in the genus. First, the section *Musa,* including the previous *Eumusa* and *Rhodochlamys* and second, the section *Callimusa* including the previous *Callimusa*, *Australimusa* and *Ingentimusa*. This new sub-division of the genus is also consistent with known cytogenetic characteristics of the different species as all the species of the section *Musa* exhibit 2n=2x=22 chromosomes, while all the species of the new *Callimusa* section have 2n=2x=20 chromosomes, with the noticeable exception of *M. beccarii* N. W. Simmonds (2n=2x=18) and of *M. ingens* (2n=2x=14).

The reappraisal of *Musa* sectional taxonomy based on molecular results is a perfect example of the essential supporting role that DNA and phylogenetic analyses can play to clarify taxa delimitation and occurrence. The systematic use of a common platform or methodology for the molecular characterization of wild *Musa*, similar to what exists for cultivated banana (Christelová et al., 2011b), could help sort out taxonomy problems such as synonymy. Nowadays, the general trend in plant research includes genomic studies based on next-generation sequencing technologies, such as the recent study of the radiation of *M. acuminata* sub-species using whole genome sequences (Rouard et al., 2018). Combined with the incentives of journals for the submission of datasets into appropriate public data repositories, it will likely prevent scientists' natural reluctance to share materials and data and will help better characterize and delineate wild *Musa* species.

4.2 Prioritizing wild *Musa* species collection

According to the passport data of 30 *ex-situ* banana collections around the world and representing a total of 6606 accessions, 31 wild *Musa* species out of the 70 are conserved in *ex-situ* facilities (Ruas et al., 2017; https://www.crop-diversity.org/mgis/). This number represents less than half of the *Musa* species still found in the wild. In addition, with the noticeable exception of the two direct ancestors of edible bananas *M. acuminata* (247 accessions conserved in genebanks) and *M. balbisiana* (169 accessions conserved in genebanks) that are conserved in 12 and 16 collections, respectively, most of the species are poorly represented in collections, with 16 of them conserved as only one accession in one collection (Fig. 4). Therefore, a priority target for wild *Musa* collecting should be those taxa and species that are not conserved yet and those that are conserved but for which no back-up exists. Taxa that are under-represented *ex-situ* should also be further collected to increase the coverage of the intra-specific diversity conserved and to minimize the risk of loss. Ideally, the conservation status of the wild banana *in-situ* should also account for prioritization of collection. A problem here is that, once again, data on wild bananas conservation status in their natural range is sparse. As an example, the International Union for Conservation of Nature (IUCN) red list database displays information on only eight *Musa* species (IUCN, 2019), which is far from

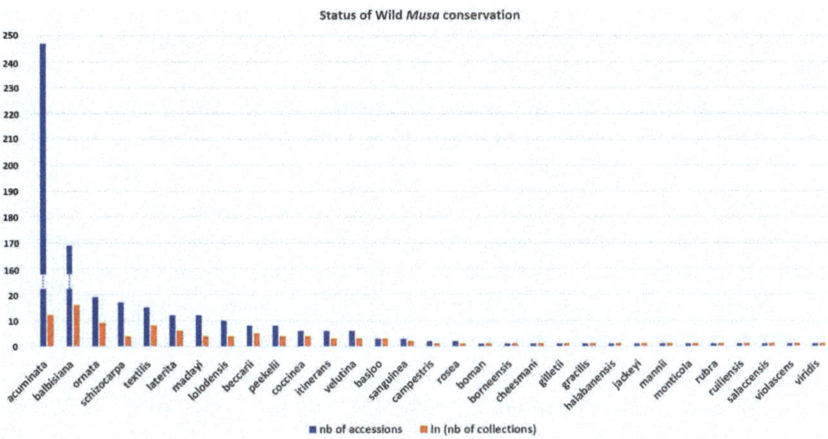

Figure 4 Current status of *ex-situ* conservation of wild banana species: out of more than 70 species of wild bananas, 31 are conserved in genebanks, of which 16 species are conserved in the form of a single accession. Data retrieved from the *Musa* Germplasm Information System (MGIS) and based on the passport data on 6606 banana accessions conserved in 30 Genebanks.

representing the entire genus *Musa*. Noticeably, out of these eight species, three are considered critically endangered: *M. argentii*, *M. arunachalensis* and *M. insularimontana* (Allen and Gogoi, 2019; Allen and Sabu, 2019; Allen, 2019). The lack of data on the conservation status of wild bananas demands a broad ranging collecting strategy in which all taxa and all populations encountered should be systematically collected.

5 How to collect wild *Musa* species

5.1 What material to collect?

The clonal nature of the banana crop has also largely disfavoured past efficient wild banana collecting. Edible seedless bananas are cultivated and propagated as clones using suckers or as *in vitro* plantlets as landraces and cultivars are highly or completely sterile. Once a clone (containing a combination of favourable genes and alleles) is identified and deemed successful (i.e. acceptable to farmers), vegetative propagation is the only way to maintain it from one cultivation cycle to another. In such context, *ex-situ* conservation facilities are designed around the idea that sterile clones rather than seeded wild populations must be conserved. One has to keep in mind that field collections of clones regularly need to be replanted, through suckers, *in-vitro* culture and more recently through cryopreservation of meristems (Van Den Houwe et al., 2000; Panis, 2009). Such a clonal approach to conservation is relevant and mandatory for the banana crop. However, many of the wild *Musa* species are difficult to conserve *ex-situ* as plants in the field because they are poorly adapted to cultivation conditions. Some may establish but do not flower, such as *M. kattuvazhana* (Jacob, 1952), others may even not establish at all, such as *M. ingens* (Argent, 1976).

Wild *Musa* species are fully fertile with many of them being allogamous. In the wild, they exist as populations that hold an inherent potential for adaptation, through the continuous reshuffling of genes and allelic combinations during sexual reproduction. Therefore, to correctly capture traits and adaptation capacities of wild bananas, populations, in addition to individuals/genotypes, should be collected. As wild bananas have been long conserved along with their cultivated counterparts, collection and conservation methods applied to wild *Musa* species were largely similar to those for cultivated bananas. As a result, most of the wild bananas collected in the past were collected as unique genotypes and then conserved clonally, for example, Paul Allen collecting missions during 1959–1961 (Rosales et al., 1999) or the expeditions to Papua New Guinea in the late 1980s (Sharrock, 1988, 1989; Sharrock et al. 1988, 1989). Seeds of wild Musa species were also collected, but not for the purpose of conserving the widest range of diversity as possible. Past collectors, even though extremely curious about the taxonomy of wild bananas, mainly targeted the discovery of specimens that would be of direct use for commercial banana breeding schemes. For example, the extensively used accession 'Calcutta 4' (*M. acuminata* ssp. *burmanica*, originating in Myanmar) was initially sent to West Indies from the Calcutta botanical garden as a seed (Cheesman, 1948). In the same way, although Simmonds (1956b) during his trip to South-East Asia and the Pacific collected seeds of new wild material, the aim was to later grow them to enable cytogenetic studies.

Recently, the need for the conservation of wild banana populations in the form of seeds rather than clonal individuals has been recognized. A few initiatives have been conducted to collect seeds, in addition to suckers, of wild bananas. It was for example the case in India between 1999 and 2005 during a National Project (Pandey et al., 2008). In 2015, wild banana seeds from Vietnam were collected in the framework of the Crop Wild Relatives project (https://www.cwrdiversity.org/project/) and sent to the Millennium Seed Bank at Kew's Royal Botanical Garden that opened in 2000 (https://www.kew.org/wakehurst/whats-at-wakehurst/millennium-seed-bank). In 2016, acknowledging the need for a global coordinated wild banana seed collection, the Bioversity International *Musa* Germplasm Transit Centre (ITC) (https://www.bioversityinternational.org/banana-genebank/) signed an agreement with the Botanic Garden of Meise (BGM) in Belgium to establish the Global Seed Bank for Banana at the BGM. In this context, wild banana seed collections occurred in the Autonomous Region of Bougainville and in mainland Papua New Guinea in 2016 and 2017, respectively (Sardos et al., 2017a,b). In 2019, seeds from other species and other populations were collected in other provinces of Papua New Guinea (Carpentier and Janssens, pers. com.; Sardos et al., 2019b). After a few years of initial research, mainly to resolve wild banana population dynamics, the wild banana seeds will be available for distribution upon request under the conditions of the International Treaty for Plant Genetic Resources for Food and Agriculture (ITPGRFA) of the FAO.

6 Theoretical and practical issues in collecting wild bananas

The shift in focus to collecting populations as seeds rather than fixed genotypes (suckers) has raised new theorical and practical issues to ensure the efficient collection of wild *Musa* diversity, such as how many seeds to collect, from how many individuals or what should be the optimal sampling design. To answer these questions, it is necessary to understand well the genetic structure of the targeted species. Given that genetic structure is correlated to the reproductive biology of a species, we review below the reproductive modes of wild *Musa* species.

6.1 Wild Musa species reproductive biology

Wild *Musa* plants are monoecious, i.e. each plant contains both female and male flowers that are usually unisexual. On the unique inflorescence of the plant, or flowering stalk, the exposure of female and male flowers is asynchronous. That is to say that the basal female flowers are exposed first, then when the ovaries develop into fruits, male flowers located on the upper nodes of the stalk start to be exposed. This behaviour prevents pollen from the male flowers' anthers to be deposited on the stigmas of female flowers of the same plant. Therefore, most of wild *Musa* species are allogamous and cross-pollination is the rule. In a few species, however, female flowers are male fertile which implies that self-pollination should be the dominant reproductive mode for those taxa. This is, for example, the case of *M. boman* Argent and *M. ingens* in the section Callimusa and of *M. schizocarpa* Simmonds in the section *Musa* (Simmonds, 1956a; Argent, 1976). If all individuals of the previously cited species are expected to bear hermaphrodite flowers, there are several cases where some level of inter-specific variability exists with regard to this feature. In *M. acuminata* (section *Musa*), all subspecies are expected to bear unisexual flowers with the exception of *M. acuminata* ssp. *errans* (Blanco) Valmayor in the Philippines, in which all basal flowers have been described as hermaphrodites (Valmayor, 2001) and *M. acuminata* ssp. *banksii* Muell. in New-Guinea, Samoa and Queensland (Australia), which exhibits various levels of male fertility in its female flowers according to their position on the stalk (Simmonds, 1956a). Recently, it was also established that *M. kattuvazhana* Jacob, native to South India and re-described as a synonym of *M. acuminata* ssp. *burmanica* Simmonds, exhibits both basal female and bisexual flowers (Joe et al., 2016). Equally, out of the four botanical varieties of *M. lawitiensis* Nasution & Supard (section *Callimusa*) described in Borneo, three exhibit basal hermaphrodite flowers while the fourth one only bears unisexual flowers (Häkkinen, 2006). In theory, in a species or taxa that is predominantly selfing, the seeds of a single individual are expected to hold a limited amount of diversity compared to what may exist in the entire taxa. On the contrary, in species or taxa that are out-crossing, seeds of a given individual plant are expected to hold a wider amount of diversity. The occurrence of hermaphrodite flowers on the stalk will direct the reproduction regime towards self-pollination while the absence of such flowers will stimulate cross-pollination. Therefore, levels of intra-specific variation mean that each taxon should be considered separately when establishing collecting protocols.

6.2 Pollinators of wild Musa species

Both pollen and seed dispersal affect the genetic structure of a population by influencing the rate and distance at which genes move across its range. The spatial structure of the population and the levels of genetic differentiation between and among populations will therefore be impacted by the nature of pollination and seed dispersal. In predominantly selfing species, i.e. those taxa bearing hermaphrodite flowers, geneflow mainly occurs through seed dispersal, but in outcrossing species it will also occur through pollination. The nature and mobility of the pollinator will therefore impact a species population genetic structure. In the case of pollinators traveling over short distances, higher levels of genetic structure are expected among populations given that geneflow is confined within a small area. On the contrary, when pollination is performed by certain birds and bats, geneflow is expected to occur over longer distances, smoothing the genetic structure among populations by connecting separated plant populations.

In banana, pollination occurs mainly through nectarivore birds and bats and to some extent by small climbing mammals such as tree shrews in Sumatra (Nur, 1976; Itino et al., 1991). It has been demonstrated that none of the numerous insects and bugs observed in banana flowers in Java and Sumatra were efficient pollinators; however, they may attract mammals that feed on them and pollinate bananas (Nur, 1976). Wild banana inflorescences show some levels of adaptation to the main pollinators and it is assumed that taxa with pendent or horizontal flowering stalks are primarily pollinated and dispersed by nectar-eating bats. These species usually flower at night-time and release a strong odour that is attractive to bats. Given that bats locate their food mainly through echolocation, the colours of the buds and flowers of these species need not be bright and visually appealing. This may explain the inconspicuous colours of the buds and the pale bell-shaped flowers observed in these taxa. The shape of the bud is also well adapted as bats can potentially hold it while feeding and the opened bracts help them to echolocate flowers (Murphy et al., 2016). Even the nectar composition, which has high levels of polysaccharides and proteins and a jelly-like consistence, has been shown to be attractive to bats. On the contrary, the species which exhibit erect inflorescences are pollinated by birds, at least for those species of high ornamental value that display bright colourful buds and flowers, and that flower during day-time. In these taxa, adaptation to bird pollination also induced the production of a watery, more dilute nectar (Nur, 1976) which is, in addition to vivid floral displays, another classic feature in plants pollinated by birds (Cronk and Ojeda, 2008). However, most of these observations have been made on a limited set of species, all in Indonesia (Nur, 1976; Itino et al., 1991). Liu et al. (2002), studying the reproductive ecology of *M. itinerans* Cheesman in the rain forests of Yunnan (China), discovered that this species was flowering twice a day, during early morning and late afternoon, and was effectively pollinated both by sunbirds and bats despite its horizontal (female phase) and pendent (male phase) flowering stalk. In Papua New Guinea, it is also noticeable that the colours of the bud and of the flowers of the erect *M.*

maclayi F. Muell. are not very different from the colours of the bud and flowers of the pendent *M. peekilii* Lauterbach (Fig. 5). Therefore, it may well be that pollination in wild *Musa* is more complex than a simple segregation between erect and horizontal/pendent flowering species.

6.3 Seed dispersers of wild Musa species

The fleshy fruits of wild bananas, even though they are full of hard seeds, are appealing for a variety of animals. Seed dispersal can occur in two ways. Either fruits are entirely removed from the bunch and carried away further, or they are consumed directly on the bunch. In the first case, in addition to the seeds that can be spat out during fruit consumption, notably by some bat species, fruits can accidentally fall during transportation, thus leading to the dispersal of seeds to potential new habitats (Tang et al., 2007). In the second case, the seeds fall on the ground around the mother plant. Once on the ground, secondary dispersers such as ants further spread the seeds. It is to be noted here that experimental ingestion of ripe fruits of *M. acuminata* and *M. balbisiana* by pigs and the subsequent digestion of their seeds affected seed behaviour with regard to germination. It altered the shape of the germination time curve in both species, but only accelerated germination in *M. balbisiana* (Simmonds, 1959). It is therefore considered that ingestion of seeds by animals,

Figure 5 Male bud of *M. maclayi* doesn't exhibit bright colors even though the species has erect inflorescences (left). An inflorescence of the pendent *M. peekilii* entering the male stage (right). Both photos taken in Papua New Guinea from Sardos et al. (2017b).

if they are not destroyed during feeding, and their expulsion via faeces is also a likely process for seed dispersion with different incidences on dispersal rates according to the animal species considered.

The primary wild banana seed dispersers are fruit bats and birds, but also civets and rodents for *M. itinerans* (Liu, 2001 *in* Liu et al., 2002). In south India, monkeys were mentioned to be consumers of fruits of *M. kattuvazhana* in addition to birds (Jacob, 1952). In China, observations of mats of *M. acuminata* established that most of the ripe fruits were consumed at night by bats which remove fruits and then fly them back to their roosts, and that some 20% of ripe fruits were eaten by birds during the day time. In birds, two behaviours were observed: one species was removing entire fruits while the two others were pecking the pulp directly on the plant (Meng et al., 2012). Personal observations in Papua New Guinea show that bananas are indeed either eaten on the bunch or taken away (Figs 6 and 7). There is no study, to my knowledge, targeting wild banana seeds dispersers but there is a profusion of birds and bats in Papua New Guinea. In addition, the erect *M. maclayi* is usually found growing along with the pendent *M. schizocarpa* and/ or M. *acuminata* ssp. *banksii* (Argent, 1976; Sardos et al., 2017b), and it is likely that seed dispersers are the same for these three species. It is probable here that the nature of the seed dispersers for a given wild banana species is linked to the fauna present in the area of study at present rather than to the target species.

Figure 6 Bunch of *M. acuminata* ssp. *banksii* with ripe fruits (left). It is noticeable from the rotten fruit pedicels at the basal part of the bunch that older fruits were totally removed from the bunch. Ripening bunch of *M. peekilii* (right). Black rotten skin of fruits eaten on the bunch is visible at the base of the bunch. Photos taken in Papua New Guinea (Sardos et al., 2017b).

Figure 7 Partially eaten fruits of *M. peekilii* (left) and of *M. schizocarpa* (right). It is noticeable that ripe fruits of *M. schizocarpa* self-peel when ripe (center). Photos taken in Papua New Guinea (Sardos et al., 2017b).

6.4 Practical implication for wild Musa species collecting

The reproduction regime, along with the pollen and seed dispersal modes, influences the genetic diversity within a geographical range and the differentiation among populations. The study conducted by Ge et al. (2005) on a wide sample from many wild populations of *M. balbisiana* in China showed that pollen and seed dispersal unequally contribute to geneflow. It appeared that seed dispersal was limited to local populations and geographical regions, while pollen was dispersed more widely, including across geographical barriers. This pattern can be explained by the capacity of fruit bats, feeding on nectar and covered by pollen, to travel over long distances. On the contrary, fruits are heavy and eaten by bats or other animals in the vicinity of the wild banana population. The same study also showed that the presence of bats in the region correlated to lower levels of genetic structuring among populations, due to a higher incidence of geneflow between them, while their absence induced higher genetic differentiation among populations. Another study conducted on *M. balbisiana* on a larger scale, found high genetic differentiation among seed collections made in different countries (Bawin et al., 2019). It also showed low genetic diversity within seeds collected on a local scale. Seeds from the same bunch are indeed half-siblings by their mother, and it also appeared here that they include alleles from a relatively small number of pollen donors.

Altogether, information on pollen and seed dispersal is of great help to develop collecting strategies. First, it confirms that, for a given taxon, collecting should occur across a wide geographical range to maximize the genetic diversity collected. Second, it suggests that great care should be taken in exhaustively collecting populations in areas with more differentiated populations. Finally, more focus should be given to the collection of within-population diversity, especially in the regions where genetic structuring is lower. However, in the field, seed collection obeys rules that often prevent teams from following theorical protocols. Experience has shown that seeds collected from unripe bunches have poor germination rates compared to those collected on ripe bunches (Simmonds, 1959; B. Panis, pers. com.). Therefore, it is of prime importance in the field to collect fruits at the appropriate stage of development. Even though wild bananas flower all year round, it is not easy to find several bunches at the right stage for a given population. In Papua New Guinea, despite the frequent occurrence of wild bananas in the regions visited, we were not able to collect many ripe bunches (Sardos et al., 2017b). It is therefore important in wild *Musa* species collection strategies to collect both suckers from parental populations and seeds. It also calls for collection strategies that would repeatedly visit the same sites rather than make only one collection. Such strategies would result in enough seeds with good germination rates to ensure the long-term conservation of the taxa collected. In addition, it would include seeds from different individuals within the same population and thus would help minimize intra-specific sampling gaps (Bawin et al., 2019).

7 On the importance of capturing passport data and making them publicly available

Data on wild *Musa* species is often incomplete and scattered in recent reports, journal articles and books. It is therefore of utmost importance to gather the right information on the taxa and populations collected. This information includes collecting codes, time and place of collection, type of material collected (e.g. seeds or suckers), the assumed taxonomy of the sample, local name if existing, a quick description of the habitat and also characterization data in the form of photographs (IPGRI, 1996). If past records of passport data linked to the accessions conserved in genebanks were of unequal quality, it is expected that modern means, such as GPS, will enable the capture of sharp data at least for locations. Equally, digital photography allows the capture of numerous good photos at low cost. Standard sets of photos for the description of bananas have been developed (TAG, 2010) and combined with the use of the appropriate colour charts (IPGRI, 1996); they can help resolve taxonomic conflicts (Fig. 8). In addition, molecular markers such as microsatellites (SSR) have been proven many times to be excellent tools for the characterization of cultivated and wild bananas (e.g. Perrier et al., 2019 Christelová et al., 2017; Ge et al., 2005; Bawin et al., 2019). Combined with the recent advances made in genomics that allow the production of invaluable sequence data, silica-dried leaf samples taken when collecting wild bananas accessions can be used to produce DNA in sufficient quality and quantity to perform high-throughput

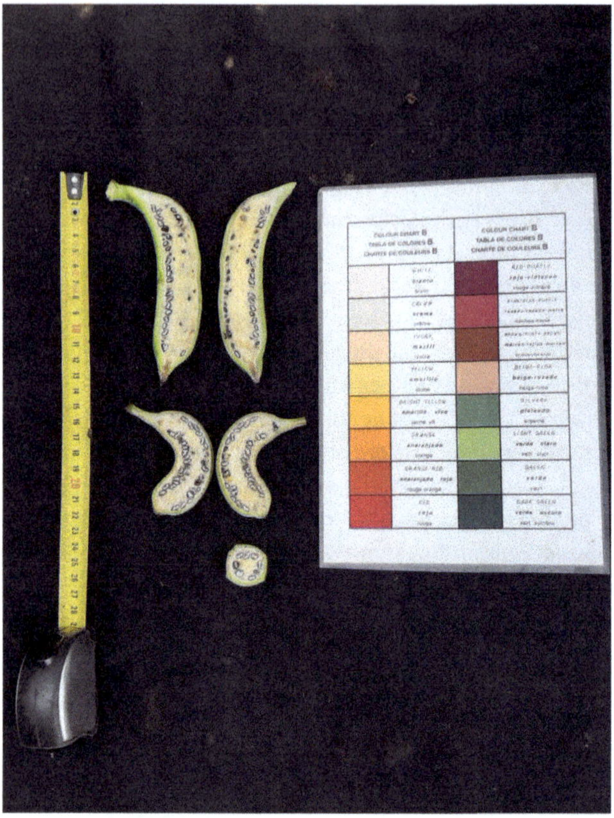

Figure 8 Descriptors and tools are used to standardize passport data collection of wild *Musa* species. Here, the color chart from the standard descriptors developed by IPGRI (1996) is used to document *M. acuminata* ssp. *banksii*. Photo taken in Samoa by Sardos et al. (2019b).

genotyping such as Genotyping-by-Sequencing or RADseq. It is important to coordinate projects concomitant to collecting expeditions to enable the production of such datasets.

Data associated with the germplasm collected and conserved in genebanks represents a mine of information for researchers, scientists and breeders, now and in the coming generations. Such data can allow the long-term monitoring of species and help direct conservation initiatives. It can also guide researchers and breeders towards unravelling the trait potential of specific taxa. It is therefore of high importance that such information is produced in a standardized format, properly stored and backed-up. Several information systems for plant germplasm collections exist, such as Genesys (www.genesys-pgr.org) and GRIN-Global (www.ars-grin.gov/ngps), and a specific database for banana has been created. The *Musa* Germplasm Information System

(MGIS) is the largest database existing for this crop. It includes passport and characterization data, voucher images and genomic-based marker information generated for a large number of accessions. Thirty banana collections have deposited the data associated with their germplasm in this system, which allows free and easy access to any information provided by the collections and features an easy portal for ordering material from the Bioversity International *Musa* Germplasm Transit Centre (ITC) (Ruas et al., 2017). In the context of wild *Musa* species collection, this system provides a reliable interface between collections in which the collected material is deposited for conservation and for any user in need of information on wild *Musa* species.

8 Conclusion

In the past several years, many projects targeting the collection and study of populations of wild *Musa* germplasm have taken place, such as in the Autonomous Region of Bougainville in 2016 (Sardos et al., 2017a) and in the Morobe and Madang provinces of Papua New Guinea in 2017 (Sardos et al., 2017b). The wide populations sampled during these campaigns constitute a remarkable basis for in-depth studies of wild bananas. As an example, the analysis by Bawin and colleagues (2019) of *M. balbisiana* populations from different sources, including Papua New Guinea, with microsatellite markers allowed for a better understanding of the geographical organization of the genetic diversity of this species and led to practical recommendations for future seed collecting. Additional samples from different species have been collected more recently in Papua New Guinea (S. Carpentier, pers. com.), in North Vietnam (Caruana et al., 2019; S. Janssens, pers. com.), in Laos and in China (M. Chabannes, pers. com.). Their upcoming genetic analysis, together with analyses of different taxa collected in Bougainville and Papua New Guinea, should also allow the development of recommendations for collecting missions adapted to the specificity of the different species studied. It will also provide information on the dynamics of genetic diversity in *Musa* at a level that was never reached before, and will emphasize the urge for collecting and conserving more than 70 species of wild bananas described from the wild.

9 Where to look for further information

To explore information and passport data of banana germplasm conserved in 30 ex-situ collections around the world and to order plant materials from the International Transit Centre (ITC): Musa Germplasm Information System (MGIS) https://www.crop-diversity.org/mgis/.

To explore scientific literature about bananas, visit MusaLit website: http://www.musalit.org/.

For general public: learn about wild and cultivated bananas on the ProMusa website http://www.promusa.org/.

As references for early banana collections:

Simmonds, N. W. (1956a), 'Botanical results of the banana collecting expedition', 1954–5. *Kew Bull.* 11(3), 463–489.

Simmonds, N. W. (1956b), 'A banana collecting expedition to South East Asia and the Pacific', *Trop. Agr.* 33(4), 251–271.

For a comprehensive study of wild bananas in Papua New Guinea:

Argent, G. (1976), 'The wild bananas of Papua New Guinea', *Notes from the Royal Botanic Garden, Edinburgh* 35(1), 77–114.

and latest collecting mission in Papua New Guinea:

Sardos, J., Paofa, J., Janssens, S., Vanden Abeele, S., Roux, N. and Panis, B. (2017b), Report from the Exploration of wild banana populations in Papua New Guinea, 7–17 June 2017. Bioversity International, Montpellier, 20p.

On the importance of molecular phylogeny to understand the history of wild bananas:

Janssens, S. B., Vandelook, F., De Langhe, E. A. L., Verstraete, B., Smets, E., Vandenhouwe, I. and Swennen, R. (2016), 'Evolutionary dynamics and biogeography of Musaceae reveal a correlation between the diversification of the banana family and the geological and climatic history of Southeast Asia', *New Phytol.* 210(4), 1453–1465. http://dx.doi.org/10.1111/nph.13856.

To understand how the assessment of the genetic diversity and the levels of structuring between populations of wild banana is important for conservation:

Bawin, Y., Panis, B., Vanden Abeele, S., Li, Z., Sardos, J., Paofa, J., Ge, X.-J., Mertens, A., Honnay, O. and Janssens, S. (2019), 'Genetic diversity and core subset selection in ex situ seed collections of the banana crop wild relative *Musa balbisiana*', *Plant. Genet. Resour.-C,* 17(6), 536–544. doi:10.1017/S1479262119000376.

Ge, X., Liu, M. H., Wang, W. K., Schaal, B. A. and Chiang, T. Y. (2005), 'Population structure of wild bananas, *Musa balbisiana*, in China determined by SSR fingerprinting and cpDNA PCR-RFLP', *Mol. Ecol.*, 14(4), 933–944. http://dx.doi.org/10.1111/j.1365-294X.2005.02467.x.

For wild banana species accepted names, but still to be complemented and updated:

Häkkinen, M. (2013), 'Reappraisal of sectional taxonomy in *Musa* (*Musaceae*)', *Taxon,* 62(4), 809-813. http://dx.doi.org/10.12705/624.3.

10 References

Allen, R. (2019). *Musa insularimontana, The IUCN Red List of Threatened Species* 2019: e.T111907086A111907091. Available at: https://dx.doi.org/10.2305/IUCN.UK.2019-3.RLTS.T111907086A111907091.en. Downloaded on 14 January 2020.

Allen, R. and Gogoi, R. (2019). Musa argentii, *The IUCN Red List of Threatened Species* 2019: e.T123707198A123802003. Available at: https://dx.doi.org/10.2305/IUCN.UK.2019-3.RLTS.T123707198A123802003.en. Downloaded on 09 January 2020.

Allen, R. and Sabu, M. (2019). *Musa arunachalensis, The IUCN Red List of Threatened Species* 2019: e.T123707215A123802008. Available at: https://dx.doi.org/10.2305/IUCN.UK.2019-3.RLTS.T123707215A123802008.en. Downloaded on 09 January 2020.

Argent, G. (1976). The wild bananas of Papua New Guinea, *Notes from the Royal Botanic Garden, Edinburgh* 35(1), 77–114.

Bakry, F., Carreel, F., Jenny, C. and Horry, J. P. (2009). Genetic improvement of banana. In: Jain, S. M. and Priyadarshan, P. M. (Eds), *Breeding Plantation Tree Crops: Tropical Species*. Springer, New York, pp. 3-50. Available at: http://aob.oxfordjournals.org/content/104/6/vii.extract.

Batte, M., Swennen, R., Uwimana, B., Akech, V., Brown, A., Tumuhimbise, R., Hovmalm, H. P., Geleta, M. and Ortiz, R. (Feb 5, 2019). Crossbreeding East African Highland bananas: lessons learnt relevant to the botany of the crop after 21 years of genetic enhancement, *Frontiers in Plant Science* 10, 81. doi: 10.3389/fpls.2019.00081.

Bawin, Y., Panis, B., Vanden Abeele, S., Li, Z., Sardos, J., Paofa, J., Ge, X.-J., Mertens, A., Honnay, O. and Janssens, S. B. (2019). Genetic diversity and core subset selection in ex situ seed collections of the banana crop wild relative *Musa balbisiana*, *Plant Genetic Resources: Characterization and Utilization* 17(6),536-544. doi: 10.1017/S1479262119000376.

Bellon, M. R., Dulloo, E., Sardos, J., Thormann, I. and Burdon, J. J. (2017). In situ conservation-harnessing natural and human-derived evolutionary forces to ensure future crop adaptation, *Evolutionary Applications* 10(10), 965-977. doi: 10.1111/eva.12521.

Caruana, M.-L., Ly, N.-S., Haevermans, T., Jenny, C., Perrier, X., Sardos, J. and Chabannes, M. (2019). Screening of wild seedy *Musa balbisiana* biodiversity to fill up phylogenetic gaps and fuel future breeding programs, *Rencontres de Virologie Végétale, Aussois* 87, 28.

Cheesman, E. E. (1947). Classification of the bananas. II. The genus Musa L, *Kew Bulletin* 2(2), 106–117.

Cheesman, E. E. (1948). Classification of the bananas. III. Critical notes on species: Musa balbisiana, *Kew Bulletin* 1,11-28.

Christelová, P., De Langhe, E. A. L., Hribova, E., Cizková, J., Sardos, J., Hušáková, M., Van Den Houwe, I., Sutanto, A., Kepler, A. K., Swennen, R., Roux, N. and Dolezel, J.

(2017). Molecular and cytological characterization of the global *Musa* germplasm collection provides insights into the treasure of banana diversity, *Biodiversity and Conservation* 26(4),801–824. Available at: http://dx.doi.org/10.1007/s10531-016-1273-9.

Christelová, P., Valarik, M., Hribova, E., De Langhe, E. A. L. and Dolezel, J. (2011a). A multi gene sequence-based phylogeny of the Musaceae (banana) family, *BMC Evolutionary Biology* 11, 103. Available at: http://dx.doi.org/10.1186/1471-2148-11-103.

Christelová, P., Valarik, M., Hribova, E., Van Den Houwe, I., Channeliere, S., Roux, N. and Dolezel, J. (2011b). A platform for efficient genotyping in Musa using microsatellite markers, *AoB PLANTS* 2011, plr024. Available at: http://dx.doi.org/10.1093/aobpla/plr024.

Cronk, Q. and Ojeda, I. (2008). Bird-pollinated flowers in an evolutionary and molecular context, *Journal of Experimental Botany* 59(4),715–727. Available at: https://dx.doi.org/10.1093/jxb/ern009.

Daniells, J. W., Sachter-Smith, G. and Taylor, M. (2016). Bananas adrift in time – a case study in the Solomons, Proceedings of the IX International Symposium on Banana: ISHS-ProMusa Symposium on Unravelling the Banana's Genomic Potential, Brisbane. *Acta Horticulturae* 1114, 27–32. ISHS, Leuven, Belgium. Available at: http://dx.doi.org/10.17660/ActaHortic.2016.1114.4.

De Langhe, E. A. L., Vrydaghs, L., De Maret, P., Perrier, X. and Denham, T. P. (2009). Why bananas matter: an introduction to the history of banana domestication, *Ethnobotany Research and Applications* 7, 165–177. Available at: http://journals.sfu.ca/era/index.php/era/article/view/356.

Dempewolf, H., Baute, G., Anderson, J., Kilian, B., Smith, C. and Guarino, L. (2017). Past and future use of wild relatives in crop breeding, *Crop Science* 57(3),1070–1082. doi: 10.2135/cropsci2016.10.0885.

Ge, X. J., Liu, M. H., Wang, W. K., Schaal, B. A. and Chiang, T. Y. (2005). Population structure of wild bananas, *Musa balbisiana*, in China determined by SSR fingerprinting and cpDNA PCR-RFLP, *Molecular Ecology* 14(4), 933–944. Available at: http://dx.doi.org/10.1111/j.1365-294X.2005.02467.x.

Gogoi, R. and Borah, S. (2014). *Musa Argentii* (Musaceae): a new species from Arunachal Pradesh, India, *Edinburgh Journal of Botany* 71(2), 181–188. Available at: http://dx.doi.org/10.1017/S0960428614000079.

Häkkinen, M. (2006). *Musa lawitiensis* Nasution and Supard. (Musaceae) and its intraspecific taxa in Borneo, *Adansonia* 28(1), 55–65. Available at: http://sciencepress.mnhn.fr/en/periodiques/adansonia/28/1/musa-lawitiensis-nasution-supard-musaceae-et-ses-taxons-intraspecifiques-borneo.

Häkkinen, M. (2013). Reappraisal of sectional taxonomy in *Musa* (*Musaceae*), *Taxon* 62(4), 809–813. Available at: http://dx.doi.org/10.12705/624.3.

Hendrickx, K. (2007). The origins of banana-fibre cloth in the Ryukyus, Japan. Leuven University Press, Leuven, 318p. Available at: muse.jhu.edu/book/24395.

IPGRI (1996). *Descriptors for Banana (Musa spp.)*. IPGRI, Rome (ITA), 55p.

Itino, T., Kato, M. and Hotta, M. (1991). Pollination ecology of the two wild bananas, *Musa acuminata* subsp. *halabanensis* and *M. salaccensis*: chiropterophily and ornithophily, *Biotropica* 23(2),151–158. doi: 10.2307/2388300.

IUCN (2019). *The IUCN Red List of Threatened Species*. Version 2019-3. Available at: http://www.iucnredlist.org. Downloaded on 10 December 2019.

Jacob, K. C. (1952). *Madras Bananas: A Monograph*. Government Press, Chennai, 228p.

Janssens, S. B., Vandelook, F., De Langhe, E. A. L., Verstraete, B., Smets, E., Vandenhouwe, I. and Swennen, R. (2016). Evolutionary dynamics and biogeography of Musaceae reveal a correlation between the diversification of the banana family and the geological and climatic history of Southeast Asia, *New Phytologist* 210(4), 1453-1465. Available at: http://dx.doi.org/10.1111/nph.13856.

Joe, A., Sreejith, P. E. and Sabu, M. (2016). The identity of *Musa kattuvazhana* (Musaceae) with reduction of *Musa acuminata* subsp. *burmannica* and *Musa banksii* var. *singampatti* as its synonyms, *Webbia: Journal of Plant Taxonomy and Geography* 71(2), 203–208. Available at: http://dx.doi.org/10.1080/00837792.2016.1200820.

Joe, A., Sreejith, P. E. and Sabu, M. (2014). *Musa cylindrica*, a new species of Musa (Musaceae) from North-East India, *Phytotaxa* 172, 137–140. Available at: 10.11646/phytotaxa.172.2.11.

Li, L. F., Häkkinen, M., Yuan, Y. M., Hao, G. and Ge, X. J. (2010). Molecular phylogeny and systematics of the banana family (Musaceae) inferred from multiple nuclear and chloroplast DNA fragments, with a special reference to the genus *Musa*, *Molecular Phylogenetics and Evolution* 57(1),1–10. Available at: http://dx.doi.org/10.1016/j.ympev.2010.06.021.

Li, W. M., Dita, M., Wu, W., Hu, G. B., Xie, J. H. and Ge, X. J. (2015). Resistance sources to *Fusarium oxysporum* f. sp. *cubense* tropical race 4 in banana wild relatives, *Plant Pathology* 64(5), 1061–1067. Available at: http://dx.doi.org/10.1111/ppa.12340.

Liu, A. Z., Liz, D. Z., Wang, H. and Kress, W. J. (2002). Ornithophilous and chiropterophilous pollination in *Musa itinerans* (Musaceae), a pioneer species in tropical rain forests of Yunnan, Southwestern China, *Biotropica* 34(2), 254–260. Available at: http://dx.doi.org/10.1111/j.1744-7429.2002.tb00536.x.

Lý, N., Lê, C., Triệu, T., Haevermans, A., Lowry, P. I. P. and Haevermans, T. (2012), A distinctive new species of wild banana (Musa, Musaceae) from northern Vietnam, *Phytotaxa* 75(1),33–42. doi: 10.11646/phytotaxa.75.1.3.

Meng, L., Gao, X., Chen, J. and Martin, K. (2012). Spatial and temporal effects on seed dispersal and seed predation of *Musa acuminata* in southern Yunnan, China, *Integrative Zoology* 7(1), 30–40. Available at: http://dx.doi.org/10.1111/j.1749-4877.2011.00275.x.

Murphy, M., Clare, E. L., Rydell, J., Yovel, Y., Bar-On, Y., Oelbaum, P. and Fenton, M. B. (2016). Opportunistic use of banana flower bracts by *Glossophaga soricina*, *Acta Chiropterologica* 18(1), 209–213.

MusaNet (2016). *Global Strategy for the Conservation and Use of Musa (Banana) Genetic Resources*. Montpellier, France, 226p.

Nur, N. (1976). Studies on pollination in Musaceae, *Annals of Botany* 40(2),167–177. Available at: http://dx.doi.org/10.1093/oxfordjournals.aob.a085120.

Ortiz, R. and Vuylsteke, D. (1992). Inheritance of black Sigatoka resistance and fruit parthenocarpy in triploid AAB plantain. In: *Agronomy Abstract*. American Society of Agronomy, Madison, WI, p. 109.

Ortiz, R. and Vuylsteke, D. (1994). Plantain breeding at IITA. In: Jones, D. R. (Ed.), Proceedings of the First Global Conference of the International Musa Testing Program, La Lima (HND). INIBAP, Montpellier, pp. 130–156.

Pandey, A., Tomer, A. K., Bhandari, D. C. and Pareek, S. K. (2008). Towards collection of wild relatives of crop plants in India, *Genetic Resources and Crop Evolution* 55(2), 187–202. doi: 10.1007/s10722-007-9227-4.

Panis, B. (2009). Cryopreservation of *Musa Germplasm*. Technical Guidelines No. 9 (2nd edn.). Bioversity International, Montpellier.

Perrier, X., Jenny, C., Bakry, F., Karamura, D. A., Kitavi, M., Dubois, C., Hervouet, C., Philippson, G. and De Langhe, E. A. L. (2019). East African diploid and triploid bananas: a genetic complex transported from South-East Asia, *Annals of Botany* 123(1), 19–36. Available at: http://dx.doi.org/10.1093/aob/mcy156.

Pollefeys, P., Sharrock, S. and Arnaud, E. (2004). *Preliminary Analysis of the Literature on the Distribution of Wild Musa species Using MGIS and DIVA-GIS*. IPGRI, Rome.

Rosales, F. E., Arnaud, E. and Coto, J. (1999). *A Tribute to the Work of Paul Allen: A Catalogue of Wild and Cultivated Bananas*. INIBAP, Montpellier.

Rouard, M., Droc, G., Martin, G., Sardos, J., Hueber, Y., Guignon, V., Cenci, A., Geigle, B., Hibbins, M. S., Yahiaoui, N., Baurens, F. C., Berry, V., Hahn, M. W., D'Hont, A. and Roux, N. (2018). Three new genome assemblies support a rapid radiation in *Musa acuminata* (wild banana), *Genome Biology and Evolution* 10(12), 3129–3140. http://dx.doi.org/10.1093/gbe/evy227

Rowe, P. R. and Richardson, D. L. (1975). Breeding bananas for disease resistance, fruit quality and yield, *SIATSA Bulletin*. SIATSA, La Lima 2017, bax046.

Ruas, M., Guignon, V., Sempere, G., Sardos, J., Hueber, Y., Duvergey, H., Andrieu, A., Chase, R., Jenny, C., Hazekamp, T., Irish, B., Jelali, K., Adeka, J., Ayala-Silva, T., Chao, C. P., Daniells, J., Dowiya, B., Effa Effa, B., Gueco, L., Herradura, L., Ibobondji, L., Kempenaers, E., Kilangi, J., Muhangi, S., Ngo Xuan, P., Paofa, J., Pavis, C., Thiemele, D., Tossou, C., Sandoval, J., Sutanto, A., Vangu Paka, G., Yi, G., Van Den Houwe, I., Roux, N. and Rouard, M. (2017). MGIS: managing banana (*Musa* spp.) genetic resources information and high-throughput genotyping data, *Database*. doi: 10.1093/database/bax046.

Sabu, M., Joe, A. and Sreejith, P. E. (2016). Brief overview of diversity of wild Indian *Musaceae*. In: Proceedings of the IX International Symposium on Banana: ISHS-ProMusa Symposium on Unravelling the Banana's Genomic Potential, Brisbane, Australia, 17–22 August 2014. ISHS, Leuven. *Acta Horticulturae* 1114. Available at: http://dx.doi.org/10.17660/ActaHortic.2016.1114.10.

Sachter-Smith, G. (2011). *Bananas of the Solomon Islands: A Catalog of the Banana Plants Seen in the Islands of Guadalcanal, Makira and Malaita*. Bioversity International, Montpellier, 54 p.

Sardos, J., Paofa, J., Janssens, S., Sachter-Smith, G., Rauka, G. and Roux, N. (2017a). *Banana Collecting Mission in the Autonomous Region of Bougainville (AROB), Papua New Guinea*. Bioversity International, Montpellier, 26 p.

Sardos, J., Paofa, J., Janssens, S., Vanden Abeele, S., Roux, N. and Panis, B. (2017b). *Report from the Exploration of Wild Banana Populations in Papua New Guinea, 7–17 June 2017*. Bioversity International, Montpellier, 20 p.

Sardos, J., Paofa, J., Sachter-Smith, G., Togabalaguguwa, E., Hribova, E., Van Den Houwe, I. and Roux, N. (2019b). *Report from the Banana Collecting Mission to the Province of West New Britain, Papua New Guinea*. Bioversity International, Montpellier, 23 p.

Sardos, J., Perrier, X., Dolezel, J., Hribova, E., Christelová, P., Van Den Houwe, I., Kilian, A. and Roux, N. (2016). DArT whole genome profiling provides insights on the evolution and taxonomy of edible Banana (*Musa* spp.), *Annals of Botany* 118(7),1269–1278. Available at: http://dx.doi.org/10.1093/aob/mcw170.

Sardos, J., Sachter-Smith, G., Shandil, A., Van Den Houwe, I., Hribova, E., Matalavea, P., Vaai, A., Iosefa, T., Hunter, D. T. and Roux, N. (2019a). *Report from the Banana Collecting Mission to Samoa*. Bioversity International, Montpellier, 21 p.

Sharrock, S. (1988). *Report on the First IBPGR-QDPI Banana Germplasm Collecting Mission to Papua New Guinea, 27 February to 22 March 1988*. IBPGR, Rome, 47 p.
Sharrock, S., Jones, D. R. and Banag, J. (1989). *Report on the Third IBPGR-QDPI Banana Germplasm Collecting Mission to Papua New Guinea, 15 February to 12 March 1989*. IBPGR, Rome, 17 p.
Sharrock, S., Daniells, J. W. and Kambuou, R. (1988). *Report on the second IBPGR-QDPI banana germplasm collecting mission to Papua New Guinea, 22 October to 27 November 1988*. IBPGR, Rome, 29 p.
Sharrock, S. L. (1989). *Report on the Fourth IBPGR-QDPI Banana Germplasm Collecting Mission to Papua New Guinea, 4 June to 9 July 1989*. IBPGR, Rome, 18 p.
Simmonds, N. W. (1956a). Botanical results of the banana collecting expedition, 1954–5, *Kew Bulletin* 11(3), 463–489.
Simmonds, N. W. (1956b). A banana collecting expedition to South East Asia and the Pacific, *Tropical Agriculture* 33(4), 251–271.
Simmonds, N. W. (1959). Experiments on the germination of banana seeds, *Tropical Agriculture* 36(4), 259-273.
Swangpol, S. C., Traiperm, P., Somana, J., Sukkaewmanee, N., Srisanga, P. and Suksathan, P. (2015). *Musa nanensis*, a new banana (Musaceae) species from northern Thailand, *Systematic Botany* 40(2), 426–432.
Tang, Z., Sheng, L., Ma, X., Cao, M., Parsons, S., Ma, J. and Zhang, S. (2007). Temporal and spatial patterns of seed dispersal of *Musa acuminata* by *Cynopterus sphinx*, *Acta Chiropterologica* 9(1), 229–235.
Taxonomy Advisory Group (TAG) (2010). *Guidelines for Documenting the Minimum Set of Photos for Bananas*. Bioversity International, Montpellier.
Valmayor, R. V. (2001). Classification and characterization of Musa exotica, *M. alinsanaya* and *M. acuminata* ssp. *errans*, *The Philippine Agricultural Scientist* 84(3),325–331.
Van Den Houwe, I., Panis, B. and Swennen, R. (2000). The *in vitro* germplasm collection at the *Musa* INIBAP Transit Centre and the importance of cryopreservation. In: Engelmann, F. and Takagi, H. (Eds), *Cryopreservation of Tropical Plant Germplasm : Current Research Progress and Application*. IPGRI/JIRCAS, Rome.
WCSP (2020). World Checklist of Selected Plant Families. Facilitated by the Royal Botanic Gardens, Kew. Published on the internet; Available at: http://wcsp.science.kew.org/. Retrieved 9 January 2020.

Chapter 6

Collection and evaluation of wild Musa species

Hugo A. Volkaert, Center for Agricultural Biotechnology – Kasetsart University, Thailand

1 Introduction

The genus *Musa* includes about 70 species distributed throughout the Asian tropics and subtropics from India, Nepal, Bhutan, Southern China and Southern Japan in the north to Northern Australia in the south and the islands in the western Pacific Ocean. The haploid chromosome number in the subgenus *Musa* is 11 while in the subgenus *Callimusa* the haploid chromosome number is 10 except for *Musa beccarii N. W. Simmonds* where 2n = 2x = 18. *Musa ingens N. W. Simmonds* has a haploid chromosome number of 7 and is classified separate from the two subgenera. Several of the *Musa* species are appreciated as ornamentals in gardens and greenhouses, but the most important wild species are the progenitors and relatives of the edible bananas and plantains (*M. acuminata Colla*, *M. balbisiana Colla* and *M. schizocarpa N. W. Simmonds* from the subgenus *Musa*) and the relatives of the Fe'I (*M. peekelii Lauterb*, *M. lolodensis Cheesman* and *M. maclayi F. Muell* in the subgenus *Callimusa*).

Though botanists had collected *Musa* specimens for herbaria and exotic gardens since early on, the first attempts to explore the *Musa* diversity for breeding and conservation purposes started only in the 1930s and 1940s. Batches of seed of wild bananas were collected from various places throughout the British Empire and sent to the Imperial College of Tropical Agriculture (ICTA)

http://dx.doi.org/10.19103/AS.2020.0070.07

in Trinidad where they were germinated and the resulting plants described (Cheesman, 1947, 1948a,b,c, 1949a,b,c,d, 1950a,b, Simmonds, 1953). Several of these plants are still being used in research and breeding programmes such as the 'Selangor2' and 'Calcutta4'. The study of wild bananas was resumed after WWII among others with a survey by N. W. Simmonds covering various areas in Southeast Asia (Simmonds, 1957). In the 1980s and later several germplasm collection missions were organized at national levels focusing mostly on banana cultivars such as in Thailand (Silayoi and Babprasert, 1983, Silayoi and Chomchalow, 1987), Malaysia (Jamaluddin, 1986), Vietnam (Danh et al., 1998) and India (Uma et al., 2001). A rare systematic study of the diversity of wild banana relatives was undertaken by Nasution (1991) in Indonesia. One multinational collection effort was made to explore the diversity of cultivated and wild bananas in Papua New Guinea and surrounding islands (Arnaud and Horry, 1997). Most of the collected germplasm has been maintained at national institutes while a number of accessions have also been transferred to the 'Musa Germplasm Transit Center (ITC)' maintained in trust by Bioversity International hosted at the Katholieke Universiteit Leuven, Belgium under the auspices of the Food and Agriculture Organization of the United Nations. The ITC currently maintains almost 1600 accessions of wild and edible bananas for conservation and distribution across the globe.

2 The importance of collecting and characterizing wild Musa germplasm

In spite of efforts for many years, the diversity of the wild relatives currently represented in *Musa* germplasm collections in Southeast Asia and the world is far from satisfactory. There are large sets of wild accessions not present in any germplasm collection and those wild relatives present have not been rigorously characterized.

There are three main reasons why a thorough and systematic study of the wild relatives of the edible bananas is urgently needed.

First of all an exploration of the diversity among the wild ancestors is a high priority to trace the origins of the various forms of edible bananas that have been brought into cultivation by farmers in Southeast Asia and the Pacific over the past hundreds or thousands of years. Though this might be a rather esoteric study, knowledge of the genetic and geographic origins of this crop can lead to better understanding of the limitations and possibilities of breeding programmes. Our present understanding of the origins of the edible bananas is far too fragmentary and biased.

The second reason is exactly to collect and characterize this diversity to discover desirable traits in the wild germplasm for introduction into new cultivars. One can observe great diversity for fruit size, shape, flavour, bunch size and plant stature in the wild populations while it can be presumed that there is diversity for drought and disease tolerance, nematode resistance and other traits.

The third reason is that the natural habitat of the wild relatives is increasingly being invaded by people. Although wild bananas seem to thrive for some time in disturbed habitats (e. g. along roads) ultimately they disappear as agricultural fields and plantation crops are introduced into their habitat. The discovery, characterization and conservation of the genetic diversity is becoming an urgent matter before it disappears forever.

I estimate that about half of the cultivars with unique genome constitutions from local germplasm collections in Southeast Asia have not been entered into the ITC though for some of them information might be available through the Musa Germplasm Information System (MGIS, https://www.crop-diversity .org/mgis/). Most of these are lesser-known cultivars of only local importance. However the situation for the wild relatives is less satisfactory. Currently (December 2019) the MGIS website holds information on 91 wild accessions of *Musa acuminata* present in the ITC. Out of the 91 wild accessions available, 12 are listed having a ploidy different from diploid, though for at least two of them (ITC1178, Buitenzorg and ITC0399 Malaccensis) that does not fit with molecular analysis which clearly indicates these accessions are diploid. Among those 91 accessions, several are potential duplicates such as Higa (ITC0378, ITC0428, ITC0464), Pahang (ITC0070, ITC0609, ITC0727), Cici/Cici Alas (ITC0415, ITC0681, ITC1079), Long Tavoy (ITC0093, ITC0283) and Hawain2/3 (ITC0602, ITC0616, ITC0617, ITC1209). Thus the effective number of wild diploid *Musa acuminata* genotypes represented in the ITC is even less, maybe as low as 72. For many of the wild accessions in the ITC the original collection information is uncertain or has been lost, except for those collected in Papua New Guinea. The ITC apparently has only a single accession of *M. acuminata* that unambiguously originated from Sumatra (a recently introduced accession of *M. acuminata* ssp. *sumatrana*, ITC1701) and only two from Borneo (ITC1520 from Kalimantan and ITC0253 'Borneo'). This is in spite of the fact that Nasution (1991) described five morphologically distinct varieties of *Musa acuminata* from Sumatra and two from Kalimantan. Regretfully one has to conclude that the representation of wild *Musa acuminata* in the ITC collection is very incomplete and in addition poorly documented and insufficiently characterised. For most of the other *Musa* species there are even far fewer accessions in the ITC collection. Although the MGIS contains information on wild *Musa* accessions in other collections, it is also scant and incompletely documented.

Similar to other important crops, the ITC collection was initiated (under INIBAP/IPGRI around 1985–1988) with the primary aim of conserving the genetic diversity in cultivated bananas. The occasional accessions of wild relatives that were transferred to the ITC were of course included. After the addition of the accessions collected in Papua New Guinea a degree of complacency may have crept into the banana conservation effort. The feeling was that representatives from all the subspecies deemed important contributors to the germplasm of the cultivars (Simmonds, 1966, Horry and Jay, 1988, Horry, 1989) were included in the ITC collection. A lack of leadership has resulted in poor coverage and insufficient characterization of the germplasm in the *Musa* wild relatives. This is in stark contrast to the germplasm collections of other major crops such as potato, rice, corn and the cereals where efforts have been made and are still ongoing (e.g. Castañeda-Álvarez et al., 2015) to ensure that crop wild relatives are well represented and documented. Further exploration of wild banana relatives is given minimal importance with some plans of possible explorations in the peripheral distribution of *Musa acuminata* including southeast Indonesia, Myanmar, northeast India and islands near East Africa (MusaNet, 2016) while the core region of diversity (triangle Java-Sumatra-Borneo) is not included.

3 An estimation of the genetic diversity missing from the Musa Germplasm Transit Centre (ITC)

To demonstrate the limited genetic diversity represented in the ITC germplasm collection, the analysis of the sequence diversity at one of the genes coding for the catalase enzyme (CAT2) can be used. The DNA sequences at single-copy loci reveal not only the diversity in a population, but also contain detailed information of their genealogy (Olsen and Schaal, 1999). The study of highly conserved 'housekeeping' genes offers the advantage that they are transferable and comparable among species and have a long tradition as isozyme variants in various plant species. Starting in 1998 I developed primers for the PCR amplification and sequencing of part of the coding regions of catalase, isocitrate dehydrogenase, alcohol dehydrogenase, granule-bound starch synthase and several other similar low-copy-number genes based on highly conserved regions in the sequences from various plant species. For each gene a fragment containing exons (coding region, conserved) and introns (non-coding region, variable) was amplified. After cloning of the fragments and sequencing a new generation of PCR primers could be designed to amplify single loci with high specificity in the various species of *Musa*.

During 2008 I used this approach to analyse the allelic diversity of about 100 accessions from the ITC, including approximately 60 cultivars and 31 wild *Musa acuminata* at four gene-coding regions (Volkaert, 2011). Later on several cultivated and wild accessions from Thailand (Toprasi et al., 2012) and in 2014 some accessions from the Indonesian Institute of Sciences (Lembaga Ilmu Pengetahuan Indonesia, LIPI) were added to the data set. A preliminary evaluation at that time indicated that out of a total of 45 haplotypes at the locus coding for a catalase gene (CAT2), 27 could be found in the ITC germplasm collection: 20 in the cultivars and 19 in the wild accessions (Fig. 1). Most importantly, the cultivars in the ITC collection contained eight haplotypes that could not be found in any of the wild accessions, neither in the ITC nor in the Thai or Indonesian (LIPI) wild accessions that had been analysed. Most of these missing haplotypes seemed to be of Western Indonesian origin (ssp. *zebrina* or ssp. *malaccensis*). In particular, one haplotype detected in three cultivars of the 'Pisang Jari Buaya' group (AA genome) was very similar to haplotypes in a single wild accession from Sumatra in the LIPI collection. Considering the incompleteness of the available accessions from the region it was decided that an exploration of the wild *Musa acuminata* of Sumatra should be undertaken. This exploration was coordinated by LIPI and supported by the 'Stichting Dioraphte' from the Netherlands in the first half of 2017. Through this effort the presence of all five varieties described by Nasution (1991) on Sumatra was confirmed and a totally new variety of *Musa acuminata* was identified (Fajarudin Ahmad et al. in prep). For 69 of the 164 accessions collected from Sumatra DNA sequence data were obtained. This information, supplemented with 19 additional accessions from the LIPI collection, expanded the network for the CAT2 locus (Fig. 1) from 45 to 71 haplotypes plus an additional 4 haplotypes that cannot be placed with certainty in the network since they have been found only in single accessions that turned out to be heterozygous at multiple SNP positions. Among those 69 accessions from a single island of Indonesia we found 33 haplotypes, more than in all of the analysed samples from ITC. Based on these findings we can conclude that the 27 haplotypes in the ITC collection constitute only just over a third of all haplotypes that have been detected thus far. In addition, of the eight haplotypes present in cultivars but absent from the ITC wild accessions, three (A01, A03, A42, see Fig. 1) were detected in Sumatran wild banana populations and one (A16) was detected in accessions from LIPI, leaving still four haplotypes (A02, A15, A17, A20) to be discovered in natural populations. Analysis in progress on other gene loci indicates that the CAT2 locus is no exception as other loci have comparable or even higher number of haplotypes with a comparable proportion missing from ITC.

As there is very likely still more diversity to be detected in the forests of Borneo, Java, Sulawesi, peninsular Malaysia, the Philippines and Thailand, for this particular locus the diversity conserved in the ITC collection may represent only about 10–20% of the total diversity within *Musa acuminata* in nature.

4 Conserving cultivars versus wild germplasm

One may question whether it is essential to detect and conserve all the genetic diversity that can be found at any one particular locus. Indeed it is not necessary to do so and practically probably even impossible to guarantee that all diversity has been discovered let alone then conserved *in situ* or *ex situ*. A random

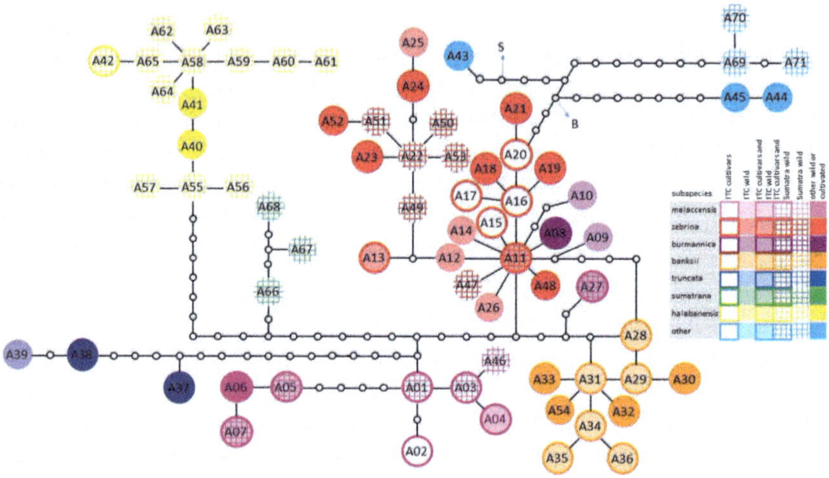

Figure 1 Haplotype diversity in *Musa acuminata*. The third intron and part of the fourth exon of the catalase2 gene was PCR amplified with specific primers and sequenced in both directions with primers internal to the fragment. The diploid sequence chromatographs were separated into their corresponding haplotypes using Phase (Stephens et al., 2001) as implemented in DnaSP version 6.12.01 (Librado and Rozas, 2009). For diploid accessions, the sequence chromatograms with indels were resolved with Indelligent (http://dmitriev.speciesfile.org/indel.asp Dmitriev and Rakitov, 2008). For triploid accessions, haplotypes were manually inferred using base substitution polymorphisms before the indel sites as anchors to select candidate haplotypes that in combination would generate the pattern observed in the sequence chromatograms. Haplotype networks were calculated with TCS version 2.1 (Templeton et al., 1992) with indels recoded as single positions and considered as fifth character state. 96% parsimony confidence was used. The different haplotypes are coloured more or less following subspecies accessions that make up the majority in each cluster. Note that haplotype A27 is closer to the cluster of *ssp. banksii* and *ssp. zebrina* but is found in 'Pahang' (ITC0609) and *ssp. malaccensis* wild accessions from Sumatra up to northern Thailand. The haplotypes A43, A44, A45, A69, A70 and A71 may belong to a different species or subspecies admixed into some wild plants and cultivars from Thailand (A43–45) or wild bananas from Sumatra (A69–71). The arrows leading to 'S' and 'B' indicate the branching point to the sequences of *M. schizocarpa* and *M. balbisiana*. Each line segment in the graph represents a single mutation.

locus such as CAT2 is meaningless as an indicator for human utilitarian use of wild germplasm since there may be very little or no correlation between its haplotypes and phenotypic traits of interest to breeding and production. However, molecular markers of this type provide insight in the population structure of wild *Musa* germplasm and can provide a scientific basis for guided collection of wild relatives and assist in the assembly of a core collection that then could be further characterized for traits that might be valuable in breeding. Contemporary banana breeding relies mostly on the germplasm included in the edible seedless cultivars supplemented with a few wild accessions that, mostly by chance events, have since long time been in the ITC collection. The genetic traits included in the edible cultivars are obviously relevant, but difficult to use since the cultivars typically all have low seed fertility and in many cases are also pollen sterile. As shown by the analysis of the CAT2 haplotypes, a part of the genetic diversity in cultivars is not available through the seeded wild accessions maintained in ITC. In addition, the breeding value of the germplasm in the cultivars is difficult to evaluate as it may be combined in an admixed form, in a triploid and even in combination with germplasm of a different species, *Musa balbisiana*. The difficulty, or in many cases inability, to make crosses that include cultivars further hinders the study of the inheritance behind traits of interest. If a wider diversity of wild accessions would be available to banana breeders, progress could be made using regular crossing schemes. Once a desirable improved seeded line has been produced, it could then be further crossed with an accession that contributes parthenocarpy and sterility traits to obtain an improved seedless banana cultivar. As an example, the cultivars of the group 'Pisang Jari Buaya' are used as resistant checks in nematode research. Presumably they could be considered as a source of resistance to burrowing nematodes (Marin et al., 1998, Ray, 2002). It is however very difficult to introgress desirable traits from cultivars since they rarely produce seed. Introgression of these desirable traits would be more feasible if the sources of resistance could be traced to the fertile seeded ancestors, the wild *Musa acuminata halabanensis* of Sumatra (Fajarudin Ahmad et al., in prep).

The 'International Musa Testing Programme' has evaluated a number of the major cultivars groups of banana (landraces, cultivar selections and some hybrids from participating breeding programmes) for agronomic performance, host reaction to pests and diseases and post-harvest characteristics in diverse environments in several countries. Though some reference clones were included in all planting sites, the inclusion of wild accessions was extremely limited. Almost nothing is known about wild accessions with regards to, for example, drought resistance. Drought tolerance in A × B hybrid cultivars is generally attributed to the contribution of the B genome (Simmonds, 1966, Thomas et al., 1998). However, the possibility that the different A genomes among the few studied cultivars might have differential drought tolerance has not been taken into consideration. Drought tolerance probably exists in several wild *M. acuminata* populations (e.g. in Northeastern Thailand), but this has never been investigated. Only a representative collection of wild accessions supplemented with progenies of crosses among them and the systematic study of their phenotypic plasticity in a range of environments (good old-fashioned 'common garden experiment' replicated in multiple locations that differ in soil, altitude and rainfall patterns) would start to shed some light how the alleles of the wild relatives have contributed to the characteristics of present-day cultivars and reveal their true potential for improved cultivars in the future.

5 Conclusions

The ITC collection has been initiated as a means to conserve banana cultivars from all over the world. However the representation of the genetic diversity in the immediate ancestors of the cultivated bananas is lacking. The exploration of the wild relatives and immediate ancestors of the cultivated bananas is urgently needed to safeguard the gene pool for future breeding and conservation purposes. The value of wild relatives has been appreciated in most other crops and also in bananas it will improve our knowledge and boost the breeding potential once it has been identified, collected and evaluated.

6 Where to look for further information

In spite of the many shortcomings, the MGIS (https://www.crop-diversity.org/mgis/) and Promusa (http://www.promusa.org/) websites maintained by Bioversity International are still useful to access information about diversity in wild and cultivated bananas. For the wild and cultivated bananas from Papua New Guinea, Musalogue (https://www.bioversityinternational.org/e-library/publications/detail/musalogue/) is a good starting point.

Häkkinen and Väre (2008) give a broad overview of the different *Musa* species that have been identified and highlight the problematic naming of species and cultivar groups that has happened in the past.

7 Acknowledgements

Yuyu S. Poerba, Fajarudin Ahmad and LIPI (Indonesia) are thanked for the invitation to assist in part of the exploration of wild bananas in Sumatra and access to some of the banana genetic resources in the LIPI collection.

Ines van den Houwe and ITC (KULeuven, Belgium) are acknowledged for sending leaf materials of banana accessions and information on the wild accessions in the ITC collection.

Thanks to all the students who have at various times helped with the work in the laboratory: Sasivimol (Pupee) Swangpol, Piyaporn (Bee) Phansak, Siriporn (Pook) Jaengsuthivorawat, Siriporn (Phrae) Cheuayjaeng, Jongkol (Nid) Jeua-ngam, Poom Predakul, Tanat (Ake) Wattanakulpakin, Nichnipha (Nich) Toprasi, Nipha (Oil) Cheunpom and Sukanya (Meaw) Jeennor.

8 References

Arnaud, E. and Horry, J. P. 1997. *Musalogue: A Catalogue of Musa Germplasm. Papua New Guinea Collecting Missions, 1988–1989*. International Network for the Improvement of Banana and Plantain, Montpellier, France.

Castañeda-Álvarez, N. P., de Haan, S., Juárez, H., Khoury, C. K., Achicanoy, H. A., Sosa, C. C., Bernau, V., Salas, A., Heider, B., Simon, R., Maxted, N. and Spooner, D. M. 2015. *Ex situ* conservation priorities for the wild relatives of potato (*Solanum* L. section Petota). *PLoS ONE* 10(4): e0122599.

Cheesman, E. E. 1947. Classification of the bananas. II. The genus *Musa* L. *Kew Bulletin* 2(2): 106–117.

Cheesman, E. E. 1948a. Classification of the bananas. III. Critical notes on species (a: *M.*, b: M. acuminata). *Kew Bulletin* 1: 11–28.

Cheesman, E. E. 1948b. Classification of the bananas. III. Critical notes on species (c: M. *paradisiaca* and M. *sapientium*, d: *Musa banksii*). *Kew Bulletin* 2: 145–157.

Cheesman, E. E. 1948c. Classification of the bananas. III. Critical notes on species (e: *Musa basjoo*, f: *Musa nagensium*). *Kew Bulletin* 3(3): 323–328.

Cheesman, E. E. 1949a. Classification of the bananas. III. Critical notes on species (g: *Musa itinerans*, h: *Musa ornata*). *Kew Bulletin* 1: 23–28.

Cheesman, E. E. 1949b. Classification of the bananas. III. Critical notes on species (i: *Musa sanguinea*, j: *Musa velutina*). *Kew Bulletin* 2: 133–137.

Cheesman, E. E. 1949c. Classification of the bananas. III. Critical notes on species (k: Musa *laterita*, l: *Musa textilis*). *Kew Bulletin* 3: 265–272.

Cheesman, E. E. 1949d. Classification of the bananas. III. Critical notes on species (m: *Musa fehi*, n: *Musa peekelii*). *Kew Bulletin* 4(4): 445–452.

Cheesman, E. E. 1950a. Classification of the bananas. III. Critical notes on species (o: *Musa lolodensis*, p: *Musa maclayi*, q: *Musa coccinea*). *Kew Bulletin* 1: 27–31.

Cheesman, E. E. 1950b. Classification of the bananas. III. Critical notes on species (r: *Musa borneensis*, s: *Musa violascens* and M. *gracilis*). *Kew Bulletin* 2: 151–155.

Danh, L. D., Nhi, H. H. and Valmayor, R. V. 1998. Banana collection, conservation and characterization in Vietnam. *Infomusa* 7(1): 10–13.

Dmitriev, D. A. and Rakitov, R. A. 2008. Decoding of superimposed traces produced by direct sequencing of heterozygous indels. *PLOS Computational Biology* 4(7): e1000113.

Häkkinen, M. and Väre, H. 2008. Typification and check-list of *Musa* L. names (Musaceae) with nomenclatural notes. Adansonia, sér 3(30): 63–112.

Horry, J. P. and Jay, M. 1988. Distribution of anthocyanins in wild and cultivated banana varieties. *Phytochemistry* 27(8): 2667–2672.

Horry, J. P. 1989. Chimiotaxonomie et organisation genetique dans le genre Musa. *Fruits* 44: 455–474+509–520+573–578.

Jamaluddin, S. H. 1986. Characterization, evaluation and utilization of the banana germplasm in Malaysia. In: Chan, Y. K. and Raveendranathan, P., Eds. Prosid Simposium Buah Buahan Kebangsaan, MARDI, Kuala Lumpur, Malaysia, pp. 315–329.

Librado, L. and Rozas, J. 2009. DnaSP v5: a software for comprehensive analysis of DNA polymorphism data. *Bioinformatics* 25: 1450–1452.

Marin, D. H., Sutton, T. B., Barker, K. R., Kaplan, D. T. and Opperman, C. H. 1998. Burrowing nematode resistanceof black sigatoka resistant hybrids. *Nematropica* 28: 241–247.

MusaNet 2016. Global Strategy for the Conservation and Use of *Musa* Genetic Resources (Laliberté, B., Compiler). Bioversity International, Montpellier, France. ISBN: 978-92-9255-050-9.

Nasution, R. E. 1991. A taxonomic study of the species *Musa acuminata* Colla with its intraspecific taxa in Indonesia. *Memoirs of the Tokyo University of Agriculture* XXXII: 2–122.

Olsen, K. M. and Schaal, B. A. 1999. Evidence on the origin of cassava: Phylogeography of *Manihot esculenta*. Proceedings of the National Academy of Sciences USA, 96: 5586–5591.

Ray, P. K. 2002. *Breeding Tropical and Subtropical Fruit*. Springer, Berlin. ISBN: 978-3-540-42855-8.

Silayoi, B. and Babprasert, C. 1983. *Banana Genetic Resources Exploration in Thailand*. Kasetsart University, Bangkok, Thailand.

Silayoi, B. and Chomchalow, N. 1987. Cytotaxonomic and morphological studies of Thai banana cultivars. In: Persley, G. J. and De Langhe, E. A., Eds. *Banana and Plantain Breeding Strategies*. Proceedings of an International Workshop Held at Cairns, Australia, 13–17 October 1986. ACIAR Proceedings No. 21, Canberra, Australia, pp. 557–560.

Simmonds, N. W. 1953. Classification of the bananas. III. Critical notes on species (t: *Musa erecta*, u: *Musa angustigemma*). *Kew Bulletin* 4: 571–574.

Simmonds, N. W. 1957. Botanical results of the banana collecting expedition, 1954–5. *Kew Bulletin* 11(3): 463–489.

Simmonds, N. W. 1966. *Bananas*. Longmans, London.

Stephens, M., Smith, N. J. and Donelly, P. 2001. A new statistical method for haplotype reconstruction from population data. *American Journal of Human Genetics* 68(4): 978–989.

Templeton, A. R., Crandall, K. A. and Sing, C. F. 1992. A cladistic analysis of phenotypic associations with haplotypes inferred from restriction endonuclease mapping and DNA sequence data. III. Cladogram estimation. *Genetics* 132(2): 619–633.

Thomas, D. S., Turner, D. W. and Eamus, D. 1998. Independent effects of the environment on the leaf gas exchange of three banana (*Musa* spp,) cultivars of different genomic constitution. *Scientia Horticulturae* 75(1–2): 41–57.

Toprasi, N., Suvittawat, K. and Volkaert, H. 2012. Relationship of molecular and morphological characteristics for classification among Thai cultivated bananas. *Agricultural Science Journal* 43: 185–193 (in Thai with English abstract).

Uma, S., Shyam, B., Selvarajan, R. and Sathiamoorthy, S. 2001. Collection and characterisation of banana and plantains of Northeastern India. In: Molina, A. B., Roa, V. N. and Maghuyop, Eds. *Advancing Banana and Plantain R&D in Asia and the Pacific*, Vol. 10. INIBAP-ASPNET, Los Baños, Philippines, pp. 103–109.

Volkaert, H. 2011. Molecular analysis reveals multiple domestications of edible bananas. *Acta Horticulturae* 897(897): 143–152.

Chapter 7

Collection and evaluation of banana and plantain landraces in Africa

D. Karamura and W. Ocimati, Bioversity International, Uganda; G. Blomme, Bioversity International, Ethiopia; J. G. Adheka, University of Kisangani (UNIKIS), Democratic Republic of the Congo; C. Sivirihauma, Université Catholique du Graben (UCG), Democratic Republic of the Congo; D. B. Dhed'a, University of Kisangani (UNIKIS), Democratic Republic of the Congo; and E. Karamura, Bioversity International, Uganda

1 Introduction

Since their introduction into Africa over two thousand years ago, bananas and plantains have multiplied, diversified and evolved into numerous landraces in two discrete zones in Africa, which are now considered as secondary centers of genetic diversity. Plantains (*Musa* AAB) predominate the lowland regions of West and Central Africa, while East African Highland green-cooking and beer bananas (*Musa* AAA) predominate the highlands of East and Central Africa. Apart from these two main subgroups, there are several locally dominating subgroups - Cavendish-dominated (AAA), Mchare-dominated (AA), Kayinja and Bluggoe (ABB), Kamaramasenge (AAB), Ibota (AAA) and Gros Michel (AAA) bananas that are widely distributed in Africa. Of these, Mchare (AA) has been in Africa for a long time while the rest are considered relatively recent introductions. These Musa genome groups are cultivated widely in banana-growing areas

http://dx.doi.org/10.19103/AS.2020.0070.08

in Africa. Recently, the productivity of bananas seems to be shrinking due to a combination of factors including declining soil fertility, pest and disease outbreaks and climate change-linked factors. Moreover, the introduction and/or re-emergence of diseases and attendant socioeconomic challenges, together with a general lack of knowledge on the landraces, increasing population pressure, limited resource allocation to crop improvement and crop landrace development, and policies that do not take into account the long-term and sustainable utilization of landraces further reduces the full exploitation of the available Musa diversity. Consequently, these genetic groups are severely threatened by genetic erosion, primarily due to their replacement by modern, genetically uniform varieties which are resistant to some diseases and pests. Several collection missions have been carried out at regional and national levels to collect plantain and highland banana landraces, and other locally dominating edible landraces in particular locations (e.g. ABBs, other AAAs and AAs) but accurate identification, characterization and evaluation studies remain limited. Most of the evaluation studies carried out have been at the plant level (i.e. morphological characterization) while molecular characterization of these landraces in these regional and national collections remains to be done. Crop improvement programs require accurately characterized and evaluated landraces and data collected at intraspecific and infraspecific levels with respect to biotic and abiotic stress responses, as well as delayed ripening, micronutrient content and yield. Characterization and evaluation of these materials need to be extended and deepened by deploying molecular tools and post-harvest characteristics determined to allow end users (breeders, entrepreneurs and growers) to make informed decisions. Much of this information is still missing and hampering the full exploitation of the potential of the African *Musa* landraces. Hence, most landraces from ex-situ and in-situ collections in Africa remain underutilized due to the lack of characterization and evaluation data. As a way forward, it is proposed that forward-looking policies and institutional frameworks be developed to ensure financial resources, trained manpower and the implementation of effective data management tools at national and regional levels, in the quest to realize the sustainable development and utilization of *Musa* landraces in Africa. This chapter provides a review of the processes involved in the collection and evaluation of banana and plantain landraces, the challenges and opportunities encountered and suggestions for the way forward.

2 Overview of banana and plantain landraces in Africa

The genus *Musa* is native to Southeast Asia while the closest related genus, *Ensete*, is native to Africa and Madagascar (Perrier et al., 2019). Since their introduction from Asia to Africa several thousand years ago (Kitavi et al., 2016;

Perrier et al., 2011) bananas and plantains have been distributed widely across the continent. Traders from the Near East and Far East are thought to have been responsible for the introduction of *Musa* germplasm into Africa via the coastal islands of Madagascar, Zanzibar and Pemba. From these islands, bananas were carried westwards across the continent by African migrations (Simmonds, 1976). Although the East African diploids and triploids belong to the same genetic complex, we postulate that they arrived in East Africa independently as AA and AAA genotypes.

Cultivation and selection of these materials over more than two millennia has resulted in the currently available landraces (Swennen and Vuylsteke, 1987) and since their introduction into Africa, bananas and plantains have contributed greatly to the socioeconomic well-being of Africa, in terms of nutrition, food security and incomes (Table 1). The continuous selection pressure applied by farmers and the environment over a long period of time has given rise to local adaptation such that secondary genetic diversity regions for *Musa* are now considered in Africa, with plantains dominating the lowland regions of West and Central Africa while the highlands of East Africa are dominated by the green-cooking and beer bananas (Ortiz and Vuylsteke, 1994) also called the East African Highland Bananas (EAHBs) (Karamura, 1999).

The introduction of new *Musa* germplasm into Africa accelerated over the last century. Following the end of World War II, many Africans returning from the war fronts in the Far East carried a diversity of cultivars including the ABB Pisang Awak, Silver Bluggoe, AAB Silk, Kamaramasenge, Prata, Mysore, AB-Ney Poovan (Kisubi) and Ibota (AAA). Similarly, in the 1950s and 1960s, several *Musa* landraces from the Latin American and Caribbean subcontinent were collected and exchanged with African *Musa* landraces by the European colonizers of Africa. From the British Caribbean research stations, Gros Michel, Dwarf and Giant Cavendish and Uganda Red were among the *Musa* diversity introduced

Table 1 Banana/plantain consumption in Africa compared with other lead producers

Country	Consumption level (kg/cap/year)
Uganda	243
Rwanda	197
Gabon	161
Cameroon	128
Papua New Guinea (Asia)	**121**
Sao Tomé & Principe	93
Ghana	92
Burundi	89
Ecuador (Latin America)	**88**

Ref: Frison and Sharrock, 2000.

into the African continent, thereby further enriching the *Musa* diversity in Africa. Plantain, the EAHBs, Cavendish and Mchare constitute the four dominating banana systems in different locations (Fig. 1). Within Africa, there has been a lot of undocumented germplasm exchange both formal and informal, across the traditional growing regions. More recently, locally organized collection expeditions have helped to move *Musa* landraces among the regions.

The movement of banana landraces crisscrossing the region has not been without challenges. Over the last 50 years or so, banana and plantain systems in the key *Musa* secondary diversity centers in Africa have been constrained by a range of biotic and abiotic stresses. The effects of new pest and disease outbreaks often favored by changes in climatic conditions, coupled with socioeconomic threats emanating from civil unrests, policies that promote replacement of landraces by bred (hybrids) cultivars have undoubtedly led to widespread extinction of some landraces and genetic erosion of the crop's

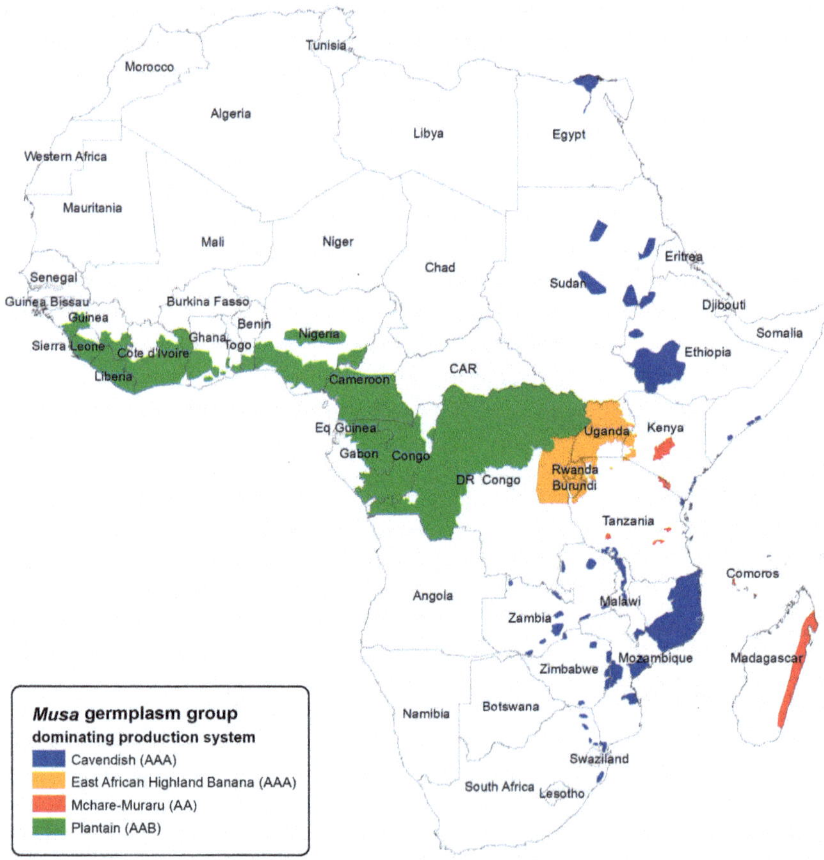

Figure 1 Musa germplasm group dominating production systems.

gene pool. Consequently, the management and maintenance of banana and plantain productivity and diversity in Africa became a major concern for agricultural scientists given the important role the banana landraces play in the livelihoods of the people in Africa. This resulted in actions by various stakeholders at the national and international level. For instance, in the quest to address productivity constraints, several countries in the East and Central African region prioritized banana as their number one research crop and accordingly initiated national banana research and development programs to promote the improvement and sustainable development of the crop. At the international level, the Global Strategy for Plant Conservation (GSPC) has underscored that reliable baseline and evaluation data are needed for better exploitation of these landraces. The GSPC recommended measuring of genetic diversity among landraces to understand the extent, distribution, basic characteristics, comparative performance and utility value of each landrace. Such recordings provide baseline information essential for the establishment of country, regional and global priorities for the conservation, management and improved utilization of these landraces. They also remain a vital resource for contemporary plant breeding. Therefore, these landraces need to be collected, evaluated and pertinent information about them communicated and/or shared to enhance utilization.

3 Banana and plantain landrace diversity and its conservation in Africa

All countries (Uganda, Kenya, Tanzania, Rwanda and Burundi) in the East and Central African region grow East African Highland Bananas with about 11.7 million metric tons produced in 2017 (FAOSTAT, 2019). It is here that bananas reach their greatest importance as a staple food crop. In countries like Uganda, Burundi and Rwanda, the annual consumption rate per capita reaches 243 kg, making the region the leading consumer of bananas per person in the world (Frison and Sharrock, 2000). The crop is not only used as food but is also important for several cultural functions (often overlooked by crop development planners), and in addition, sales in local markets provide an important source of household income. The EAHBs, also known as the Lujugira-Mutika subgroup (Shepherd, 1957; Simmonds, 1966) or just Mutika (Perrier et al., 2019), though limited to the east African plateau, are the dominant bananas in the region. In Uganda, the crop is grown in all parts of the country not experiencing a pronounced dry season, that is, the area within 80 km of the shore of Lake Victoria, the South-Western highlands, and the slopes of Mt. Elgon in the east and the high rainfall areas of the western part of the country. In Tanzania, banana landraces are a staple food in the high rainfall areas, located at higher altitude and low-lying places around Lake Victoria in addition to valley bottoms, where they are a major constituent of the local diet (Ngeze, 1994; Karamura, 1999).

In Kenya, *Musa* landraces are widely grown in the western part, in areas surrounding Lake Victoria, on the slopes of Mt. Kenya and along the coast. Banana is also a dominant feature in the landscapes of Rwanda, Burundi and eastern DR-Congo. In Rwanda, the crop is produced between 800 m and 2000 m above sea level and occupies more than 23% of the area under cultivation. There are three banana-production zones in Rwanda; the Kivu lake border, Kigali-Butare axis and Kigali rural-Kibungo (Nsabimana et al., 2004). Inventory and survey work in East Africa (De Langhe et al., 2001; Byabachwezi et al., 2005; Karamura et al., 2006) revealed some AAA types which were not confirmed as AAA-Mutika, scattered in the Kenya-Tanzania-Malawi highland axis.

There are three types of triploid East African banana landraces in the Pare-Usambara region in Tanzania: the EAHBs, the Ilalyi (AAA genome) and 10–15 'ill-defined' astringent triploids (Karamura et al., 2016; Perrier et al., 2019). There are also two groups of diploid East African banana landraces: the Mchare-Muraru found in the Kilimanjaro region in Tanzania and in the Gikuyu highlands of Kenya and 8–10 'ill-defined' diploids found in Tanzania (Perrier et al., 2019). Ecologically, bananas in highland regions of East Africa grow within the range of 1000–3000 m above sea level (masl). The Lujugira-Mutika cultivars along the East African plateau tend to favor the 1400–1800 masl range with a mean annual temperature range of about 20–26°C and evenly distributed high rainfall. This temperature range reduces pest and disease pressure. On Mount Kilimanjaro, however the Ilalyi subpopulation is grown between 1600 masl and 3000 masl. Partly because of reduced pest and disease pressures, and partly because of their importance to communities, both the Ilalyi and the rest of the Lujugira-Mutika banana stands may last 50–200 years. Representative samples of the East African highland banana varieties are being preserved in all national collections of each country whereas representative samples of Ilalyi, Mchare, ill-defined triploid and diploids landraces, are being preserved in the East and Central African regional banana field gene bank at Mbarara, Kisii gene bank in Kenya, the Tengeru field gene bank in Arusha and the Maruku collection in Bukoba, Tanzania (Tables 2 and 3).

Plantains are a major staple in Central and Western Africa with a production estimated at approximately 8 million tons (Lescot, 2007). About two-thirds of *Musa* cultivated and produced in these two regions are plantains. The other third consists of dessert and other cooking bananas (FAOSTAT, 2019). The majority (82%) of plantains are produced in the area stretching from the lowlands of Guinea and Liberia to the central basin of the Democratic Republic of Congo. These areas represent the secondary centers of diversification of plantain (*Musa* AAB) (Swennen et al., 1995). The major plantain-producing

Table 2 National Musa collections in Africa (*Musa*Net Survey report, 2011)

Region	Country	Name of the Institute	Role of collection	Maintenance type	Station	Total no. of accessions	Cultivated	Wild species	Breeding lines	Other types
Western and central Africa	Benin	Institut National de Recherches Agricoles du Bénin	National	Field	INRAB	20	20	0	0	0
Western and central Africa	Cameroon	Centre africain de recherche sur bananes et plantains	Regional	Field	Njombe, Station	650	450	50	40	115
Western and central Africa	DR Congo	Université de Kisangani, Faculté des Sciences de Kisangani	National	Field	Kisangani	128	120	2	6	0
Western and central Africa	DR Congo	Institut National pour l'Etude et la Recherche Agronomiques (INERA)	National	Field	Mvuazi, Station	88	87	1	0	0
Western and central Africa	DR Congo	Universite Catholique du Graben (UCG), Butembo, North Kivu	National	Field	Butembo, UCG, Station	112	105	0	7	0
Western and central Africa	Ivory Coast	Centre National de Recherche Agronomique (CNRA)	National	Field	CNRA, Azaguié	92	37	2	13	40

(Continued)

Table 2 National Musa collections in Africa (*Musa*Net Survey report, 2011) (*Continued*)

Region	Country	Name of the Institute	Role of collection	Maintenance type	Station	Total no. of accessions	Cultivated	Wild species	Breeding lines	Other types
Western and central Africa	Nigeria	International Institute of Tropical Agriculture	Regional	Field	Ibadan	371	193	18	160	180
Western and central Africa	Togo	Institut Togolais de la Recherche Agronomique (ITRA)	National	Field	ITRA	13	13	0	0	0
Eastern and Southern Africa	Burundi	Institut de recherches agronomiques et zootechniques (IRAZ) de la CEPGL	Regional	Field	IRAZ, Gitega Station	297	234	31	30	2
Eastern and Southern Africa	Ethiopia	Ethiopian Agricultural Research Organization	National	Field	Melkasa Agriculture Research Center (in vivo)	88	12	0	64	12
Eastern and Southern Africa	Ethiopia	Horticultural Crops Research, Areka Agricultural Research Center	National	Field	Areka Agricultural Research Center	197	100	0	7	0
Eastern and Southern Africa	Kenya	Kenya Agricultural Research Institute	National	Field	Kisii Regional Research Center	176	0	0	0	0

Eastern and Southern Africa	Kenya	Kenya Agricultural Research Institute	National	In vitro	Thika	63	60	0	3	0
Eastern and Southern Africa	Madagascar	Centre National de Recherche Appliquée au	National	Field	Fofifa, Toamasina	23	20	0	3	0
Eastern and Southern Africa	Malawi	Department of Agricultural Research and Technical Services	National	Field	Bvumbwe Research Station	70	69	1	0	0
Eastern and Southern Africa	Mauritius	Agricultural Research and Extension Unit, Mauritius	National	Field	Richelieu Research Station	66	50	1	0	0
Eastern and Southern Africa	Rwanda	Institut des Sciences Agronomiques du Rwanda	National	Field	Ngoma Station/ Rukira site	117	107	2	8	0
Eastern and Southern Africa	South Africa	Institute for tropical and sub-tropical crops	National	Field	Burgershall research Station	114	37	23	0	54
Eastern and Southern Africa	Sudan	Agricultural Research Corporation (ARC)	National	Field	ARC, Wad Medani	355	351	0	4	0

(Continued)

Table 2 National Musa collections in Africa (*Musa*Net Survey report, 2011) (*Continued*)

Region	Country	Name of the Institute	Role of collection	Maintenance type	Station	Total no. of accessions	Cultivated	Wild species	Breeding lines	Other types
Eastern and Southern Africa	Tanzania	Agricultural Research and Development Institute	National	Field	Maruku Research Station	120	118	1	1	0
Eastern and Southern Africa	Uganda	National Agricultural Research Organization	Regional	Field	East and central African Musa collection, Mbarara	478	430	16	59	0
Eastern and Southern Africa	Zanzibar	Kizimbani Agricultural Research Institute	National	Field	Kizimbani Research station	37	36	1	0	0

Table 3 Proportion of banana and plantain subgroups in Africa

Musa subgroup	Genome group	Proportion (%)
Eastern and Southern Africa (ESA region)		
Highland bananas	AAA	71
Plantains	AAB	11
Gros Michel and other dessert	AAA/AAB	05
Cavendish for local consumption	AAA	12
Cavendish for export	AAA	01
West and Central Africa (WCA)		
Plantain	AAB	73
Gros Michel and other dessert	AAA/AAB	06
Cavendish for local consumption	AAA	11
ABB cooking	ABB	06
Cavendish for export	AAA	04

Source: Adheka (2014).

countries are Cameroon, Ghana, Nigeria and the Democratic Republic of Congo (DRC). It is estimated that about 70 million people in West and Central Africa derive more than 25% of their carbohydrates from plantains, making them one of the most important sources of food energy throughout the humid forest zone of African lowland (Swennen, 1990a,b). In Eastern Democratic Republic of Congo (DR Congo), the per capita consumption of *Musa* spp. is about 200 kg/year (Ndungo, 2004) and it contributes nearly 70% of total crop production, with 23.5% of the production realized in the North Kivu Province (Bakelana and Ndungo, 2004). Eastern DR Congo constitutes an exceptional area where the EABHs (AAA genome) and plantains (AAB genome) meet in an area characterized by the presence of three agroecological zones that are high (1400 masl), medium (800–1300 masl) and low (below 700 masl) altitude sites resulting in the cultivation of many different banana and plantain varieties (Ndungo, 2002; Ocimati et al., 2013; Sivirihauma et al., 2017) (Tables 4 and 5).

4 Access to and utilization of characterization and evaluation data for banana and plantain landraces

Information on characterization and evaluation concerning landraces is very useful to a diversity of stakeholders – farmers, entrepreneurs, nutritionists, educators, breeders and other applied scientists and policy makers. Characterization and evaluation increase the value of landraces to all these stakeholders. Landrace characterization and evaluation, in the broad sense and in the context of genetic resources, is describing collected materials. It covers the whole range of activities starting from the receipt of the new samples by

Table 4 Distribution of plantain[1] clone sets[2] in DR Congo

Province	French	False Horn	Horn	Total
Oriental	45	18	6	69
South Kivu	10	2	2	14
North Kivu	36	17	5	58
Maniema	21	6	4	31
Katanga	2	2	4	8
Eastern Kasai	10	8	3	21
Western Kasai	11	7	4	22
Bandundu	14	7	4	25
Equateur	36	18	6	60

Source: Adheka (2014).

[1]Representative samples of different plantain landraces have been preserved in the national collections of Cameroon, Nigeria, DR Congo, Ivory Coast and others. The plantains do much better in low lands (the West African coastal regions and the Congo basin) with higher temperatures. These regions are however also very prone to pest and disease attacks/outbreaks. Hence, in many plantain agroecologies, the plantation life of plantain stands may last only 3-5 years, when the land is rendered fallow to regain fertility but also to reduce pest pressure.

[2]A series of collecting missions was carried out in the Democratic Republic of Congo (DR Congo) from 2005 to 2012 which revealed that at least 119 plantain landraces exist in the country (Adheka, et al., 2018ab). The plantain diversity in DR Congo was shown to decrease and change in composition from rainforest to the savanna regions (Adheka et al., 2013, 2018b; Losimba et al., 2018). Adheka et al. (2018b) have classified plantains in the DR Congo into three clone sets - French, False Horn and Horn plantains. This is the first time these plantain groupings are named and classified as clone sets.

the curator and growing these to increase the sample for characterization and preliminary evaluation, and for further or detailed evaluation and documentation, using a standard descriptor list. Based on these descriptors, the data recorded for each accession fall into the following sections: passport data, characterization, preliminary evaluation and further evaluation, the latter in multilocation studies. Such information is usually kept by the national banana research programs and horticultural research institutes (Table 2).

4.1 Passport data

Passport data as part of the collection process include all basic information recorded during the time of collecting samples or the information provided by the collector regarding, for example, source and origin. It is the identity of the landraces as determined at the point and time of collection. Passport data should include a description of the site in terms of the microecological conditions, the landscape and the ecological systems in which the banana landrace is naturally growing. Records on topography or soil characteristics can be valuable for plant breeders as well as entrepreneurs for improving adaptation to conditions or tolerance of edaphic or climatic stresses. The important passport descriptors

Table 5 *Musa* landrace groups and subgroups in Africa

Landrace	Genome group	Subgroup	Distribution in Africa	Possible origin
Kisubi, Sukari	AB	Ney poovan	East Africa	South India, Sri Lanka, West Indies
Gonja, Nkonjwa, Mzuzu	AAB	Plantain	West and Central Africa, East Africa	Not known if it came from SE Asia as a triploid or is a product of diploid hybridization in Africa
Kikonde	AAB	Mysore	Zanzibar, East Africa	SE Asia islands
Kamaramasenge/ Sukali Ndizi	AAB	Kamaramasenge	East Africa	Unknown
Kipungusa/ Manzano/ Kipukusa /Sukari	AAB	Silk, Figue Pomme	Kenya, Mozambique, Tanzania and Zanzibar	Malaysia, Papua New Guinea and India
Kijakazi, Kibungara, Bakweri	AAB	Pome	Kenya, Tanzania, Zanzibar, west Africa	India
Kivuvu/ Mkojosi/ Bokoboko/	ABB	Bluggoe	East, central, west Africa	India
Kayinja/ Kijivu, Nyeupe,	ABB	Pisang Awak	East, central, west and South Africa	Main land SE Asia
Yangambi (KM 5), Mufunyankoni	AAA	Ibota	Democratic Republic of Congo, Rwanda, Uganda	Islands of SE Asia
Matooke	EAHB-AAA	Lujugira-Mutika	Eastern Africa	Not known if it came from SE Asia as a triploid or is a product of diploid hybridization in Africa
Ilalyi	EAB-AAA	Ilalyi	East Africa	Not known
Kitombo	EAB-AAA	Kitombo	Tanzania	Not known
Paka	EAB-AA		Tanzania, Zanzibar	Unknown
Mchare, Mlali, Muraru	EAB-AA	Mchare	Tanzania, Kenya, Comoros, Madagascar	Unknown
Mlambichi, Ntindi 1	AA	Sucrier	Tanzania	SE Asia
Mwitupemba	AA	Not defined	Tanzania, Zanzibar	Unknown
Bogoya	AAA	Gros Michel	East, central, west and south Africa	Main land SE Asia
Red Bogoya, Mzungu Mwekundu,	AAA	Red	East, central and west Africa	SE Asia
Bogoya omumpi, Mtwishe	AAA	Cavendish	Africa	Mainland SE Asia

Source: Musanet Report (2016).

are the site of collection (village/state/country), collector's number, type of material (population or pure line), date of collection, altitude, latitude and longitude for site of collection, status (wild, weedy, landrace, cultivar), growing condition and source (such as field, farm store or institute). Written notes need to be supplemented by photographs and maps. The curators of the national and regional banana collections in Table 2 are charged with the responsibility of keeping custody of these data but quite often, the collectors do not provide a comprehensive set of data and it is usually not possible to re-collect such information.

4.2 Characterization data

According to a recent International Plant Genetic Resources (IPGR) descriptor-list definition (1996), characterization consists of recording those characters which are highly heritable, can be easily assessed by eye and are expressed in natural environments. Characterization should provide a standardized record of readily assessable plant characters which, together with passport data, go a long way to identify an accession (Frankel, 1986). These are mainly morphological characters indicating the precise qualitative description of the plant when growing in optimal conditions and whereby its characteristics are maximally expressed. The different possible states of each descriptor are morphological markers. In case of cultivated bananas and plantains, which are clonally propagated, these markers are highly stable and only change due to the occurrence of somatic variants. Characterization data may also include a limited number of additional agronomic and quantitative traits thought desirable by a consensus of users of the crop. Characterization descriptors include pseudostem, bunch and rachis appearances, rachis and bunch positions and other characters which are generally used for taxonomic purposes. Recorded characterization data, along with the passport data, provides an overall picture of the range of diversity in the collections. This descriptor system has been adopted by the curators of field collections. But since many descriptors deal with colors or forms with sometimes delicate gradation (e.g. 'light green' versus 'green') there is a risk of different interpretations by different observers, that effectively can lead to 'artificial' differences between described cultivars or accessions. Some definitions of descriptor-states may also have to be made more explicit or more adapted to intra-subgroup diversity and would benefit from specific illustrations to improve accurate documentation. Constructing well-defined and user-friendly descriptor lists, as well as the exchange of description data among curators, goes some way to help clarify differences in interpretation. Molecular characterization can follow morphological characterization to gain a better insight into the underlying genetic relatedness among the landraces. The molecular-marker techniques currently used for *Musa* description globally are the simple sequence repeats (SSRs), the single-nucleotide polymorphisms (SNPs) and diversity arrays technology (DArT). In Africa, such information is found in research institutes and universities along with the CGIAR centers - IITA and Bioversity International.

4.3 Evaluation data

While passport and characterization data determine the identity of a particular landrace, evaluation data such as agronomic traits, host reaction to pests and diseases and abiotic constraints, post-harvest characteristics and quality are crucial to help scientists and potential end users to select the right materials. Germplasm evaluation refers to the agronomic description of the material in a gene bank, for traits that are generally important to breeders, researches in crop improvement and to business entrepreneurs in the industry. Lack of such evaluation data for a given *Musa* landrace limits its utilization to only cultural or traditional uses (FAO, 2010). Breeders and other end users require readily available and comprehensive datasets to select germplasm for further studies and/or development and subsequent testing and promotion to farmers and other end users. Evaluations can be carried out on landraces at the national and regional levels in their areas of cultivation. There are different stages and levels of *Musa* germplasm evaluation (MusaNet, 2016), including:

- Preliminary evaluation that consists of recording a limited number of agronomic traits thought desirable by users of the crop. Curators and breeders can provide this data, such as cycle time under evaluation, time of planting to shooting, flowering to maturity, and general performance of yield, pseudostem height, girth and height following ratoon, bunch weight, number of hands, fruits, fruit weight, length and diameter, which would help to narrow the selection of potential breeding stocks.
- On-farm evaluation of landraces includes the compilation of indigenous traditional knowledge, cultural uses, ease with mixed cropping and mixed-farming systems to guide early selection and acquisition of new accessions with relevant traits during collection.
- Phenotyping and genotyping. Using mass-production techniques such as tissue culture technology, a wide range of landraces can be evaluated in a screenhouse for specific traits of interest through phenotyping and genotyping, including high-throughput mass-screening for pest and disease resistance such as for, example, Sigatoka leaf spot and Fusarium wilt resistance under controlled conditions.
- Global evaluation. Evaluation of landraces introduced from regional or global collections (such as the International Transit Center, ITC, in Belgium) alone and/or in combination with hybrids to assess the performance of

accessions for specific traits under field conditions is commonly carried out in more than one country by different teams of varying experiences. It is therefore critical that key participants discuss and agree on the methods (study designs, replications, local reference cultivars/checks), evaluation protocols and which variables will be measured, following agreed protocols and reporting procedures. These evaluations will not only aid our understanding of the influence of the environment and growing conditions on the performance of promising accessions in multilocation trials, but also provide insights into the potential utilization options by the communities.

Further evaluation of introduced landraces consists of observations and/or recording potential agronomic characters, which will determine the usefulness of an accession under specific circumstances. It is the follow-up observations after selections that have been introduced onto farm. Typically, these include stress tolerance, disease and pest resistance, agronomic quality characters. Detailed evaluation of large collections requires capturing of data using a range of approaches involving the disciplines of cytogenetics and genetics, physiology, pathology, entomology, biochemistry, agronomy and post-harvest technologies. They all contribute information to form a complete picture of the identity and agronomic utility of landraces for the benefit of the banana sector.

Since evaluation needs may vary over time in response to changes in agricultural practices, farm economics, breeding strategies and the occurrence and impact of pest and diseases it should be kept in mind that evaluation work should always be open-ended and not necessarily be confined to characters on the descriptor list. In general, characterization and evaluation should be done by various discipline experts, initially led by the curator/germplasm scientists. Further evaluation or detailed evaluation is mostly done by the breeders in close collaboration with agronomists, pathologists, entomologists and post-harvest technologists.

4.4 Data access and availability

Availability and access of data on *Musa* landraces in Africa remain problematic. First, only a limited number of studies has been carried out on *Musa* landraces and of those that were carried out, datasets remain incomplete. In some cases information of evaluations is only available in student thesis as, for example, in the NARO-Uganda Library as is the case for evaluations for weevils and nematodes (Rukazambuga, 1996; Kashaija et al., 1994), black Sigatoka (Tushemereiwe et al., 1993) and Fusarium wilt (Kangire et al., 2000) conducted in landraces within the EAHBs. Africa has also participated in global evaluations of *Musa* accessions that included both landraces and hybrids. The international Musa testing program (IMTP) initiated by the International Network for Banana and Plantain (INIBAP) was carried out in two phases. The IMTP1 was conducted during 1989–1995 and focused on black Sigatoka while IMTP2 spanned 1996–2000 and also comprised Fusarium wilt in Nigeria, Cameroun, Canary Islands, Uganda and South Africa. For black Sigatoka, the report gives disease development time (DDT) and the youngest leaf spotted (YLS). Calcutta 4 (a wild AA diploid) was used as the resistant reference whereas Pisang Ceylan (AAB) and Pisang Berlin (AA) were used as checks for partial resistance and susceptibility, respectively. For Fusarium wilt, Bluggoe was found to be most susceptible, followed by Gros Michel and Saba to Fusarium wilt race 1. Yangambi Km 5 was resistant to race 1 strains in Africa, but was very susceptible to tropical race 4 strains when tested in Asia in disease-garden trials.

At the regional level, the genotyping with SSR markers of the East African accessions collected during past missions (Karamura et al., 2006) showed that the East African diploids are morphologically diverse, but genetically homogenous and are part of the same genetic complex as EAHBs but are not their direct progenitors. This enabled inferences on the origin of the East African banana complex to a restricted area of Southeast Asia, thereby providing useful information on the genetic makeup and history of these important landraces (Perrier et al., 2019).

5 Challenges for the maintenance and conservation of collected landraces in Africa

It is extremely important to realize the importance of well-characterized and evaluated *Musa* landraces for research and distribution to farmers. Large-field collections of banana and plantain landraces have been established at various African countries (Table 2). In East and Central Africa as well as in West Africa *Musa* landrace collections have been identified as core units of commodity research programs. However, ongoing changes in programs and projects, funding and locations have resulted in loss of landraces during successive phases of re-establishment. In general, *Musa* landrace collections face the following challenges.

5.1 Poor documentation and poor-quality data

Frequently, one encounters recording of different datasets in different excel spread sheet files. Data collected include, for example, the total number of accessions, cultivated (and/or wild materials) landraces, bunch weights, number of fingers/hands/bunches and fruit length, but for most collections these data are often of poor quality and has not been collected in a systematic manner. Some of these data are presented in poorly accessible government and institutional reports or publications in national journals. Such information at times is very difficult to locate and cannot easily be found through standard literature searches. Hence, for quite some time there has been a lack of adequate knowledge on the existing banana and plantain landrace diversity both in Africa and globally (De Langhe, 2004). Regular training to improve skills coupled with supervision of data capture and storage may help to eliminate this challenge.

5.2 Reduction in genetic diversity

Field collections have been too expensive to maintain by different national programs, leading to a tremendous loss of diversity, particularly if that diversity no longer exists on farms. While every national banana program has and needs a collection and will endeavor to keep it, the size and its role in research and development will depend on the priorities and resources of each *Musa* research program. Crop germplasm collections, whether large or small or haphazardly acquired and maintained, may contain some unique and potentially valuable landraces and particular phenotypes worthy of characterization, evaluation and maintenance, despite the costs involved. When these landraces are identified and described, their peculiarities, value, potential and limitations become known, hence making it easier for plant breeders and farmers to exploit these all-important resources.

5.3 Limited investment for effective management

Field gene banks need re-establishment after a period, the length of which depends on a range of factors. For instance, pest and disease buildup in the gene bank can begin to overwhelm the collection, making it difficult to carry out routine research work. Similarly, the original filed layout of the collection can be very limiting to some research designs/objectives so that it is necessary to re-establish the gene bank. By far the biggest limitation for the maintenance of these collections comprises (a) maintenance costs; (b) lack of trained manpower; (c) lack of a clear policy framework for their use; and (d) the problem of pests and diseases. Especially viruses, bacteria and fungi are important constraints. Currently all the banana germplasm collections in Africa are under threat due to Xanthomonas wilt and banana bunchy top. In vitro collections have recently been preferred because of the little space they use in conserving a large amount of germplasm protected from pest and disease. However, the in vitro collections are not immediately available for evaluation, study and use. Moreover, the cost and continuous availability of electric power in many parts of Africa remains a challenge. Overall, the in vitro conservation capacity is very limited and will require revamping in terms of resources to make it effective.

There are two general methods used in maintaining and conserving *Musa* landraces, ex-situ and on-farm in situ methods. Ex-situ techniques suitable for conservation include field gene banks which are considered the most suitable for plant breeding (Pillay et al., 2004). Conservation activities need to be targeted so as to preserve most of the available genetic diversity present in the crop and its wild relatives. Conservation efforts only focused on domesticated landraces, which share the same genetic background while showing some somatic variation, benefit from genetic analysis to be able to conserve much of the diversity in the smallest number of genotypes. The largest array of *Musa* landrace germplasm has been, and continues to be, in farmers' fields where farmers continue to source and maintain a diverse supply of bananas (Karamura and Mgenzi, 2004).

5.4 Lack of supportive policies

On-farm conservation can be a valuable and sustainable way to conserve landraces. However, the lack of supportive government policies and institutional frameworks has been an important drawback to the development of effective strategies in this area. There are several reasons why on-farm conservation is important. First, landraces in on-farm gene banks can be evaluated at a minimum cost, concurrently with farmer selection. Secondly, a relatively wide diversity of landraces can be targeted for evaluation in a short time and without the limitations of on-station research. Reductions in genetic diversity among old, phenotypically diverse landraces occur when they are being replaced by improved, modern cultivars which generally represent a narrow genetic base (Paroda and Arora, 2018). Awareness-raising through local media such as radio and television talk shows, newspapers, targeted field demonstrations and generally engaging the stakeholders is the key to influence stakeholders with regard to the preservation of landraces.

5.5 Political and social upheavals

Quite often, political and social upheavals arise in Africa, including regions where *Musa* collections are located. Civil wars, natural calamities such as drought and/or floods may lead to unprecedented movement of people in the vicinity of a large *Musa* collection (i.e. a source of food) which are occasionally appropriate sites for settling of victims of such upheavals. During the civil war in Uganda, during the early 1980s, the Kawanda Research Station (today NARO-Uganda) had several field crop collections including a banana that was used by rebels and the internally displaced refugees as a source of food. Within three years the crop collections vanished and several genotypes disappeared completely. As a risk management strategy all landraces need to be present in at least two or more different separate locations.

6 Recommendations for addressing challenges and exploiting opportunities for sustainable development of Musa landraces in Africa

- Networking the collection and evaluation data and subsequently sharing the data and genotypes among different areas in Africa and beyond;
- Strengthening the capacity for characterization and evaluation, gathering and storage of quality data in a standardized manner;
- Raise awareness and influence policy and increase investment for capacity building and infrastructure development to safeguard collections of landraces; and
- Strengthen on-farm conservation of *Musa* landraces by giving incentives to farmers to conserve landrace diversity to minimize genetic losses.

7 Where to look for further information

7.1 Student thesis studies

Adheka, G. (2014). Contribution to the characterization and classification of the Congo basin African plantains (Musa AAB) in the Democratic Republic of Congo. PhD theses, University of Kisangani, DR Congo.

Barekye, A. (1999). Relative damage caused by *Radopholus similis* and *Helicotylenchus multicinctus* on bananas in Uganda. MSc Thesis, Makerere University, Kampala, Uganda.

Kangire, A. (1998). Fusarium Wilt (Panama disease) of exotic bananas and wilt of East African Highland bananas (Musa, AAA-EA) in Uganda. PhD Thesis, Department of Agriculture, University of Reading, UK.

Kashaija, I. N. (1996). Factors influencing nematode population densities and root damage on banana cultivars in Uganda. PhD Thesis, University of Reading, UK

Namaganda, J. M. (1996). Nematode parasites of banana-root crop systems in Uganda. PhD Thesis, University of Reading.

Nankinga, C. M. (1994). The potential of indigenous fungal pathogens for the biological control of the banana weevil, (Cosmopolites sordidus) in Uganda. MSc Thesis, Makerere University, Faculty of Agriculture and Forestry, Uganda.

Nankinga, C. M. (1999). Characterization of entomopathogenic fungi and evaluation of delivery systems of Beauveria bassiana for the biological control of the banana weevil, Cosmopolites sordidus. PhD Thesis, University of Reading, UK.

Tushemereirwe, W. K. (1997). Factors influencing the expression of leaf spot diseases of highland bananas in Uganda. PhD Thesis, University of Reading, UK.

7.2 Field studies

Adheka, J. G., Dhed'a, D. B., Sivirihauma, C., Karamura, D., Langhe, E. de., Swennen, R. and Blomme, G. (2013). Plantain collection and morphological characterization in Democratic Republic of Congo: past and present activities and prospects, Banana systems in the humid highlands of sub-Saharan Africa: enhancing resilience and productivity. In: Blomme, G. et al. (Eds). CAB International, UK, p. 1-7.

Dhed'a, D. B., Nzawele, B. D., Roux, N., Ngezahayo, F., Vigheri, N., De Langhe, E., Karamura, D., Picq, C., Mobambo, P., Swennen, R. and Blomme, G. (2011). *Musa* collection and characterization in central and eastern DR Congo: a chronological overview. *Acta Horticulturae* 897, 87-94.

Karamura, D., Karamura, E. and Tinzaara, W. (2012). *Banana Cultivar Names, Synonyms and Their Usage in East Africa*. Bioversity International, Kampala.

MusaNet (2016). *Report on the East and Southern African Regional Workshop on East African Highland Banana Characterization and Documentation*. Bioversity International.

Ocimati, W., Karamura, D., Rutikanga, A., Sivirihauma, C., Ndungo, V., Adheka, J., Dhed'a, B., Muhindo, H., Ntamwira, J., Hakizimana, S. and Ngezahayo, F. (2013). *Musa* germplasm diversity status across a wide range of agro-ecological zones in Rwanda, Burundi and eastern Democratic Republic of Congo. Banana Systems in the Humid Highlands of Sub-Saharan Africa: In: *Enhancing Resilience and Productivity*. CAB International, Wallingford, UK, pp. 8-21.

Ocimati, W., Karamura, D., Sivirihauma, C., Ndungo, V., De Langhe, E., Adheka, J., Dhed'a, B., Ntamwira, J., Muhindo, H., Ragama, P. and Blomme, G. (2014). On-farm banana (*Musa*) cultivar diversity status across different altitudes in North and South Kivu provinces of eastern Democratic Republic of Congo. In: XXIX International Horticultural Congress on Horticulture: Sustaining Lives, Livelihoods and Landscapes (IHC2014): IX 1114 (pp. 35-44).

Onyango, M., Karamura, D., Keely, S., Manshardt, R., Haymar, D. (2011). Morphological characterization of East African AAB and AA dessert bananas (*Musa* spp.). *Acta Horticulture* 897, 95-105.

Sivirihauma, C., Ocimati, W., Valimuzigha, K., Karamura, D., Adheka, J., Ibanda, B., Dhed'a, B., Kamira, M. and Blomme, G. (2017). Diversity and morphological characterization of *Musa* spp. in North Kivu and Ituri provinces, Eastern Democratic Republic of Congo. *International Journal of Biodiversity and Conservation* (IJBC), 9(10): 292-305.

Talwana, H., Speijer, P. R., Gold, C. S., Swennen, R., and De Waele, D. (1996). Effect of nematode infection and damage on the root system and plant growth of three Musa cultivars commonly grown in Uganda. *Nematology* 8(2):177-189.

Tushemereirwe, W. K. and Waller, J. M. (1993). Black leaf streak (*Mycosphaerella fijiensis*) and associated diseases of bananas in Uganda. *Plant Pathology* 42:471–472.

Tushemereirwe, W. K., Gold, C. S., Speijer, P. R. and Holderness, M. (1994). Banana foliar disease in Uganda: preliminary results from a diagnostic survey. African *Crop Science Journal* 2: 145–149.

8 References

Adheka, J. G., Komoy, J., Tamaru, C., Sivirahauma, C., Dhed'a, D. B., Karamura, D., De Langhe, E., Swennen, R. and Blomme, G. 2018a. Banana diversity in the Oriental provinces, northeastern Democratic Republic of Congo. *Acta Horticulturae*. ISHS 1196:255-264. DOI 10.17660/ActaHortic.2018.1196.31.

Adheka, J. G., Dhed'a, D. B., Karamura, D., Blomme, G., Swennen, R. and De Langhe, E. 2018b. The morphological diversity of plantain in the Democratic Republic of Congo. *Scientia Horticulturae* 234:126–133.

Bakelana, K. and Ndungo, V. 2004. *La maladie de Bwere: une bactériose dévastatrice de la culture de la banane dans la province du Nord Kivu en République Démocratique du Congo*, Rapport de mission FAO, 1 pp.

Byabachwezi, M., Shaban, M. and Karamura, D. 2005. *Tanzania Musa Expedition 2005 Morogoro-Mbeya Region*. INIBAP, Kampala, Uganda.

De Langhe, E. 2004. Project: Plantain in the Eastern Congo Basin (Democratic Republic of Congo). Report of a Consultancy Mission – 18th to 24th of January 2004 - at the University of Kisangani with the Objective to Organise the Implementation of the Project. INIBAP (International Network for the Improvement of Banana and Plantain), Montpellier, France.

De Langhe, E., Karamura, D. A. and Mbwana, A. 2001. *Tanzania Musa Expedition 2001*. INIBAP, Montpellier, France.

FAO 2010. *Global Forest Resources Assessment: Country Report for Uganda Forestry Department*. Food and Agriculture Organization of the United Nations, Rome.

FAOSTAT 2019. *Online Statistical Database*. Food and Agriculture Organization of the United Nations, Rome. Available at: http://faostat.fao.org/.

Frankel, O. H. 1986. Genetic resources - museum or utility. In: Proceedings of Plant Breeding Symposium, Department of Scientific and Industrial Research (DSIR), Wellington.

Frison, E. and Sharrock, S. 2000. The economic, social and nutritional importance of banana in the world. In: Frison, E. and Sharrock, S. (Eds.), *Bananas and Food Security*. Aniane Louma Productions, France, pp. 21-35.

IPGRI, INIBAP and CIRAD. 1996. *Descriptors for Banana (Musa spp.) Montpellier*. CIRAD, France.

Kangire, A., Karamura, E. B., Gold, C. S. and Rutherford, M. A. 2000. Fusarium wilt of bananas in Uganda, with special emphasis on wilt-like symptoms observed on East African highland cooking cultivars. In: Craenen, K., Ortiz, R., Karamura, E. B. and Vuylsteke, D. (Eds.), Proceedings of the International Conference on Banana and Plantain for Africa. *Acta Horticulture* 540:343-353.

Karamura, D. 1999. *Numerical Taxonomic Studies of the East African Highland Bananas (Musa AAA-EA) in Uganda*. INIBAP, France.

Karamura, D., Kitavi, M., Nyine, M., Ochola, D., Ocimati, W., Muhangi, S., Talengera, D. and Karamura, E. B. 2016. Genotyping the local banana landrace groups of East Africa. *Acta Horticulturae* 1114:67–74.

Karamura, D., Mgenzi, B., Karamura, E. and Sharrock, S. 2004. Exploiting indigenous knowledge for the management and maintenance of Musa biodiversity on farm. *African Crop Science Journal* 12(1):67–74.

Karamura, D., Njuguna, J. and Nyamongo, D. 2006. *A Report on the Kenya Musa Expedition*. Musalit, Bioversity International, Montpellier.

Kashaija, I. N., Speijer, P. R., Gold, C. S. and Gowen, S. R. 1994. Occurrence, distribution and abundance of plant parasitic nematodes on bananas in Uganda. Preliminary results of a diagnostic survey. *African Journal of Crop Science* 2(1):99–104.

Kitavi, M., Downing, T., Lorenzen, J., Karamura, D., Onyango, M., Nyine, M., Ferguson, M. and Spillane, C. 2016. The triploid East African Highland Banana (EAHB) genepool is genetically uniform arising from a single ancestral clone that underwent population expansion by vegetative propagation. *Theoretical and Applied Genetics* 129(3):547–561. Available at: http://hdl.handle.net/10568/70223.

Lescot, T. 2007. Les dossiers de Fruitrop: la diversite´ ge´ne´tique des bananiers. Fruitrop n8145, Mai, p. 28.

Losimba, J. K., Giria, J. A., Djailo, B. D. A., Karamura, D., Blomme, G., Swennen, R. and Langhe, E. D. 2018. The complex distribution of plantain cultivars (Musa sp., AAB subgroup) in the Bas-Uele province of the Democratic Republic of Congo. *African Journal of Agricultural Research* 13(26):1358–1373. Available at: https://doi.org/10.5897/AJAR2018.13202.

MusaNet [online] 2011. Homepage of the MusaNet portal. Available at: https://sites.google.com/a/cgxchange.org/musanet/home. Accessed 25 July 2011.

MusaNet 2016. Global strategy for the conservation and use of Musa (Banana) Genetic resources (B. Laliberté, compiler). Bioversity International, Montpellier, France. 226p. Available at: https://cdn.croptrust.org/wp/wp-content/uploads/2017/06/Musa_Strategy_2016.pdf.

Ndungo, V. 2002. Analyse de la sécurité alimentaire menacée en période de guerre dans les territoires de Beni et de Lubero. Parcours et initiatives no 01, CRIG/UCG, pp. 89–95.

Ndungo, V. 2004. La menace de la fusariose sur la production des bananiers et plantains dans les pays des grands lacs Africains (Burundi, RDC, Rwanda). Parcours et initiatives n°3, CRIG/UCG, Butembo, pp. 30–34.

Ngeze, P. B. 1994. *Bananas and Their Management*. Kagera Writers and Publishers Co-Operatives Society, Tanzania.

Nsabimana, A., van Staden, J. and Gaidashova, S. V. 2004. Importance of the East African Highland bananas and their local nomenclature in Rwanda. In: Karamura, D. A., Karamura, E. and Tinzaara, W. (Eds.) *Banana Cultivar Names, Synonyms and Their Usage in East Africa*. Musalit Document ID: IN130120. Bioversity International, Uganda.

Ortiz, R. and Vuylsteke, D. 1994. Inheritance of black sigatoka disease resistance in plantain-banana (Musa spp.) hybrids. *Theoretical and Applied Genetics* 89(2–3): 146–152.

Paroda and Arora 2018. Plant genetic resources: general perspective. Available at: http://bioversityinternational.org/fileadmin/bioversity/publications/Web_version/174/ch05.htm#TopOfPage.

Perrier, X., Jenny, C., Bakry, F., Karamura, D., Kitavi, M., Dubois, C., Hervouet, C., Philippson, G. and De Langhe, E. (2019). East African diploid and triploid bananas: a genetic complex transported from South-East Asia. *Annals of Botany* 123(1):19–36. ISSN: 0305-7364.

Perrier, X., De Langhe, E., Donohue, M., Lentfer, C., Vrydaghs, L., Bakry, F., Carreel, F., Hippolyte, I., Horry, J. P., Jenny, C., Lebot, V., Risterucci, A. M., Tomekpe, K., Doutrelepont, H., Ball, T., Manwaring, J., de Maret, P. and Denham, T. 2011. Multidisciplinary perspectives on banana (*Musa* spp.) domestication. *Proceedings of the National Academy of Sciences of the United States of America* 108(28):11311–11318.

Pillay, M., Ssebuliba, R., Hartman, J., Vuylsteke, D., Talengera, D. and Tushemereirwe, W. K. 2004. Conventional breeding strategies to enhance the sustainability of *Musa* biodiversity conservation for endemic cultivars. *African Crop Science Journal* 12:59–65.

Rukazambuga, N. D. T. M. 1996. The effects of banana weevil (*Cosmopolites sordidus* germar) on the growth and productivity of bananas (*Musa* AAA EA) and the influence of host vigour on weevil attack. Available at: https://ethos.bl.uk/OrderDetails.do?uin=uk.bl.ethos.319250.

Shepherd, K. 1957. Banana cultivars in East Africa. *Tropical Agriculture* 34:277–286.

Simmonds, N. W. 1966. *Bananas* (2nd edn.). Tropical Agriculture Series. Longman, London, 512 pp.

Simmonds, N. W. 1976. Bananas (*Musaceae*). In: Simmonds, N. W. (Ed.) *Evolution of Crop Plants*. Tropical Agriculture Series. Longman, London, pp. 211–215.

Swennen, R. 1990a. Limits of morphotaxonomy. Names and synonyms of plantains in Africa and elsewhere. In: Jarret, R. L. (Ed.) *The Identification of Genetic Diversity in the Genus Musa*. Proceedings of the an International Workshop, Los Bãnos, Philippines, 5–10 September 1988. International Network for the Improvement of Banana and Plantain (INIBAP), Montpellier, France, pp. 172–210.

Swennen, R. 1990b. *Plantain Cultivation under West African Conditions: A Reference Manual*. IITA, Ibadan, Nigeria, 24 p.

Swennen, R. and Vuylsteke, D. 1987. Morphological taxonomy of plantains (Musa cultivars AAB) in West Africa. *Banana and Plantain Breeding Strategies*. In: Persley, G. I., and De Langhe, E. A. (Eds), Banana and plantain breeding strategies: proceedings of an international workshop held at Cairns, Australia, 13–17 October, 1986. ACIAR Proceedings, Brown Prior Anderson Pty. Ltd, Burwood, Victoria 3125. Vol 21, pp.165–171. ISBN: 0949511366.

Swennen, R., Vuylsteke, D. and Ortiz, R. 1995. Phenotypic diversity and patterns of variation in West and Central African plantains (*Musa* spp., AAB group, *Musaceae*). *Economic Botany* 49(3):320–327.

Tushemereirwe, W. K., Gold, W. K., Speijer, P. R. and Holderness, M. 1993. Foliar diseases of banana in Uganda: results from a diagnostic survey. *African Crop Science Journal* 1(2):145–149.

Chapter 8

Seed germination, preservation and population genetics of wild Musa germplasm

Bart Panis, Bioversity International and Katholieke University of Leuven (KUL), Belgium; Simon Kallow, Royal Botanical Gardens Kew, UK and Katholieke University of Leuven (KUL), Belgium; and Steven B. Janssens, Meise Botanic Garden, Belgium

1 Introduction

Bananas and plantains (*Musa* spp.) are amongst the most important tropical and subtropical food crops in the world. The genus Musa contains about 70 species grouped in 2 sections, Section Callimusa (incorporating Australimusa) containing 20 chromosomes (n = 10) and Section Musa (incorporating Rhodochlamys) containing 22 chromosomes (n = 11) (Wong et al., 2001). Most of the cultivated varieties are derived from one of, or are hybrids among, the *Musa acuminata* subspecies (e.g. *M. acuminata* ssp. *banksii*, or *M. acuminata* ssp. *malaccensis*) and *Musa balbisiana*, both belonging to the section Musa (FAO 2018). There exists another group, the Fe'i-type varieties, containing 18 chromosomes (n = 9) originating from the south pacific region that is probably mainly derived from the wild *Musa maclayi*, belonging to the section *Callimusa* (Simmonds, 1962). All Musa species are native to tropical Indomalaya, Asia and Australia, and are likely to have been first domesticated in New Guinea and surrounding islands (De Langhe et al., 2009).

http://dx.doi.org/10.19103/AS.2020.0070.09

Banana cultivation is under continuous attack from pests and diseases which are threatening banana cultivation (bacteria, fungi, viruses, nematodes and insects), especially, fungal and viral diseases such as the Sigatoka disease complex (*Pseudocercospora* spp.), Panama disease (*Fusarium oxysporum* f.sp. *cubense*), and banana bunchy top disease (banana bunchy top virus (BBTV)) and banana streak disease (banana streak virus (BSV)) are among the most devastating ones (Remans et al., 2007; Niyongere et al., 2015; Ordonez et al., 2015; Arango Isaza et al., 2016). Recently a new strain of *Fusarium oxysporum* f.sp. *cubense*, Tropical race 4, is threatening the world banana industry that almost completely relies on just one single variety, the Cavendish. Also the impact on banana cultivation of bacterial diseases such as *Xanthomonas* bacterial wilt (*Xanthomonas campestris* pv. *musacearum*), Moko/Bugtok disease (*Ralstonia solanacearum*) and Erwinia-associated diseases (*Erwinia* spp.) should not be underestimated (Blomme et al., 2017). As with other crops, the rapid spread of pests and diseases in banana is exacerbated by the large scale monoculture of Cavendish varieties in plantations the world over. Therefore, in order to reduce the impact of biotic and abiotic factors on banana cultivation, it is important to allow for genetic enrichment of the currently cultivated gene pool (Ordonez et al., 2015).

Crop Wild Relatives play an important role in modern plant breeding as they contain important traits for agriculture. Their conservation is therefore essential for future sustainable agriculture and food security. This review focuses on the different aspects and techniques that are required to establish seed collections of wild bananas. This covers physiology and morphology of the banana seeds, difficulties encountered during their germination and putative solutions, possibilities for medium- and long-term storage and population genetics. The number of publications and reports dealing with germination, seed physiology, population genetics and especially seed storage of wild *Musa* is very limited. This is caused by the fact that seed storage experiments rely heavily on (i) the availability of good quality seed batches and (ii) efficient seed germination protocols. Both issues are related since the ultimate way to screen whether a seed batch is of good quality is to test its germination rate. Until now, seed germination rates of wild bananas are unpredictable and mostly very low irrespective of the species.

2 The importance of conserving crop wild relatives (CWRs)

Most crops commonly grown in agriculture today have been domesticated from only one or two wild ancestors. Since being separated from the wild by early farmers the process of these ancient crops has been under specific selection pressure for often hundreds or even thousands of years, generally resulting in a huge diversity of landraces often well adapted to their local environments.

Seed and planting materials were collected and stored by farmers in the past but this has changed dramatically in the modern world due to industrialization of the seed business and intensified globalization. From an agricultural point of view, the increased industrialization of agriculture has resulted in a dramatic, over 90%, loss of landraces of different crops that were once selected and maintained by local farming communities and formed the basis of their livelihoods (FAO, 1999). Unfortunately, the importance of these landraces - in many cases adapted to specific local conditions - has too often been neglected and with their demise also the unique genetic alleles encoding for specific adaptations have been perpetually lost. The modern agricultural community relies on only a few genetic lineages to support the growing demand for food. This shift to less genetic diversity in our crops has resulted in erosion of crop genetic diversity and significantly increased our genetic vulnerability which is seriously limiting our ability to respond to emerging agricultural problems such as pests and diseases, drought, changes in climate and soil degradation. In the long term this may result in a decline of the overall food production. Former agricultural practices induced and accelerated evolutionary selection and increased the diversity among landraces as large numbers of farmers selected a wide range of different desirable traits for several millennia (Thrall et al., 2010). With the onset of modern industrial agriculture and its intensive and large-scale cultivation of monocultures, farmers no longer control their own seed but typically all purchase the same seed from a private company in order to achieve high yields with the aid of high levels of input such as irrigation, fertilizers and pesticides. This massive modernization of agricultural practices has led to genetic erosion and created a high level of genetic vulnerability. Hence, the need for novel and superior alleles from crop wild relatives (CWRs) and landraces is more critical than ever.

CWRs play an important role in modern plant breeding as they contain important traits for agriculture (Pistorius, 1997). Their conservation is therefore essential for future sustainable agriculture and food security. Moreover, modern crops contain a specific, but limited, selection of the genetic material of their ancestors (and close relatives). Despite the fact that several beneficial alleles were selected and captured into landraces by early farmers, a significant proportion of these alleles present in crop wild relatives were not incorporated into landraces during domestication (Heywood et al., 2007). It is now widely understood that CWRs contain a wealth of beneficial alleles and nowadays, CWRs are once again incorporated into modern breeding programs as they have the potential to contribute several novel and beneficial traits to existing crops (Hajjar and Hodgkin, 2007; Zhang et al., 2017). Of these traits, most are associated with pest and disease resistance, yet other traits such as drought or salt tolerance have also been successfully introgressed into modern cultivars (Hajjar and Hodgkin, 2007). Cultivars specifically tailored for pest and disease resistance are considered to become ineffective due to changes in the pathogen populations after 5 years for some crops (Heywood et al., 2007). As a result, understanding the evolution of alleles of economic and agricultural importance as well as their identification and population genetic structure is crucial for future cultivation of economically important crops. Moreover, a large amount of untapped genetic variation present in crop wild relatives is still unexploited. Unfortunately, increased biodiversity loss caused by habitat destruction and climate change pose a major threat to the inter- and intraspecific diversity of CWRs (Hammer and Teklu, 2008; Ford-Lloyd et al., 2011). In addition, many CWRs are found in areas that are used for the cultivation of related crops, allowing the introgression of crop-specific alleles into the wild gene pool, and jeopardizing the unique diversity present in CWRs (Ellstrand et al., 1999; Jarvis and Hodgkin, 1999; Aerts et al., 2013).

3 Seed morphology

The berry-type fruit of wild *Musa* contain whitish, sometimes yellowish, pulp and hard dark seeds (Fig. 1) which are dispersed by mammals and birds (Liu et al., 2004; Tang et al., 2005; Meng et al., 2008; Kulavmode et al., 2015). They can vbe cylindrical, subglobular, angular or flattened and vary within a seed batch depending on the placement in the berry. Seed weight in the genus *Musa* varies from 27 mg (*Musa ornata*) to around 70 mg (*Musa violascens).* Seeds in the sister genus *Ensete,* have a smooth seed coat and can be much larger than *Musa*

Figure 1 *Musa* balbisiana fruit containing seeds (Image: S. Kallow).

seeds measuring up to 2 cm in diameter; *Musella* (also sister genus to *Musa*) seeds are similar in size to *Musa* yet differ in having a smooth seed coat. Despite these differences, seed morphology is largely similar across the family Musaceae (Humphrey 1896; McGahan 1961; Graven et al., 1996; Benedict et al., 2016).

Musa seeds have two chambers. The first and largest chamber contains an undifferentiated capitate embryo. The embryo is small (1–2mm) and wedged into the micropylar collar, below which extends the cotyledonary haustorium of the embryo (Fig. 2). The first chamber also contains powdery starchy endosperm (McGahan, 1961). The second chamber is the chalazal mass, a large cylindrical or conical chamber containing a mass of gelatinous cells that are automorphic to the family (Benedict et al., 2016). These chambers are surrounded by tegmen, the outer of which is characterized by three layers: the silicified exotesta, the sclerotic mesotesta and the endotesta made of

Figure 2 Longitudinal section of a *Musa acuminata* subsp. *banksii* seed (top), showing micropyle (mi), embryo (em), chalazal mass (ch) endosperm (en); and detail (bottom) showing micropyle cap (mi-cap), micropyle channel (mi-ch), haustorium (hau), outer integument (out) and inner integument (in) (Images: S. Kallow).

thickened 'U shaped' cells. The inner integument has two layers, a thick inner cuticle which encloses the larger chamber only, and the outer which encloses the whole seed (Graven et al., 1996). The micropyle is situated above the embryo and is filled with a micropyle plug thereby forming an operculum. On the outer surface of the micropyle plug is the hilum. Such morphology is found in many monocots, but more developed in some of the Zingiberales and Commelinales (Danilova et al., 1995).

4 Seed germination

During germination the micropyle plug is forced off, facilitated by a predetermined rupture layer, as cell elongation of the radicle-hypocotly axis followed by the root primordium increases pressure in the seed (Fig. 3). Following this, the primary root becomes visible followed by the adventitious roots and the epicotyl. The primary root is short-lived as the adventitious roots take over. The seed coat often remains attached to the growing seedling for several weeks.

Various attempts to germinate *Musa* seeds have been made (reviewed by Ellis et al., 1985). Methods have included: soaking in water (in the light and dark), addition of gibberellic acid, H_2SO_4, HCl, HNO_3 and NaOH, chipping with a knife, removal of the micropyle lid, splitting the seed longitudinally, removing the chalazal mass, scorching with fire, cold and heat treatment, leaving seeds uncleaned, use of different containers, growth substrates and watering regimes and even passing through the intestinal canal of pigs and birds (Simmonds, 1952; Simmonds, 1959; Stotzky et al., 1962; Bhat et al., 1994; Chin, 1996; Burgos-Hernández et al., 2014; Vineesh et al., 2015; Wilson and Tenkouano, 2019). Regrettably, none of these methods have resulted in reliable or reproducible germination protocols. There are, no doubt, further factors which require exploration and consideration. Factors may be related to the seed itself, for instance, seed dormancy or poor understanding of germination

Figure 3 *Musa balbisiana* seed following germination, (left) showing epicotyl (ep), primary root (pr) and micropyle cap (mi-cap); (middle) showing adventitious roots (ad), epicotyle (ep) and primary root (pr); (right) *Musa balbisiana* seedling, showing shoot (sh), seed coat (sc) and adventitious roots (ad) (Images: S. Kallow).

requirements; or to do with the way in which seeds are collected and stored, which may have consequences on germination and viability

Some of the parameters shown to be important for seed germination are described in the next section.

4.1 Seed collecting

The many variables involved in seed collecting from wild populations make the immediate post-harvest handling of seeds difficult to control; these could, consequently, have confounding effects on the quality and germination potential of collected seed (Probert et al., 2007). Such variables can be difficult to unpick, for example, when collecting in the field in challenging terrain or travelling long distances, it is often impossible to know the timing of pollination and maturity of the bunch during harvest of the fruit for seed. Additionally, if seeds have been collected by different people from different institutions, it is hard to fully understand the variables associated with harvest and post-harvest if one was not present (e.g. seed extraction techniques, the method and extent of drying after extraction, the length of time in different post-harvest and storage conditions, or transportation methods). Often the viability of seed from collecting missions is low but data on why this might be the case is lacking. Experimentation using controlled material from field collections will likely provide useful insights on how to maximize viability, some examples of which are described below; however, it is not straight forward how to apply these findings directly to collecting missions.

The effect of specific climatic conditions on seed harvesting in field experiments was investigated by Ortiz and Vuylsteke (1995); they pollinated triploid bananas with diploid 'Calcutta 4' pollen (AA) and assessed seed number, presence of embryo, and germination of embryo, each month over five years and correlated this with climatic data (from Onne, Nigeria - outside of *Musa's* native range). Their results demonstrate strong seasonal effects including higher seed production in Jan-Feb and Aug–Sept, with a decline to May; peaks in seed production were positively correlated with high relative humidity, low solar radiation, low temperature and high rainfall suggesting that collecting from wild populations would be best following the wet season. Interestingly, the percentage of germination (from embryo rescue) was higher in May when seed production was lower. These findings, however, have to be interpreted with care since the mother plant was an edible triploid (and thus highly sterile).

The influence of maturity and ripeness of the fruit on the germination of *Musa acuminata* and *M. balbisiana* was examined in several experiments by (Simmonds, 1952). In one of these experiments, seeds from 6 weeks before to 4 weeks after maturity were collected and germinated (days after pollination were not defined). Simmonds (1952) managed to germinate seeds from across the entire period, but the highest germination rate (97%) was achieved at maturity or up to 3 weeks before full maturity (92%). Germination was still high up to 2 weeks after maturity (82%), after which it reduced to 46%. Whilst not controlling for some factors in this study, (e.g. he ripened fruit after collection, during which time seeds may have been able to continue maturing), these results seem to suggest that maturity may be important for seed germination but the exact timing does not need to be strictly controlled – there is a 5-week window to collect fruit containing seeds that still germinate well. Uma et al. (2011) found that during ripening of *Musa acuminata* (Pisang Jajee), the endosperm of seeds converted from a transparent fluid into a semi-solid state which was then followed by visible embryo development commencing after 70% seed maturity (based on full maturity at 110 days after pollination). Embryo rescued immature embryos (<90% seed maturity) required plant growth regulators in the culture media for successful regeneration, whereas >90% mature embryos regenerated just as well without.

From our experience during collecting missions, fully ripe or mature bunches are rarely, if ever, to be found in wild *Musa* populations; more often than not the bunches possible to find and collect are green and the fruit is firm. Ripe bananas are often eaten by bats, birds or mammals or they fall to the ground where they decompose rapidly. During collecting trips seeds can be superficially visually assessed for the developmental stage: bunches with seeds that are hard, with black or brown coats, powdery rather than milky or clear endosperms and with observable embryos should be selected. In fact, if the aim of the seed collection is long-term storage, collecting at late maturation is likely important for traits such as desiccation tolerance and potential longevity (Leprince et al., 2017).

4.2 Alternating temperature requirements

An ecological perspective may help better understand *Musa* seed germination and dormancy. In their native environment *Musa* plants are readily found on steep mountain gullies, rocky cliffs or in forest patches; as well as human-modified ecosystems, farmland edges, roadsides or recently cleared land (Simmonds, 1962; Chin, 1996). Such open-canopy niches suggest that bananas display pioneer or secondary-successional characteristics. This means that seeds may be able to survive in the soil between dispersal and disturbance and then respond to environmental changes when disturbance occurs. Such triggers in the natural environment can provide clues to then help understand germination in the lab or field.

Stotzky and Cox (1962) found that *Musa balbisiana* seeds require alternating, as opposed to constant, temperatures to germinate. They observed that when seeds were incubated at all combinations of low (4°C, 12°C, 15°C,

18°C and 27°C) and high temperatures (27°C, 32°C and 35°C) for either 5 hours or 19 hours, germination occurred only at alternating temperatures. Additionally, across all combinations the high temperature that stimulated the highest germination was 35°C for 5 hours (mean germination rate 59%), and temperature at the low side of the diurnal cycle that had the highest germination was 15°C for 19 hours (mean germination 70%). As a combination, germination was optimal (80%) at 5 hours of 35°C and 19 hours of 12°C. This response has also been observed in *Musa acuminata* subspecies (Kallow et al., 2019). In the dormancy classification developed by Baskin and Baskin (2004) this would be considered non-dormancy, as the seeds germinate given suitable environmental conditions.

Such a germination response to alternating temperatures can be a mechanism for canopy gap detection (Vázquez-Yanes and Orozco-Segovia, 1982), whereby a new niche can be exploited from the soil seed bank. Following disturbance, the soil seed bank will likely be exposed to increased temperature differentials in diurnal fluctuation as the removal of shade from the forest canopy exposes the soil directly to the sun (Hardwick et al., 2015). Additional cues associated with forest disturbance include changes in light and red: far-red light ratios (Xia et al., 2016) and changes in exogenous soil compounds (Yoshioka et al., 1998). It is advantageous for plants to quickly colonize such an ecological niche, and after a banana seedling becomes established they form a corm that can be very long lived as they continue to regenerate vegetatively by suckering while producing seed for more long-distance dispersal.

4.3 The role of the seed coat and other structures

The hard nature of the seed coat and the cuticle layer of the inner integument (McGahan, 1961) have led to a view that the seed coat limits moisture imbibition and therefore germination i.e, physical dormancy (Stotzky et al., 1962; Nagano et al., 2010; Fortescue and Turner, 2011). Graven et al. (1996) examined the structure and macromolecular composition of *Musaceae* seed coats using mass spectrometry, scanning and light microscopy and histochemical techniques. Their observations show that the seed coat could play an important role in seed dormancy. There are several structures which could limit gas and water exchange, most notably: the silica layer on the exotesta; the thick cuticle on the innermost layer which contains wax esters; the polyphenols in the tegmen; and the tightly packed thickened walled cells of the mesotesta which contain water-repellent lignin and unknown phenols.

The barrier function of the seed coat has been examined by Wattanachaiyingcharoen and Turner (1990) who showed that when the inner integument is pierced through - as compared to the scarification of the exotesta only with glass paper - the rate and amount of water absorption increased. They suggest this is because it is the inner integument that is water impermeable. Puteh et al. (2011) examined *Musa acuminata* seeds during imbibition and found that seeds rapidly imbibe water during the first hour. They propose that following drying the shrinking of the operculum and the surrounding tissues in the hilum area resulted in the formation and widening of water channels. Such channels could also be established during the warming and cooling cycles described above. However, when only the seed coat was mechanically scarified by Stotzky et al. (1962) to expose the endosperm, an alternating temperature requirement was still necessary. Interestingly, during embryo rescue, whereby embryos are excised and germinated on growth media (Cox et al., 1960; Afele and De Langhe, 1991; Uma et al., 2011), there is no requirement for alternating temperatures. These observations suggest that the alternating temperature mechanism lies in the relationship between the embryo, endosperm and possibly the aleurone layer, and not in making the seed coat more permeable to water.

4.4 Embryo development

Observations that seed batches germinate with greater success and speed following a period of warm temperature prior to other treatments (known as warm stratification) (Stotzky and Cox, 1962) suggest that *Musa* seeds may in fact display morphological dormancy, whereby the embryo requires time and suitable conditions to develop within the seed. For example, Nagano et al. (2008) sectioned *Musa velutina* embryos during developmental warm stratification in 75% moisture content vermiculite at 25°C. They distinguished four stages of development: immature undifferentiated mushroom-shaped embryo at harvest, first leaf primordia, second leaf primordia and third leaf primordia. Development to third leaf primordia took 8 months post-harvest and embryo length increased from 2.2mm to 3.5mm during this period. Furthermore, similar morphological differentiation of the embryo during warm stratification for *Musella lasiocarpa* was observed by Tang (2014), and germination to 82% following 6 months of stratification at 23°C on river sand at 60% moisture content compared to 0% germination without stratification. An important question is whether this behavior is typical only for *M. velutina* and *Musella* seeds or if this is in fact a physiological rather than morphological process, since it has not been observed in other *Musa* species?

4.5 Conclusions

Embryo development in the seed, alongside an alternating temperature requirement, means that Musaceae seeds have been described as having morpho-physiological dormancy (Baskin and Baskin, 2014). Morphological dormancy is when seeds have undifferentiated embryos which can take several weeks or months to develop prior to germination, often during warm moist stratification. Physiological dormancy is when there is a physiological inhibiting mechanism in the embryo that limits germination when a suitable environment is available. Germination and physiological dormancy are often controlled by plant hormones such as gibberellins (GA) and abscisic acid (ABA) or sensitivity towards these compounds (Kucera et al., 2005). The amount and speed of banana seed germination increases following treatment with GA (Nagano et al., 2010; Arun et al., 2013) which supports the assertion that banana seeds are physiologically dormant. However, as seeds are able to germinate in suitable temperatures (even if they must be alternating), without the need for any dormancy alleviation means, strictly speaking, they are non-dormant.

Our ability to consistently germinate banana seeds is still some way off, hampered by harvest and post-harvest factors, the effects of which are as yet poorly understood. Banana seed germination ecology seems to include a mechanism for germination following environmental disturbance - defining where seeds germinate - by responding to alternating temperatures rather than constant temperatures. Additionally, a period of warm stratification seems to be beneficial for germination, potentially including embryo development or germination inhibitor loss - defining when seeds can germinate and thus avoiding germination immediately following full dispersal. A better understanding of seed germination ecology will assist in the conservation and use of wild banana seed germplasm.

5 Storage of wild Musa germplasm

Since all edible bananas are virtually sterile and thus obligatory vegetatively propagated, storage of *Musa* germplasm is generally associated with field collections, *in vitro* collections and cryopreservation. Each of these three conservation methods have their pros and cons. Field collections are prone to pests and diseases and adverse weather conditions, are land and labor demanding to maintain but can also be used for screening and phenotyping. *In vitro* collections can be subject to somaclonal variation, at risk of fungal and bacterial infection, but require little space compared to field collections and are a source of clean planting material. Finally, cryopreserved collections store plant material safely for the long term (hundreds of years) but require a substantial investment of labor to transfer all material into liquid nitrogen.

These three conservation methods developed for edible bananas can also be easily applied to wild *Musa* germplasm but the main limitation is that very few, mostly only one, individuals of a population are conserved. For example, the Bioversity International ITC collection in Belgium, currently holds a total of 1452 banana accessions as *in vitro* shoot cultures of which 1121 are cryopreserved for the long term. The number of wild bananas accessions *in vitro* and cryopreserved are, respectively, 227 and 166.

The aim, as for most crop wild relatives, is to conserve the widest possible genetic diversity of wild *Musa* germplasm found within a species as well as among populations. Moreover, we need to be sure that when supplying wild germplasm to users (often breeders), that the diversity within the population provided is maximized; and this also determines how many individuals (seeds) should be provided. For this, in-depth population genetic studies are required. Seeds derived from field collections of wild bananas are not suitable for population genetic studies or other activities that require a wide genetic diversity since (i) due to the space requirement of banana plants only a limited amount of individuals can be stored in the field and (ii) the seeds of these field collections has limited value as they are derived from a small number of individuals that might be cross pollinated by individuals from neighboring plants that belong to other populations or even other species. The ultimate storage method for wild bananas is thus the storage of banana seeds derived from original wild populations.

One way to deal with the problematic germination of banana seeds is to apply an alternative plant regeneration method, for seeds this is embryo rescue. Embryos are extracted from previously surface-sterilized seeds and cultured on a predefined tissue culture medium (Bakry, 2008). This is a suitable way of recovering the highest possible number of plants from seeds derived from open-pollinated wild seedy bananas and controlled hybridizations.

Two kinds of tissues can be stored: the complete seed or embryos derived therefrom.

5.1 Conservation of embryos

In 1992, Abdelnour-Esquivel and coworkers reported successful liquid nitrogen storage of zygotic embryos of the two wild ancestors of edible bananas: *Musa acuminata* and *Musa balbisiana*. After excision from sterilized seeds, embryos were cultured for 3–5 hours on an *in vitro* medium that contains 60 g/l sucrose after which they were dried under the laminar flow cabinet for 1.5–2 hours resulting in a moisture content of 11–14%. Dried embryos were enclosed in cryotubes that were immersed in liquid nitrogen for storage. After subsequent rapid warming in a water bath embryos were placed on an *in vitro* medium. This resulted in high survival rates of 80% and 92% for *M. acuminata* and *M. balbisiana,* respectively. Vineesh and coworkers (2015) also described successful cryopreservation of embryos excised from seeds extracted from fully ripened bunches of *M. acuminata* subsp *burmannica*. Fifty-five to seventy-five percent recovered irrespective of whether they were precultured or not on 6% sucrose.

We recently developed a simplified protocol where embryos of *M. acuminata* and *M. balbisiana* were extracted from seeds that were first dried for one week in a desiccator until a moisture content below 10 % is reached. Then the embryos are enclosed in a 2 ml cryotube and directly exposed to liquid nitrogen. After rapid warming, embryos were cultured on hormone-free tissue culture medium. Regeneration rates of cryopreserved and non-cryopreserved embryos on *in vitro* media proved to be identical and ranged between 51% and 92% depending on the quality of the seed batch (Fig. 4).

Although cryopreservation of Musa embryos results in high regeneration rates, it is probably not the method of choice for the storage of wild Musa genetic resources because (i) not all institutes dealing with banana genetic resource have access to cryopreservation facilities and (ii) the excision of embryos prior to storage is extremely labor-intensive; a skilled person can excise around 40 embryos per hour. The conservation of a population containing thousands of seeds is thus not a valid option.

5.2 Conservation of seeds

According to the Seed Information Database from the Royal Botanic Gardens, Kew (2019), most of the members of the Musaceae family are considered to have an intermediate storage behavior. This conclusion is mainly based on their own research and the fact that some published work reports loss of viability after drying (Darjo and Bakry, 1990; Chin and Krishnapillay, 1989). This reported loss ranges between 0% and 100% and depends on species and seed batch and research institute (Kallow et al., 2020).

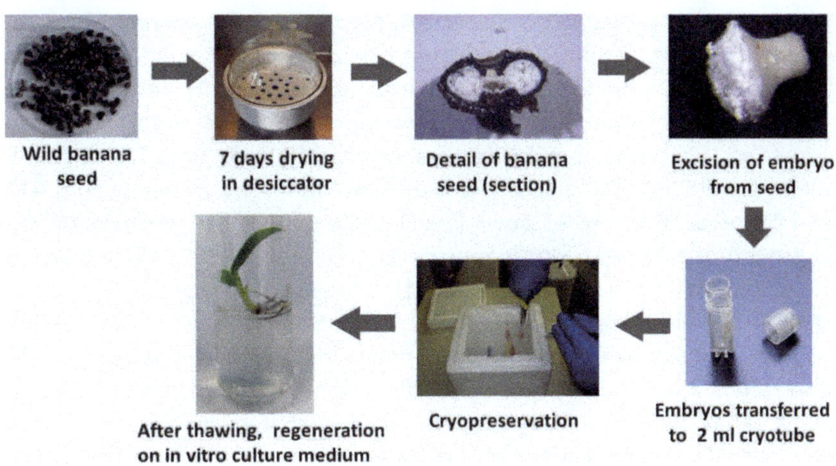

Figure 4 Workflow for cryopreservation of embryos.

Below is a short review of several reports that have dealt with the conservation of Musa seeds followed by germination in soil, germination *in vitro* and/or embryo rescue.

Back in 1952, Simmonds reported that sun-dried seeds of *M. balbisiana* could survive two years in a desiccator without losing their germination ability in soil; germination rates remained more than 90%. On the other hand, when the seeds were stored in cloth bags in the laboratory at ambient temperatures, the regeneration rate dropped to zero after one year of storage. Darjo and Bakry (1990) investigated the storage capacity of seeds that were extracted from artificially ripened fruits of *M. acuminata* and *M. balbisiana* and dried for 7 days in the air. Artificial ripening was obtained by soaking the fruits (at least 105 days after flowering) in a 0.1 % Ethrel solution. Extracted seeds were stored for 3 months at different temperatures; high temperatures +25/30°C and low temperatures -8/-9°C gave around 60% regeneration after 2 days of imbibition in water, while the regeneration rate for intermediate storage temperatures was around 80%.

Bhat and coworkers (1994) performed the first cryopreservation experiments on complete banana seeds and showed that around 20% of the *M. balbisiana* seed, when dried to around 13% moisture content, could germinate after storage into liquid nitrogen. This germination percentage was similar to that of non-cryopreserved seed. However, when seed cryopreservation was followed by embryo rescue, more than 90% of the embryos gave rise to plants, showing that the main problem in this experimental setup was not the storage but the germination in the soil. Kaya (2016) dried seeds of *M. acuminata* subsp *burmanica* and *M. velutina* with sterile air of a laminar flow bench for different periods of time. Subsequently seeds were enclosed in 2 ml cryotubes and plunged in liquid nitrogen. After storage, tubes containing the seeds were rapidly rewarmed followed by embryo rescue. He observed that the seed moisture content needs to be below 20% for the embryo extracted from the seed to survive exposure to liquid Nitrogen.

Chin (1996) conducted a conservation experiment with *M. violascens* seeds using different storage conditions and showed that seeds dried to moisture contents of 10% (original moisture content was 31%) and stored for 1 year in sealed bags at 22°C, -20°C and -196°C retained germination rates of 50%, 41% and 44%, respectively, when sown in soil, while the original germination rate was around 90%. Seeds with higher moisture contents (29% MC) could not survive liquid nitrogen exposure while 31% of the seeds that were dried down to 19% MC survived storage at -20°C for 1 year. Both moisture content as well as storage temperature are thus of importance for storing *M. violascens* seeds. These results led Dr. Chin to conclude that banana seeds have an orthodox storage behavior. Also for *M. velutina* it was proven that the inclusion of a drying step is critical to store its seed for periods surpassing one year (Pancholi et al., 1995; Nagano et al., 2008).

The reports cited above confirm recent unpublished results obtained at the Bioversity International Transit Center (ITC). It was observed that when seeds from different Musa species were sufficiently dried in a desiccator (to a moisture content below 10%) and subsequently enclosed in air-tight aluminum bags, they can be stored for prolonged periods at different storage temperatures. The following storage temperatures were tested: 25°C (room temperature), 5°C (fridge), -20°C (freezer) and -196°C (liquid nitrogen). All storage condition proved to be equal with respect to retaining viability up to 5 years. Even seeds stored at room temperature retained their regeneration capacity after 5 years of storage provided they are dried correctly and embryo rescue is applied. Longer storage duration is currently under investigation.

5.3 Conservation of pollen

An alternative method of conserving germplasm of wild *Musa* species can be found in the storage of pollen. However, reports on germination, viability and especially on storage of Musa pollen are extremely scarce. Youmbi and coworkers (2011) described storage experiments using pollen of three *M. acuminata* accessions: Calcutta 4, M53 and Zebrina, and observed that pollen drying in the desiccator for one week resulted in a complete loss of germination ability, indicating that these species produce desiccation-sensitive pollen. Avoiding desiccation damage by storing non-dried pollen at 10°C, they observed a pollen germination percentage decrease from 45% (non-stored) to 25%, 10% and 7% after 30, 60 and 90 days of storage, respectively. Storage at -20°C for 30 days reduced germination percentage to below 1%. Uma and Arun (2016) stored whole anthers of *M. acuminata* and *M. balbisiana* at different conditions: room temperature, 4°C and 0°C and observed that with storage time, germination of pollen reduced considerably. In their experiments the pollen failed to germinate at storage durations beyond 21 days, irrespective of the temperature. Pollen storage is thus currently not an option to preserve wild banana germplasm for the long term. The possibility of cryopreservation should thus be examined.

5.4 Conclusion

Like their edible counterparts, field preservation of wild banana species is a possibility. However due to space and maintenance constraints only a restricted number of individuals can be stored this way and therefore field preservation is not useful for storing wild populations. Cryopreserving embryos, excised from seeds, followed by *in vitro* culture is another option. However due to the labor-intensive nature of the embryo excision procedure prior to storage, this option is also extremely expensive. Storage of *Musa* seeds is the most suitable conservation methods since it is proven that they show orthodox storage behavior. Therefore, two main bottlenecks remain; (i) the availability of high-quality (meaning viable and regenerable) seed and (ii) the unpredictable (and often low) germination rate. The latter could be overcome by using embryo rescue. Finally, there is the option of storing pollen but before this can be put into practice more research is needed.

6 Population genetics of wild bananas

Several papers have been published about the wide range of morphological variation that is present in edible *Musa* varieties (e.g. plantains (Brown et al., 2017; Adheka et al., 2018; De Langhe et al., 2005), Cavendish-group (Robinson and Nel, 1985; Robinson et al., 1993; Chen et al., 2011)). In contrast, the morphological variation within wild *Musa* species is less well documented (Simmonds and Shepherd, 1955; Putra et al., 2010) and our current knowledge on the intraspecific genetic diversity within species of the genus *Musa* is rather limited. Most of the population genetic and phylogeographic studies that have been carried out within the genus *Musa* are focused on *M. balbisiana* (Ge et al., 2005; Uma et al., 2006; Wang et al., 2007; Rotchanapreeda et al., 2016; Jayaweera and Samarsinghe 2016; Bawin et al., 2019) and *M. acuminata* (Wong et al., 2001; Padmesh et al., 2012; Mukunthakumar et al., 2013), the two direct ancestors of most of our edible banana varieties. Only a few wild accessions such as *M. basjoo* var. *formosana* in Taiwan (Chen et al., 2017) and naturalized populations of *M. ornata* in Mexico (Burgos-Hernández et al., 2013) have been subjected to population genetic analysis. Within Musaceae, only *Musella lasiocarpa* has been intensively studied (Pan et al., 2007; Ma et al., 2019) and *Ensete ventricosum* populations have been analysed from a phylogeographic point of view by Borrell et al. (2019). In contrast to wild Musaceae populations, cultivated lineages have been more thoroughly investigated for their population genetic structure, either targeting specific clones or varieties (e.g. Rekha et al., 2001; Němečková et al., 2018) or by looking into more general evolutionary genetic relationships (de Jesus et al., 2013; Sardos et al., 2016; Christelova et al., 2016).

Population genetic approaches that are used to investigate cultivated bananas are often applied to infer genetic relationships between and within cultivated clones or varieties, to position the varieties among their wild ancestors, or to detect their putative ancestors (e.g. de Jesus et al., 2013; Němečková et al., 2018). In contrast, population genetic studies published on wild Musa accessions have addressed different scientific research questions. These range from obtaining more in-depth insights into the evolution and ecology of different species (Ge et al., 2005; Wang et al., 2007; Burgos-Hernández et al., 2013), to assessing the climatological and geological history of a specific geographic region using species-specific data (Ma et al., 2019). Additionally, inferring the conservation strategy of *Musa* species which may involve the collection of seed and clonal material (Bawin et al., 2019), and exploring the wild genetic resources in a given area (Padmesh et al., 2012), have been considered. One of the main questions often addressed for most of the different population genetic studies on wild species is the degree of genetic diversity among and within the different populations. This parameter shows the different responses of distinct wild species on their environment and how recent and past events have shaped the current degree of genetic diversity of the *Musa* species. In fact, each *Musa* species in its original distribution range responds differently to external events which will result in different population genetic parameters such as fixation index (FST), observed and expected heterozygosity (H_o and H_e, respectively) and inbreeding coefficient (FIS). For example, based on the current knowledge about ancestral distribution ranges of the different *Musa* species (Mertens et al., submitted; Janssens et al., 2016), population genetic analyses of wild species can provide interesting insights in what the true native distribution is of a species and how recent human-induced dispersal events have increased the distribution range to its current outline (Bawin et al., 2019). As a result, population genetic studies provide additional insights on where to collect new wild *Musa* accessions in order to find interesting alleles that could potentially be introgressed into new cultivars that can better respond to biotic and abiotic stresses.

Since the onset of molecular genetic techniques, three decades ago, the fields of phylogeography and population genetics have undergone dramatic change. Morphological, anatomical and cytogenetic results could now be expanded with large amounts of population genetic data concerning the presence of loci and alleles that allow for testing of different hypotheses about the biogeography, ecology and evolution of species. For bananas, also the aspect of cultivation and domestication could be more thoroughly addressed and the identification of specific alleles in wild *Musa* germplasm which may be useful for further crop improvement. One of the first techniques applied for molecular genetic analyses in *Musa* was Restriction Fragment Length Polymorphism (RFLP); a technique that is used to detect differences in the length of selected DNA fragments using enzyme digestion. Although initially applied to infer interspecific relationships within the *Musa* genus (Gawel et al., 1992), RFLPs were used in 2006 for a population genetic study on Indian *M. balbisiana* populations (Uma et al., 2006). However, since this method is considered to comprise some substantial drawbacks (e.g. time-consuming, inefficient resolution) it is in many cases replaced by SSR (Single Sequence Repeat; microsatellites) and AFLP (Amplified Fragment Length Polymorphism) analyses. The latter technique uses PCR amplification techniques of restricted fragments obtained from the digestion of total genomic DNA and has been successfully used by Wang et al. (2007) to infer the genetic diversity of *M. balbisiana* populations in China. However, already in 2001, Wong and coauthors used the

AFLP methodology to investigate inter- and intraspecific relationships within Malaysian *Musa acuminata* subspecies (*M. acuminata* ssp. *errans*, *M. acuminata* ssp. microcarpa and *M. acuminata* ssp. malaccensis). To date, AFLPs are still used to infer population genetic relationships as is shown by the study of Chen et al. (2017) on Taiwanese populations of *Musa basjoo* var. *formosana*. The most commonly used technique for population genetics in Musa are microsatellite markers (or SSR). This PCR-based technique, in which short tandem repeats are targeted, has been successfully used in several studies to investigate the genetic diversity and population structure of different *Musa* species (Ge et al., 2005; Padmesh et al., 2012; Mukunthakumar et al., 2013; Burgos-Hernández et al., 2013; Jayaweera and Samarsinghe 2016; Rotchanapreeda et al., 2016; Bawin et al., 2019). Except for putative homology issues on the obtained length of the bands, SSRs have shown to produce reliable results (Christelova *et al.*, 2016; Sardos et al., 2018). A less frequently used and less robust technique for population genetics in *Musa* is Random Amplified Polymorphic DNA (RAPD). This low-cost, time-effective technique randomly amplifies short DNA fragments throughout the genome, which were used by Mukunthakumar et al. (2013), to infer the genetic variation of *M. acuminata* ssp. *burmannica*. Other than the above-mentioned techniques, which are based on differences in DNA fragment lengths also single nucleotide polymorphisms (SNP) can be used to address population genetic questions. One of those recently developed techniques is DArT (Diversity Arrays Technology). DArT is a DNA hybridization-based genotyping technology that allows to detect large numbers of SNPs. The method is recently successfully applied to provide an onset in integrating wild accessions among the large number of known banana cultivars (Sardos et al., 2016).

7 Future trends and conclusion

In this chapter we show that although considerable progress has been made on different aspects related to germination, preservation and population genetics of wild bananas, some gaps still remain before the establishment of a global Musa seedbank can start. The first bottleneck is the restricted access to wild germplasm. This because wild bananas often grow in remote, inaccessible areas and germplasm exchange between countries remains challenging. Initiation of more collaborative projects involving partners with different expertise and resources are therefore essential. An important issue remains also the low and erratic germination rates of *Musa* seed. Further research on defining optimal harvesting times and post-harvest treatments, physiological studies linked to dormancy and determination of optimal germination conditions is thus necessary. Related to the seed storage, enough evidence is provided in literature to conclude that *Musa* seed has orthodox storage behavior, provided a good initial quality of seeds.

Classical seed storage at -20°C is thus an option, but cryopreservation of core collections for the long term should surely also be considered.

An important addition to classical seed banking is the genetic screening of the seed batches that are present in seed collections. An improved understanding of the genetic diversity of ex-situ collections such as seed collections will become an important element of future collection management. With the development of novel high-throughput sequencing techniques it will become possible to apply low-cost and time-efficient methods among a broad range of species within the Musa genus. Despite the continuous advance of modern molecular techniques, currently obtained, insights in population genetics and interactions at low taxonomic level remain very important to design an optimized collection strategy for future banana seed banking.

8 Acknowledgements

The authors gratefully acknowledge funding from a sub-grant from the University of Queensland from the Bill & Melinda Gates Foundation project 'BBTV mitigation: Community management in Nigeria, and screening wild banana progenitors for resistance' [OPP1130226]. The authors also thank all donors who supported this work through their contributions to the CGIAR Fund (http://www.cgiar.org/funders) and in particular to the CGIAR Research Program Roots, Tubers and Bananas (RTB-CRP). The authors thank the Research Foundation Flanders (FWO) for making this research possible through the Flanders Bilateral Research Cooperation with Vietnam project with external reference 'G0D9318N'.

9 Where to look for further information

Royal Botanic Gardens, Kew, UK (https://www.kew.org/science/collections-and-resources/collections/seed-collection) and Meise Botanic Garden, Belgium (https://www.plantentuinmeise.be/en/) are involved in preservation of wild bananas through seed collection. In particular, seed collections were made under the Crop Wild Relative Project (https://www.cwrdiversity.org).

Information on some of the banana seed accessions are included in the Genesys database (https://www.genesys-pgr.org/). Bioversity International (now the Alliance of Bioversity International and CIAT) hosts the world's largest banana genebank as in vitro proliferating shoots and as a cryopreserved collection. More information about this collection can be found at:

- https://www.bioversityinternational.org/research-portfolio/banana-genetic-resources-and-management-systems/.
- https://www.bioversityinternational.org/fileadmin/_migrated/uploads/tx_news/Cryopreservation_of_Musa_germplasm__2nd_edition_1383.pdf.

How and where to collect seeds of CRWs (Crop Wild Relatives), population genetics studies:

- Blair M.W., Díaz L.M., Buendía H.F., Duque M.C. 2009. Genetic diversity, seed size associations and population structure of a core collection of common beans (Phaseolus vulgaris L.). Theoretical and Applied Genetics. 119: 955–972.
- Prada D. 2009. Molecular population genetics and agronomic alleles in seed banks: searching for a needle in a haystack? *Journal of Experimental Botany*. 60: 2541–2552.
- Vander Mijnsbrugge K., Bischoff A., Smith B. 2009. Question of origin: Where and how to collect seed for ecological restoration? Basic and Applied Ecology. 11: 300-311.
- https://www.cwrdiversity.org/.

10 References

Abdelnour-Esquivel, A., Mora, A. and Villalobos, V. (1992). Cryopreservation of zygotic embryos of *Musa acuminata* (AA) and *Musa balbisiana* (BB), *Cryo-Letters* 13, 159–164.

Adheka, J. G., Dheda, D. B., Karamura, D., Blomme, G., Swennen, R. and De Langhe, E. (2018). The morphological diversity of plantain in the Democratic Republic of Congo, *Scientia Horticulturae* 234, 126–133.

Aerts, R., Berecha, G., Gijbels, P., Hundera, K., Van Glabeke, S., Vandepitte, K., Muys, B., Roldán-Ruiz, I. and Honnay, O. (2013). Genetic variation and risks of introgression in the wild *Coffea arabica* gene pool in southwestern Ethiopian montane rainforests, *Evolutionary Applications* 6(2), 243–252.

Afele, J. C. and De Langhe, E. (1991). Increasing *in vitro* germination of Musa balbisiana seed, *Plant Cell, Tissue and Organ Culture* 27(1), 33–36.

Arango Isaza, R. E., Diaz-Trujillo, C., Dhillon, B., Aerts, A., Carlier, J., Crane, C. F., de Jong, T. V., de Vries, I., Dietrich, R., Farmer, A. D., Fortes Fereira, C., Garcia, S., Guzman, M., Hamelin, R. C., Lindquist, E. A., Mehrabi, R., Quiros, O., Schmutz, J., Shapiro, H., Reynolds, E., Scalliet, G., Souza, M., Stergiopoulos, J. I., Van der Lee, T. A. J., De Wit, P. J. G. M., Zapater, M.-F., Zwiers, L.-H., Grigoriev, I. V., Goodwin, S. B. and Kema, G. H. J. (2016). Combating a global threat to a clonal crop: banana black Sigatoka pathogen *Pseudocercospora fijiensis* (synonym *Mycosphaerella fijiensis*) genomes reveal clues for disease control, *PLoS Genetics* 18, 1-36.

Arun, K., Uma, S., Saraswathi, S., Backiyarani, S. and Dubai, P. (2013). Effects of whole seed priming on the in vitro germination of hybrid banana embryos (*Musa* spp.), *Seed Science and Technology* 41(3), 439–451.

Bakry, F. (2008). Zygotic embryo rescue in bananas, *Fruits* 63(2), 111–115.

Baskin, C. C. and Baskin, J. M. (2014). *Seeds: Ecology, Biogeography, and, Evolution of Dormancy and Germination*, 2nd edn. Academic Press, Cambridge, MA.

Baskin, J. M. and Baskin, C. C. (2004). A classification system for seed dormancy, *Seed Science Research* 14(1), 1–16.

Bawin, Y., Panis, B., Vanden Abeele, S., Li, Z., Sardos, J., Paofa, J., Ge, X.-J., Mertens, A., Honnay, O. and Janssens, S. B. (2019). An assessment of the genetic diversity in seed batches of *Musa balbisiana* for conservation purposes', *Plant Genetic Resources* 17(6), 536–544.

Benedict, J. C., Smith, S. Y., Specht, C. D., Collinson, M. E., Leong-Škorničková, J., Parkinson, D. Y. and Marone, F. (2016). Species diversity driven by morphological and ecological disparity: a case study of comparative seed morphology and anatomy across a large monocot order, *AoB Plants* 8, 1–27.

Bhat, S. R., Bhat, K. H. and Chandel, K. P. S. (1994). Studies on germination and cryopreservation of Musa balbisiana seed, *Seed Science and Technology* 22(3), 637–640.

Blomme, G., Dita, M., Jacobsen, K. S., Pérez Vicente, L., Molina, A., Ocimati, W., Poussier, S. and Prior, P. (2017). Bacterial diseases of bananas and enset: current state of knowledge and integrated approaches Toward sustainable management, *Frontiers in Plant Science* 8, 1290.

Borrell, J. S., Biswas, M. K., Goodwin, M., Blomme, G., Schwarzacher, T., Heslop-Harrison, J. S. P., Wendawek, A. M., Berhanu, A., Kallow, S., Janssens, S., Molla, E. L., Davis, A. P., Woldeyes, F., Willis, K., Demissew, S. and Wilkin, P. (2019). Enset in Ethiopia: a poorly characterized but resilient starch staple, *Annals of Botany* 123(5), 747–766.

Brown, A., Tumuhimbise, R., Amah, D., Uwimana, B., Nyine, M., Mduma, H., Talengera, D., Karamura, D., Kuriba, J., Swennen, R. (2017). Bananas and plantains (*Musa* spp.). In: Campos, H. and Caligari, P. D. S. (Eds), *Genetic Improvement of Tropical Crops*, Cham, Switzerland: Springer International Publishing AG, pp. 219–240.

Burgos-Hernández, M., Castillo-Campos, G., Mata-Rosas, M., Gonzalez, D., Vovides, A. P. and Murguia-Gonzalez, J. (2014). Seed germination of the wild banana *Musa ornata* (*Musaceae*), *Seed Science and Technology* 42(1), 16–27.

Burgos-Hernández, M., González, H. D. and Castillo-Campos, G. (2013). Genetic diversity and genetic structure population of *Musa ornata* (Musaceae) in México, *Plant Systematic and Evolution* 299, 1899–1910.

Chen, H. B., Wang, Q. J., Qin, Y. H., Zhang, G. A. and Hu, G. B. (2011). Study on the genetic relationship of 35 Cavendish banana (*Musa* AAA) varieties by RAPD molecular marker, *Acta Horticulturae* 894, 97–104.

Chen, J. H., Huang, C. L., Lai, Y. L., Chang, C. T., Liao, P. C., Hwang, S. Y. and Sun, C. W. (2017). Postglacial range expansion and the role of ecological factors in driving adaptive evolution of *Musa basjoo* var. *formosana*, *Scientific Reports*, *Scientific Reports* 7(1), 5341.

Chin, H. F. (1996). Germination and storage of banana seeds. In: Frison, E. A., Horry, J. P. and De Waele, D. (Eds), Proceedings of the Workshop on New Frontiers in Resistance Breeding for Nematodes, Fusarium and Sigatoka, Kuala Lumpur, Malaysia, October 1995 (1996), pp. 218–227.

Chin, H. F. and Krishnapillay, B. (1989). Cryogenic storage of some horticultural species, *Acta Horticulturae* 253, 107–112.

Christelova, P., De Langhe, E., Hribova, E., Cizkova, J., Sardos, J., Husakova, M., van den Houwe, I., Sutanto, A., Kay Kepler, A., Swennen, R., Roux, N. and Dolezel, J. (2016). Molecular and cytological characterization of the global Musa germplasm collection provides insights into the treasure of banana diversity, *Biodiversity and Conservation* 26, 801–824.

Cox, E. A., Stotzky, G. and Goos, R. D. (1960). In vitro culture of Musa balbisiana Colla embryos, *Nature* 185(4710), 403–404.
Danilova, M., Nemirovich Danchenko, E., Komar, G. and Lodkina, M. (1995). The seed structure of monocotyledons, *Monocotyledons: Systematics and Evolution* 2, 461–472.
Darjo, P. and Bakry, F. (1990). Conservation et germination des grains de bananiers (*Musa* Sp.), *Fruits* 45(2), 103–113.
de Jesus, O. N., Silva, Sde O., Amorim, E. P., Ferreira, C. F., de Campos, J. M., Silva, G. de G. and Figueira, A. (2013). Genetic diversity and population structure of *Musa* accessions in ex situ conservation, *BMC Plant Biology* 13, 41.
De Langhe, E., Vrydaghs, L., de Maret, P., Perrier, X. and Denham, T. (2009). Why Bananas Matter: an introduction to the history of banana domestication, *Ethnobotany Research and Applications* 7, 165–177.
De Langhe, E. D., Pillay, M., Tenkouano, A. and Swennen, R. (2005). Integrating morphological and molecular taxonomy in *Musa*: the African plantains (*Musa* spp. AAB group), *Plant Systematics and Evolution* 255(3–4), 225–236.
Ellis, R. H., Hong, T. D. and Roberts, E. H. (1985). *Handbook of Seed Technology for Genebanks*, Rome, Italy: International Board for Plant Genetic Resources (IBPGR).
Ellstrand, N. C., Prentice, H. C. and Hancock, J. F. (1999). Gene flow and introgression from domesticated plants into their wild relatives, *Annual Review of Ecology and Systematics* 30(1), 539–563.
FAO (1999). Conference on the Multifunctional Character of Agriculture and Land, Background Paper 1 is on Agricultural Biodiversity, Maastricht, Netherlands.
FAO (2018). FAOSTAT, Retrieved July 20, 2018, from http://www.fao.org/faostat/en/#data/QC.
Ford-Lloyd, B. V., Schmidt, M., Armstrong, S. J., Barzani, O., Engels, J., Hadas, R., Hammer, K., Kell, S. P., Kang, D., Khoshbakht, K., Li, Y., Long, C., Lu, B.-R., Ma, K., Nguyen, V. T., Qiu, L., Ge, S., Wei, W., Zhang, Z. and Maxted, N. (2011). Crop wild relatives - undervalued, underutilized and under threat?, *BioScience* 61(7), 559–565.
Fortescue, J. A. and Turner, D. (2011). Reproductive biology. In: Pillay, M. and Tenkouano, A. (Eds), *Banana Breeding: Progress and Challenges*, Boca Raton, FL: Taylor & Francis, pp. 145–179.
Gawel, N. J., Jarret, R. L. and Whittemore, A. P. (1992). Restriction fragment length polymorphism (RFLP)-based phylogenetic analysis of *Musa*, *Theoretical and Applied Genetics* 84(3–4), 286–290.
Ge, X. J., Liu, M. H., Wang, W. K., Schaal, B. A. and Chiang, T. Y. (2005). Population structure of wild bananas, *Musa balbisiana*, in China determined by SSR fingerprinting and cpDNA, *Molecular Ecology* 14(4), 933–944.
Graven, P., De Koster, C. G., Boon, J. J. and Bouman, F. (1996). Structure and macromolecular composition of the seed coat of the Musaceae, *Annals of Botany* 77(2), 105–122.
Hajjar, R. and Hodgkin, T. (2007). The use of wild relatives in crop improvement: a survey of developments over the last 20 years, *Euphytica* 156(1–2), 1–13.
Hammer, K. and Teklu, Y. (2008). Plant genetic resources: selected issues from genetic erosion to genetic engineering, *Journal of Agriculture and Rural Development in the Tropics and Subtropics* 109, 15–50.
Hardwick, S. R., Toumi, R., Pfeifer, M., Turner, E. C., Nilus, R. and Ewers, R. M. (2015). The relationship between leaf area index and microclimate in tropical forest and oil

palm plantation: forest disturbance drives changes in microclimate, *Agricultural and Forest Meteorology* 201, 187–195.

Heywood, V., Casas, A., Ford-Lloyd, B. V., Kell, S. P. and Maxted, N. (2007). Conservation and sustainable use of crop wild relatives, *Agriculture, Ecosystems and Environment* 121(3), 245–255.

Humphrey, J. E. (1896). The development of the seed in the Scitamineae, *Annals of Botany* 10(37), 1–40.

Janssens, S. B., Vandelook, F., De Langhe, E., Verstraete, B., Smets, E., Vandenhouwe, I. and Swennen, R. (2016). Evolutionary dynamics and biogeography of Musaceae reveal a correlation between the diversification of the banana family and the geological and climatic history of Southeast Asia, *New Phytologist* 210(4), 1453–1465.

Jarvis, D. I. and Hodgkin, T. (1999). Wild relatives and crop cultivars: detecting natural introgression and farmer'selection of new genetic combinations in agroecosystems, *Molecular Ecology* 8 (12, suppl.1), S159–S173.

Jayaweera, S. L. D. and Samarsinghe, W. L. G. (2016). Genetic diversity and population structure of wild banana (*Musa balbisiana*) populations in Sri Lanka, *Acta Horticulturae* 1114, 53–60.

Kallow, S., Davies, R., Janssens, S., Vandelook, F., Panis, B. and Dickie, J. (2019). Wild bananas detect forest gaps. In: Proceedings of Seed Ecology VI Regensburg, July 29–August 2, 2019, Regensburg, Germany.

Kallow, S., Longin, K., Sleziak, N. F., Janssens, S. B., Vandelook, F., Dickie, J., Swennen, R., Paofa, J., Carpentier, S. and Panis, B. (2020). Challenges for Ex Situ Conservation of Wild Bananas: Seeds Collected in Papua New Guinea Have Variable Levels of Desiccation Tolerance. *Plants* 9 (9), 1243.

Kaya, E. (2016). Long-term germplasm conservation of two economically important *Musa* species via cryopreservation dehydration technique, *Biodiversity and Conservation.* 9(3), 178–182.

Kucera, B., Cohn, M. A. and Leubner-Metzger, G. (2005). Plant hormone interactions during seed dormancy release and germination, *Seed Science Research* 15(4), 281–307.

Kulavmode, A. R., Hegde, V. M. and Kambale, A. A. (2015). A note on seed dispersal of rock banana *Ensete superbum* by Asian palm civet *Paradoxurus hermaphroditus* at Sinhgad Fort, Maharashtra, India, *Journal of the Bombay Natural History Society* 112(1), 25–26.

Leprince, O., Pellizzaro, A., Berriri, S. and Buitink, J. (2017). Late seed maturation: drying without dying, *Journal of Experimental Botany* 68(4), 827–841.

Liu, M., Ge, X., Wang, W., Hsu, T., Schaal, B. and Chiang, T. (2004). Pollen and seed dispersal of Musa balbisiana in South China, *Conservative Quarterly* 47, 9–24.

Ma, H., Wang, D.-X., Li, T.-Q. and Li, Z.-H. (2019). Phylogeographic study of *Musella lasiocarpa* (Musaceae): providing insight into the historical river capture events, *Pakistan Journal of Botany* 51(1), 191–199.

McGahan, M. W. (1961). Studies on the seed of banana. I. Anatomy of the seed and embryo of Musa balbisiana, *American Journal of Botany* 48(3), 230–238.

Meng, L., Gao, X. and Chen, J. (2008). Seed dispersal and spatial-temporal variation of seed predation of Musa acuminata in Xishuangbanna, Southwest China, *Journal of Plant Ecology (Chinese Version)* 32(1), 133–142.

Mertens, A., Swennen, R., Rønsted, N., Vandelook, F., Panis, B., Sachter-Smith, G., Vu, D. and Janssens, S. B. (submitted to Diversity and Distributions, major revisions)

Conservation status assessment of banana crop wild relatives using species distribution modelling.

Mukunthakumar, S., Padmesh, P., Vineesh, P. S., Reby, S., Harikumar, K. and Krishnan, P. N. (2013). Genetic diversity and differentiation analysis among wild antecedents of banana (*Musa acuminata* Colla) using RAPD markers, *Indian Journal of Biotechnology* 12, 493–498.

Nagano, S., Mori, G. and Oda, M. (2008). Effects of temperature and moisture content of the substrate during storage on embryo development and germination in seeds of Musa velutina Wendl. & Drude, *Journal of Horticultural Science and Biotechnology* 83(1), 33–36.

Nagano, S., Mori, G. and Oda, M. (2010). Promotion of seed germination in Musa velutina Wendl. & Drude by scarification and GA3, *Journal of Horticultural Science and Biotechnology* 85(4), 267–270.

Němečková, A., Christelová, P., Čížková, J., Nyine, M., Van den Houwe, I., Svačina, R., Uwimana, B., Swennen, R., Doležel, J. and Hřibová, E. (2018). Molecular and cytogenetic study of East African highland banana, *Frontiers in Plant Science* 9, 1371.

Niyongere, C., Omondi, A. B. and Blomme, G. (2015). Banana bunchy top. In: Tennant and Fermin (Eds), *Virus Diseases of Tropical and Subtropical Crops* (1st edn, pp. 17–26), Boston, MA: CABI Publishing.

Ordonez, N., Seidl, M. F., Waalwijk, C., Drenth, A., Kilian, A., Thomma, B. P. H. J., Ploetz, R. C. and Kema, G. H. (2015). Worse comes to worst: bananas and panama disease–when plant and pathogen clones meet, *PLoS Pathogens* 11, e1005197.

Ortiz, R. and Vuylsteke, D. (1995). Factors influencing seed set in triploid *Musa* spp. L. and production of euploid hybrids, *Annals of Botany* 75(2), 151–155.

Padmesh, P., Mukunthakumar, S., Vineesh, P. S., Skaria, R., Hari Kumar, K. and Krishnan, P. N. (2012). Exploring wild genetic resources of *Musa acuminata* Colla distributed in the humid forests of southern Western Ghats of peninsular India using ISSR markers, *Plant Cell Reports* 31(9), 1591–1601.

Pan, Q. J., Li, Z. H., Wang, Y., Tian, J., Gu, Y. and Liu, X. X. (2007). Rapid analysis on the genetic diversity of wild and cultivated populations of *Musella lasiocarpa*, *Forest Research* 20, 668–672.

Pancholi, N., Wetten, A. and Caligari, P. D. S. (1995). Germination of *Musa velutina* seed - comparison of in vivo and in vitro systems, *In Vitro Cellular and Developmental Biology - Plant* 31(3), 127–130.

Pistorius, R. (1997). *Scientists, Plants and Politics*, Rome: IPGRI.

Probert, R., Adams, J., Coneybeer, J., Crawford, A. and Hay, F. (2007). Seed quality for conservation is critically affected by pre-storage factors, *Australian Journal of Botany* 55(3), 326–335.

Puteh, A. B., Aris, E. M., Sinniah, U. R., Rahman, M., Mohamad, R. B. and Abdullah, N. A. (2011). Seed anatomy, moisture content and scarification influence on imbibition in wild banana (*Musa acuminata* Colla) ecotypes, *African Journal of Biotechnology* 10(65), 14373–14379.

Putra, E. T. S., Abdullah, N. A. P., Saleh, G. and Zakaria, W. (2010). Morphological variation and geographical distribution of *Musa* sp. cv. Rastali in Peninsular Malaysia, *Journal of Cell and Plant Sciences* 1, 23–32.

Rekha, A., Ravishankar, K. V., Anand, L. and Hiremath, S. C. (2001). Genetic and genomic diversity in banana (*Musa* species and cultivars) based on D2 analysis and RAPD markers, *Infomusa* 10, 29–34.

Remans, T., Iram, S., Shuey, L., Jaufeerally-Fakim, Y. and Schenk, P. (2007). Banana streak virus: a highly diverse plant pararetrovirus, *Plant Viruses* 1, 33–38.

Robinson, J. C. and Nel, D. J. (1985). Comparative morphology, phenology and product potential of banana cultivars 'Dwarf Cavendish and Williams' in the Eastern Transvaal lowland, *Scientia Horticulturae* 25(2), 149-161.

Robinson, J. C., Nel, D. J. and Eckstein, K. (1993). A field comparison of ten Cavendish sub-group banana cultivars and selections (*Musa* AAA) over four crop cycles in the subtropics, *Journal of Horticultural Science* 68(4), 511-521.

Rotchanapreeda, T., Wongniam, S., Swangpol, S. C., Chareonsap, P. P., Sukkaewmanee, N. and Somana, J. (2016). Development of SSR markers from *Musa balbisiana* for genetic diversity analysis among Thai bananas, *Plant Systematics and Evolution* 302(7), 739–761.

Royal Botanic Gardens, Kew (2019). *Seed Information Database (SID)*. Version 7.1, Retrieved November 2019 from: http://data.kew.org/sid/.

Sardos, J., Christelová, P., Čížková, J., Paofa, J., Sachter-Smith, G. L., Janssens, S. B., Rauka, G., Ruas, M., Daniells, J. W., Dolezel, J.. and Roux, N. (2018). Collection of new diversity of wild and cultivated bananas (*Musa* spp.) in the Autonomous Region of Bougainville, Papua New Guinea, *Genetic Resources and Crop Evolution* 65(8), 2267–2286.

Sardos, J., Perrier, X., Dolezel, J., Hribova, E., Christelová, P., Van den Houwe, I., Kilian, A. and Roux, N. (2016). DArT whole genome profiling provides insights on the evolution and taxonomy of edible Banana (*Musa* spp.), *Annals of Botany* 118(7), 1269–1278.

Simmonds, N. W. (1952). The germination of banana seeds, *Tropical Agriculture* 19, 35–49.

Simmonds, N. W. (1959). Experiments on the germination of banana seeds, *Tropical Agriculture* 36, 259–273.

Simmonds, N. W. (1962). *The Evolution of the Bananas*, London: Longmans, Green and co Ltd.

Simmonds, N. W. and Shepherd, K. (1955). The taxonomy and origins of the cultivated bananas, *Journal of the Linnean Society of London, Botany* 55(359), 302–312.

Stotzky, G. and Cox, E. A. (1962). Seed germination studies in Musa. II. Alternating temperature requirement for the germination of Musa balbisiana, *American Journal of Botany* 49(7), 763–770.

Stotzky, G., Cox, E. A. and Goos, R. D. (1962). Seed germination studies in Musa. I. Scarification and aseptic germination of Musa balbisiana, *American Journal of Botany* 49(5), 515–520.

Tang, A. (2014). In vitro embryo culture of rarely endangered Musella lasiocarpa (Musaceae) with embryo dormancy, *Pakistan Journal of Botany* 46(6), 2173-2177.

Tang, Z., Cao, M., Sheng, L., Liang, B. and Zhang, S. (2005). Fruit consumption and seed dispersal of wild banana Musa acuminata by short-nosed fruit bat Cynopterus sphinx, *Acta Zoologica Sinica* 51(4), 608–615.

Thrall, P. H., Bever, J. D. and Burdon, J. J. (2010). Evolutionary change in agriculture: the past, present and future, *Evolutionary Applications* 3(5–6), 405–408.

Uma, S. and Arun, K. (2016). Understanding the diversity and reproductive biology of banana - for improvement through basic research, *Acta Horticulturae* 1114, 1–12.

Uma, S., Lakshmi, S., Saraswathi, M. S., Akbar, A. and Mustaffa, M. M. (2011). Embryo rescue and plant regeneration in banana (Musa spp.), *Plant Cell, Tissue and Organ Culture (PCTOC)* 105(1), 105–111.

Uma, S., Siva, S. A., Saraswathi, M. S., Manickavasagam, M., Durai, P., Selvarajan, R. and Sathiamoorthy, S. (2006). Variation and intraspecific relationships in Indian wild *Musa balbisiana* (BB) population as evidenced by random amplified polymorphic DNA, *Genetic Resources and Crop Evolution* 53(2), 349–355.

Vázquez-Yanes, C. and Orozco-Segovia, A. (1982). Seed germination of a tropical rain forest pioneer tree (Heliocarpus donnell-smithii) in response to diurnal fluctuation of temperature, *Physiologia Plantarum* 56(3), 295–298.

Vineesh, P. S., Skaria, R., Mukunthakumar, S., Padmesh, P. and Decruse, S. W. (2015). Seed germination and cryostorage of *Musa acuminata* subsp burmannica from Western Ghats, *South African Journal of Botany* 100, 158–163.

Wang, X. L., Chiang, T. Y., Roux, N., Hao, G. and Ge, X. (2007). Genetic diversity of wild banana (*Musa balbisiana* Colla) in China as revealed by AFLP markers, *Genetic Resources and Crop Evolution* 54(5), 1125–1132.

Wattanachaiyingcharoen, D. and Turner, D. (1990). Viability, germination and dormancy of banana see (*Musa acuminata* ssp.). *Banana Newssletter* 13, 16.

Wilson, V. and Tenkouano, A. (2019). Effects of storage and ppriming on seed emergence in soild and embryo culture of Musa acuminata Calcutta 4, *Asian Plant Research Journal* 2(1), 1–9.

Wong, C., Kiew, R., Loh, J. P., Gan, L. H., Lee, S. K., Ohn, S., Lum, S. and Gan, Y. Y. (2001). Genetic diversity of the wild banana *Musa acuminata* Colla in Malaysia as evidenced by AFLP, *Annals of Botany* 88(6), 1017–1025.

Xia, Q. Q., Ando, M. and Seiwa, K. (2016). Interaction of seed size with light quality and temperature regimes as germination cues in 10 temperate pioneer tree species, *Functional Ecology* 30(6), 866–874.

Yoshioka, T., Satoh, S. and Yamasue, Y. (1998). Effect of increased concentration of soil CO_2 on intermittent flushes of seed germination in Echinochloa crus-galli var. crus-galli, *Plant, Cell and Environment* 21(12), 1301–1306.

Youmbi, E., Fonkam, N. J. P., Tomekpe, K. and Fonbah, C. (2011). In vitro germination and pollen conservation of some *Musa* species, *Asian Journal of Biotechnology* 3(6), 554–563.

Zhang, H., Mittal, N., Leamy, L. J., Barzani, O. and Song, B. H. (2017). Back into the wild–apply untapped genetic diversity of wild relatives for crop improvement, *Evolutionary Applications* 10(1), 5–24.

Chapter 9

Safe dissemination of germplasm resources of banana

John Thomas, The University of Queensland, Queensland Alliance for Agriculture and Food Innovation, Ecosciences Precinct, Australia; Sébastien Massart, Integrated and Urban Plant Pathology Laboratory, Gembloux Agro-Bio Tech, University of Liège, Belgium; Ines Van den Houwe, Bioversity International Transit Centre, KU Leuven, Division of Crop Biotechnics – Laboratory of Tropical Crop Improvement, Belgium; Nicolas Roux, Bioversity International, France; and Kathy Crew, Queensland Department of Agriculture and Fisheries, Ecosciences Precinct, Australia

1 Introduction

The progenitors of the edible bananas we use today are a number of wild seeded species and subspecies, predominantly of *Musa acuminata* and *M. balbisiana*. The resultant selections and hybrids now comprise a collection

http://dx.doi.org/10.19103/AS.2020.0070.11

of cultivars and landraces that are essentially seedless and are propagated vegetatively (Perrier et al., 2011). Concomitant with this form of propagation goes the propensity over time to harbour and accumulate pests and diseases.

In the early phases of domestication, the movement of preferred selections would have been primarily locally between interacting human populations and, to a lesser extent, with migrating human populations. Fruit, flowers and buds for food; leaves for packing; and corms, bits and suckers for planting could all carry pests and pathogens.

Spread of most of the serious banana pests and diseases seems to have started relatively recently, with the opportunity for increased and faster long-distance transport, which has dramatically increased in modern times. For example, Fusarium wilt of banana was first described in the late 1800s and the pathogen most likely evolved in Southeast Asia. During the succeeding decades it spread worldwide (Ploetz, 2019). Similarly, banana bunchy top disease was first noted in Fiji, also in the late 1800s (Darnell-Smith, 1924), and the causal virus (banana bunchy top virus (BBTV)) also most likely evolved in Southeast Asia. Subsequent spread to sub-Saharan Africa, for example did not occur until the 1960s (Wardlaw, 1961).

The relentless increase in human populations and the increasing consumption of bananas as a dessert fruit and as staple food source in many tropical countries, coupled with increasing production pressures due to climate change, have increased the demand for the evaluation and exploitation of the new germplasm. This, in combination with increased levels of travel and trade, has increased the risk of widespread distribution of pests and pathogens. Adherence to strict guidelines for regional and international movement of germplasm is essential if we are to minimise the risk of further movement and associated impact on banana production due to pests and pathogens. This chapter addresses the biotic risks associated with germplasm exchange and the regulations and requirements in place to minimise these risks.

2 General recommendations for international *Musa* germplasm exchange

Bioversity International has facilitated the preparation of recommended guidelines for the safe movement of *Musa* germplasm (Thomas, 2015), with the focus on small-scale transfers for research, breeding, conservation and cultivar evaluation purposes. However, the same risks for spread of pests and pathogens apply to larger-scale commercial transfers of planting material. The following general recommendations have been extracted from the Bioversity International Technical Guidelines for the Safe Movement of *Musa* Germplasm (Thomas, 2015).

The following general recommendations apply to all international *Musa* germplasm exchanges:

- Germplasm should be obtained from the safest source possible and should be free of any visible disease symptoms. There is, for example, a pathogen-tested *in vitro Musa* germplasm collection available at the International Transit Center, from which safe germplasm can be ordered online (http://www.crop-diversity.org/banana).
- All germplasm should ideally be moved in the form of tissue culture. If this is not possible, full quarantine measures, including containment in secure post-entry quarantine facilities, must be taken until the vegetative material or seed is cultured *in vitro*.
- Germplasm should be tested for all viruses known to affect *Musa*, except if there is reliable evidence that particular viruses are not present in the country of origin. Reliable protocols are described in the Technical Guidelines.
- Indexing procedures and results should be documented, e.g. in a germplasm health statement.
- The movement of germplasm should be carefully planned in consultation with quarantine authorities and the relevant indexing laboratory and should comply with the regulatory requirements of the importing country.

The International Standards for Phytosanitary Measures (ISPM) as published by the Secretariat of the International Plant Protection Convention (IPPC) should be followed during movement of germplasm. In accordance with IPPC regulations, any material being transferred internationally must be accompanied by a phytosanitary certificate.

Germplasm of edible bananas has traditionally been maintained as living, vegetative material in farmers' fields (*in situ*) or in field germplasm collections (*ex situ*). In the latter case, these have normally been managed by government agencies or research organizations. Since the 1980s, tissue culture has become an increasingly important method for both propagating banana planting material and for maintaining germplasm *in vitro* (Cronauer and Krikorian, 1986, Benson et al., 2011).

3 Germplasm from field collections

Field collections are valuable components in a global *ex situ* conservation effort, particularly because of their importance for germplasm characterization, breeding (Pillay et al., 2004) and training purposes. Unlike the wild types, cultivated bananas are seedless and vegetatively propagated. Traditionally the wild and cultivated diversity is conserved in field genebanks. Around the world there are some 60 banana field collections with a total of over 9050 accessions (Laliberte, 2016) (See chapters 4 and 8 in this book). Bananas are perennial giant herbs that need to be re-planted periodically, every two to five years, in order to keep the plants vigorous. Field collections require extensive space, and expensive and time-consuming on-going maintenance if their integrity and value are to be preserved. Factors that may hamper the secure long-term preservation of the diversity in field collections are the continuous exposure to biotic threats, such as pests and diseases, and abiotic stresses including drought, flooding and other natural calamities. Even if the risks are strictly minimized, some cultivated or wild germplasm accessions may be lost from *ex situ* collections because the field sites do not always provide suitable ecological conditions. This is often the case for Australimusa species and Eumusa species requiring unique environmental conditions (mountainous and forest climates) to thrive. Field collections have a further disadvantage that plant materials are not readily available for exchange and over time may become infested with pathogens and pests. The supply of samples from a field genebank is not only time-consuming because of the low propagation rate but, with each movement of planting materials, also involves the risk of transmitting harmful pests and diseases.

4 Germplasm from *in vitro* collections

Often genebanks maintain, in addition to their field collection, all or a number of their accessions *in vitro* as tissue culture plantlets, overcoming these challenges posed by the field collections.

In the 1970s, tissue culture, and in particular shoot-tip culture, was successfully applied for the rapid multiplication of 'Cavendish' bananas (Ma and Shii, 1972) and a wide range of *Musa* genotypes have since been found amenable to this *in vitro* storage and multiplication technique. The technique offers the ability to clean and to store pathogen-free clones for a longer time, serving as a safety duplicate for field collections and facilitating lower-risk international distribution of planting material.

According to a *MusaNet* Global Survey of *Ex Situ* Collections (Laliberte, 2016) an estimated total of 4507 accessions are conserved *in vitro* in 32 institutes around the world. According to this survey, 14 institutes hold larger *in vitro* collections of over 100 banana accessions (Table 1). With its more than 1600 banana accessions, sourced from 38 countries, the ITC (Bioversity International Musa germplasm Transit Centre) at the Katholieke Universiteit Leuven, Belgium, is the most comprehensive repository of *Musa* in the world. The *in vitro* collection holds germplasm of cultivated sweet and starchy bananas belonging to 17 genome groups and 52 subgroups, a representation of the wild genepool with specimen of 34 species, and a range of high-yielding and

Table 1 Institutes holding more than 100 *in vitro Musa* accessions

Country	Institute	Number of accessions held
Australia	DAF (Department of Agriculture, Forestry and Fisheries), Queensland	417
Brazil	EMBRAPA (Empresa Brasileira de Pesquisa Agropecuária)	250
Belgium	Bioversity International Musa germplasm Transit Centre at the Katholieke Universiteit Leuven	1617
Burundi	IRAZ (Institut de recherches agronomiques et zootechniques de la CEPGL)	155
China	IFTR/GAAS (Institute of Fruit Tree Research, Guangdong Academy of Agricultural Sciences)	215
China	TSFR (Tropical and Subtropical Fruit Research Laboratory)	110
Colombia	CORPOICA (Corporacíon Colombiana de Investigacíon Agropecuaria)	164
Cuba	INIVIT (Instituto de Investigaciones de Viandas Tropicales)	137
Fiji	(SPC) Secretariat of the Pacific Community,	185
India	NBPGR (National Bureau of Plant Genetic Resources)	415
Malaysia	MARDI (Malaysian Agricultural Research and Development Institute)	180
Nigeria	IITA (International Institute of Tropical Agriculture)	180
Taiwan	TBRI (Taiwan Banana Research Institute)	220
USA (Puerto Rico)	USDA (United States Department of Agriculture)	150

disease-resistant advanced cultivars from the major banana breeding programs around the world.

In addition to the labour-intensive nature of continued *in vitro* propagation of accessions, the technique also carries the risk of somaclonal variation in the cultures over time, resulting in loss of genetic integrity of the original genotype stored. Molecular markers (e.g. SSR and DArTs) are routinely used to verify the taxonomic classification and identity of existing and new accessions (Sardos et al., 2016, Christelová et al., 2017) but are not effective to assess any genetic instability in tissue culture plants. Moreover the problem is complex as not only genetic but also epigenetic factors may be involved (Tarafdar et al., 2017). To date, trueness-to-type verification of *in vitro* conserved plant materials relies on morphological description, periodical field assessment and cytological characterization to assess and describe somaclonal variants. Research on genetics and epigenetics for a better understanding of somaclonal variation in banana and the development of a system for the early assessment and control this phenomenon in slow growth storage is needed.

To address the problem of somaclonal variation during tissue culture, two types of *in vitro* conservation methods are used for the longer-term preservation of plant germplasm. These are the slow growth method and cryopreservation (Razdan and Cocking, 2000).

4.1 Slow growth storage

Suppression of growth can be achieved by changing the physical, nutritional and osmotic environment in the tissue culture environment. Low temperature is the growth-retarding factor (Razdan and Cocking, 2000) which has been successfully and most widely adopted in *in vitro* active collections of banana. Cultures are usually pre-incubated under normal growth conditions at an optimal growth temperature of 26±2°C for 2–3 weeks. Once established, they are transferred to minimal growth conditions at a temperature between 16±1°C and 22°C, delaying the need for subculturing by several months up to one year on average (Oliveira et al., 2000). However, the frequency of subculture varies for the different genomic (sub)groups. An analysis of the storage duration of accessions maintained in ITC collection showed that effective storage duration averaged 334 days across all genotypes, but that East-African highland bananas and ABB non-plantain bananas could be stored significantly longer, up to a period of 18–22 months, under the same conditions. Wild *Musa balbisiana* (BBw) deteriorate more rapidly and can only be stored on average for a period of 9 months (Van den Houwe et al., 1995). The recommended transfer interval for diploid AA accessions at 17°C was 450 days (Oliveira et al., 2000). Periodic checks by scoring parameters for viability and deterioration are critically important to determine the need for subculturing of the plant material and to keep the collection alive and in good shape.

5 Cryopreservation of *Musa* germplasm

To ensure the secure long-term preservation of *Musa* genetic resources, cryopreservation, or storage at ultralow temperatures, is the method of choice. In liquid nitrogen (-196°C), cell division and metabolic activities are suspended and no alterations occur during extended storage periods, allowing safe storage of germplasm in perpetuity. With the development of cryopreservation protocols that are widely applicable to the different banana groups and species, safe long-term storage of banana regenerative tissues has been successfully achieved since the turn of the century. However, examples of routine use of cryopreservation for long-term conservation for plant collections are still limited to a few genebanks (Ellis et al., 2018). At the ITC, a cryopreservation protocol has been established for banana shoot tips (meristems) using the 'droplet vitrification' method (Panis et al., 2005). Today, more than 1100 banana accessions (nearly 70% of the *in vitro* collection) are safely stored in liquid nitrogen. Three replicate sets of frozen meristems are kept, each with a 95% probability that at least one plant can be regenerated, and with one duplicated set transferred to the Institut pour la recherche en développement (IRD), France, where it is stored for security reasons in a 'black box' under identical conditions as that of the ITC.

The application of cryopreservation has an initial large cost due to the investments to set up a dedicated facility and to process the germplasm for storage in liquid nitrogen. However, once cryopreserved, very little input for maintenance is required. A costing study of the ITC genebank showed that the cumulative costs of maintaining banana germplasm in medium-term storage and long-term storage become equal after a time span of 15 years (Garming et al., 2010), supporting the value of cryopreservation as a cost-effective approach for the long-term conservation of banana.

6 Seed banking

An alternative approach for long-term preservation of the seeded wild banana species, including *Musa acuminata* and *Musa balbisiana*, the ancestors of the edible bananas, could be through seed conservation. To capture and secure the broader diversity present in wild *Musa* populations and broaden the accessibility of this diversity for breeding, the potential of conserving wild species through seeds is currently under investigation. Seeds of *Musa* species have been regarded as orthodox (Bohra et al., 2020) but the germination of banana seeds is unpredictable and low. Efficient seed germination protocols and seed preservation methods for -20°C storage and cryopreservation need to be developed.

7 DNA banking

An approach complementary to the traditional conservation approaches is DNA banking by constructing genomic libraries. In recent decades, the use of molecular techniques has gained importance in biodiversity studies including molecular phylogenetics and breeding, leading to an increasing interest from researchers to have access to DNA of the widest range of banana genetic diversity. Although the method does not allow the regeneration of entire plants, it is valuable for the storage of genetic information and to facilitate the exchange of these resources among molecular scientists. The Bioversity genebank opted for the establishment of a collection of leaf tissues from greenhouse plants as source materials for DNA/RNA extraction. The preservation of these leaf materials by freeze-drying takes very little time and is relatively cheap compared to pure DNA conservation methods. Samples of some 850 diverse banana accessions are stored in a freezer at -20°C for secure long-term storage and direct use. The use of dried tissue can avoid the degradation risks with long-distance shipment of fresh material, and quarantine considerations can be reduced but not totally eliminated.

8 The need for virus indexing to ensure high health status

Many pests and pathogens can be moved around the world as contaminants with banana vegetative planting material (i.e. corms, bits, suckers), though most of these can be avoided through the use of tissue culture. The major exception is viruses, which are readily transmitted through tissue culture but can sometimes be eliminated using meristem tip culture, thermotherapy or chemotherapy techniques (Helliot et al., 2002, Ramos and Zamora, 1999, Thomas et al., 1995).

So far, eight viral diseases of banana plants and their causal agents have been described (Kumar et al., 2015, Thomas, 2019) (Table 2).

Banana streak disease is caused by banana streak viruses (BSVs), a group of closely related, mealybug-transmitted viruses with DNA genomes belonging to the genus *Badnavirus* and the family *Caulimoviridae* (Kumar et al., 2015, Iskra-Caruana et al., 2019). So far, nine species of banana streak viruses (BSV) have been recognised by the International Committee on Virus Taxonomy

Table 2 Viruses infecting banana and the diseases caused

Virus	Acronym	Disease caused	Vector type
banana bunchy top virus	BBTV	bunchy top	aphid
abaca bunchy top virus	ABTV	bunchy top	aphid
banana bract mosaic virus	BBrMV	bract mosaic	aphid
sugarcane mosaic virus	SCMV	abaca mosaic	aphid
banana mild mosaic virus	BanMMV	mild mosaic	unknown
cucumber mosaic virus	CMV	banana mosaic	aphid
banana streak viruses[a]	BSGFV, BSIMV, BSMYV, BSOLV, BSUAV, BSUIV, BSULV, BSUMV, BSVNM	banana streak	mealybug
banana virus x	BVX	unknown	unknown

[a] Banana streak disease is caused by banana streak viruses (BSVs), a group of closely related viruses belonging to the genus *Badnavirus*). So far, nine species of banana streak viruses have been recognised by the International Committee on Virus Taxonomy (ICTV, 2018): BSGF(Goldfinger)V, BSIM(Imové) V, BSMY(Mysore)V, BSOL(Obino L'Ewai)V, BSUA(Uganda A)V, BSUI(Uganda I)V, BSUL(Uganda L)V, BSUM(Uganda M)V, BSVN(Vietnam)V.

(ICTV, 2018): BSGFV, BSIMV, BSMYV, BSOLV, BSUAV, BSUIV, BSULV, BSUMV and BSVNM.

Some viruses, like BanMMV and CMV, have a worldwide distribution but, for example, two of the most important viruses, BBTV and BBrMV, have a restricted distribution, i.e. BBTV is absent from Central and South America, and BBrMV is absent from Africa and much of Central and South America (Thomas, 2019). The virus incidence in *Musa* plants sent to germplasm collection centres can be high. For example, among 434 *Musa* accessions collected to lodge in the Bioversity International germplasm collection, 68% were infected by at least one virus, mainly BSV or BanMMV (De Clerck et al., 2017). Therefore, guaranteeing the movement of pathogen-free germplasm is an important task to minimize the risk of pest introduction to new countries.

Another development which needs further attention is the recent finding that seeds of banana could transmit BBrMV (Selvarajan et al., 2020). To minimize the risk of introduction of viruses and to ensure that distributed material is free of pests of potential phytosanitary importance, effective testing procedures must be applied. Determination of the complete phytosanitary status of collected banana germplasm is therefore a prerequisite before releasing the material (De Clerck et al., 2017). The importance of distributing virus-tested material is illustrated by the absence of BBrMV and BBTV in South America underlining the necessity of virus indexing to avoid their introduction through infected germplasm.

9 Virus indexing of *Musa* germplasm

It is essential that laboratory assays are employed to ensure freedom from viruses as these pathogens can sometimes be symptomless, in field plants and especially in tissue culture plantlets. Virus indexing of banana germplasm relies on the availability of reliable diagnostic protocols for banana viruses at affordable cost. The past 40 years have seen tremendous progress in this area of virology, with the successive introduction of simple serological assays like the ELISA test, of molecular hybridization, of DNA amplification by PCR, real-time PCR and its other forms: Loop Mediated Isothermal Amplification (LAMP) and Rolling Circle Amplification (RCA)(Massart et al., 2014). For the last 30 years, technical guidelines for the safe movement of *Musa* germplasm have been published by Bioversity International, with the most recent edition of these guidelines published in 2015 (Thomas, 2015). These guidelines propose general and technical recommendations for the safe movement of *Musa* germplasm, including virus indexing protocols. For example, as the virus concentration in *in vitro* plants can be very low and barely detectable, it is recommended to grow the plants during 6 months in greenhouse and to test for the presence of viruses twice, after 3 months and 6 months. In addition, as the virus distribution in the plant can be heterogeneous, it is recommended to grow at least four plants and to sample both leaf midrib and lamina of the three youngest leaves of each plant, to minimize the chance of false negatives and to ensure appropriate diagnostic sensitivity. The detection protocols described in the guidelines rely on several techniques: electron microscopy, ELISA, (RT)-PCR, real-time PCR, immunocapture PCR and LAMP.

Banana has been relatively poorly studied compared with many temperate crops, and many viruses have only recently been discovered, and indeed continue to be revealed (Jones, 2019). The results of the full indexing of 243 accessions from the Bioversity International germplasm collection, including IC-(RT)-PCR, electron microscopy and symptom observation, have been summarized (De Clerck et al., 2017). Intriguingly, viral particles were observed by transmission electron microscopy while testing negative with the molecular test for known viruses. This highlights the need for the observation of plants for potential symptom expression, and the use of other non-specific methods such as electron microscopy or high throughput sequencing to detect previously unrecognised viruses.

10 Viruses integrated into the *Musa* genome

A particularly difficult problem is posed by BSVs, some of which (BSOLV, BSIMV, BSGFV) can be integrated in the banana B genome (endogenous BSVs; eBSVs) and subsequently be activated through stresses, including tissue culture, hybridization or water stress, to form episomal infections (Côte et al., 2010, Chabannes et al., 2013, Duroy et al., 2016). These activated BSV infections can arise in the progeny of parents that have been indexed and shown to be free of episomal BSVs (virus particles). In addition to these activatable, infectious eBSVs, there are partial, non-activatable sequences of many other BSV species present in both *Musa* A and B genomes (Geering et al., 2005). All these integrated sequences can result in false-positive PCR indexing due to contaminating *Musa* genomic DNA, necessitating the use of specific techniques such as immunocapture PCR followed by DNase treatment, to distinguish episomal BSV infections from integrated viral DNA. Strategies to deal with the presence of integrated, activatable eBSVs in germplasm distribution have been developed (Thomas et al., 2015).

11 Physical shipment of germplasm

To minimize the risk of moving pest and diseases internationally or inter-regionally along with physical plant germplasm, it is of critical importance that preventive ISPM measures developed by the International Plant Protection Convention (IPPC, 2020) are followed and effective testing of the plant material is carried out to meet the latest phytosanitary requirements of the country that imports the germplasm. The Guidelines for the Safe Movement of Musa Germplasm (Thomas, 2015) recommend that the movement of vegetative material happens in the form of tissue cultures (meristem or shoot tip), derived from conventional vegetative propagation material (suckers), and details these procedures. For wild species, if no vegetative material is available, tissue cultures can also be initiated from seeds, through embryo rescue (Uma et al., 2011). In both cases, the source material for tissue culture should be collected from a mother plant that is free from pest and disease symptoms. The *in vitro* germplasm should be tested for all known banana viruses, described above, and any other disease for compliance with the requirements of the importing country. The documents that accompany the plant materials are a health statement describing the testing procedures (Fig. 1), the official plant importation permit from the importing country and phytosanitary certificate issued by the National Plant Protection Organisation (NPPO) of the source country, declaring the absence of pathogens that may be of quarantine significance to the importing country, and a list of passport data documenting the germplasm. Before shipping, the tissue cultures should be produced in an aseptic environment, tested for endogenous bacteria (Van den Houwe and Swennen, 2000), and grown in sterile transparent culture vials. The growth medium should be translucent and antibiotic-free, allowing the detection of contamination by the plant quarantine authority at the point of entering the importing country. The individual containers with plant material should be packed in shock-absorbent packing material and sent in a solid cardboard box. Plant samples should be sent by a reliable courier to keep the period in transit as short as possible and ensuring that the germplasm arrives at its destination in good condition. ITC is an important facilitator in the international exchange of germplasm, by making disease-free, certified *in vitro* samples of accessions from its virus-indexed *in vitro* collection available for users worldwide. In more than 30 years of its existence, the ITC banana collection has distributed over 18 000 banana germplasm samples to users in 113 countries. Yearly, about 30% of the available accessions in the collection are accessed for use, indicating the great interest in diverse and pathogen- and pest-free germplasm. Two types of *in vitro* materials can be supplied: aseptic *in vitro* rooted plants suitable for growing in a nursery or multiple shoot clusters for further *in vitro* propagation or studies. To further enhance the use of germplasm, freeze-dried leaf tissue derived from indexed plants is distributed as a low-cost alternative for users interested in using the DNA of the banana plant only.

12 Regulations surrounding transfer

Several international conventions and protocols govern the legal aspects of international movement of germplasm.

BIOVERSITY INTERNATIONAL GERMPLASM HEALTH STATEMENT

ITC Accession Number:

Accession Name:

Origin of Accession:

The material designated above was obtained from a shoot-tip cultured *in vitro*. Shoot-tip culturing is used to eliminate the risk of the germplasm carrying fungal bacterial and nematode pathogens and insect pests of *Musa*. However, shoot-tip cultures could still carry virus pathogens.

SCREENING FOR VIRUS PATHOGENS

A representative sample of four plants derived from the same shoot-tip as the germplasm designated above has been grown under quarantine conditions for at least 6 months, regularly observed for disease symptoms and tested for virus pathogens as indicated below following methods recommended in the Bioversity International Technical Guidelines for the Safe Movement of *Musa* Germplasm (2015) for the diagnosis of virus diseases.

PCR-based methods
[ ] BBTV – Banana bunchy top virus
[ ] CMV – Cucumber mosaic virus
[ ] BBrMV – Banana bract mosaic virus
[ ] BSV – Banana streak viruses
[ ] BanMMV – Banana mild mosaic virus

Electron microscopy
[ ] isometric virus particles - includes CMV and unknown viruses
[ ] bacilliform virus particles – includes unknown BSVs
[ ] filamentous virus particles – includes BBrMV, BanMMV and unknown viruses

[P] = test positive, [N] = test negative, [ ] = test not undertaken.

DISTRIBUTION OF VIRUS PATHOGENS AND OTHER INFORMATION

(Example: BBTV and BBrMV are not known to occur in country of origin)

eBSVs are present in the B genome of *Musa* (banana). Consequently, almost all accessions containing the B genome may develop BSV infection and may express symptoms during any stage of growth.

The information provided in this germplasm statement is based on the results of tests undertaken at Bioversity International's Virus Indexing Centre by competent virologists following protocols current at the time of the test and on present knowledge of virus disease distribution. However, neither Bioversity International nor its Virus Indexing Centre staff assume any legal responsibility in relation to this statement.

Signature Date

This statement provides additional information on the phytosanitary status of the plant germplasm described herein. It should not be considered as a substitute for the official "Phytosanitary Certificate" issued by the plant quarantine authorities of Belgium.

Figure 1 An example of a germplasm health statement (from Bioversity International)

The Convention on Biological Diversity (https://www.cbd.int/) is a multilateral treaty with 193 participating parties, and which came into force in 1993. It has three main aims:

1 The conservation of biological diversity.
2 The sustainable use of the components of biological diversity.
3 The fair and equitable sharing of the benefits arising out of the utilization of genetic resources.

To implement the third objective, a supplementary agreement, the Nagoya Protocol (on Access to Genetic Resources and the Fair and Equitable Sharing of Benefits Arising from their Utilization to the Convention on Biological Diversity) was drafted and came into force in 2014 (https://www.cbd.int/abs/about/). It provides a transparent legal framework to guard against bio-piracy and ensures the fair and equitable sharing of benefits. Germplasm transfer is conducted under Standard Materials Transfer Agreements (SMTA).

In harmony with the Convention on Biological Diversity is the legally binding International Treaty on Plant Genetic Resources for Food and Agriculture, which came into force in 2004 (http://www.fao.org/plant-treaty/overview/en/). The Treaty aims to facilitate access for participating countries to germplasm of 64 major food crops for the purposes only of research, breeding and training, while at the same time ensuring that recipients share the benefits with the country of origin. It also recognizes the large contribution farmers make to the development of these genetic resources, and the rights they are due. There are 145 member countries plus the European Union as signatories to the Treaty.

Additionally, individual countries may have specific quarantine and phytosanitary regulations governing the importation of germplasm.

CGIAR centres (formerly the Consultative Group for International Agricultural Research) operate under FAO standards, with the underlying principles:

- Availability and use of germplasm: 'Conserved material must be available for current and future use. There will be a need to maintain sufficient quantities of seed and related information on the accessions... the genebank should have a strategy in place to multiply quickly any germplasm for distribution.'
- Availability of Information: 'In order to ensure communication of information and accountability, essential, detailed, accurate, and up-to-date information...should be recorded in electronic databases. Access, availability and sharing of this information should be treated with high priority.'

- The criteria for availability include:
  - Sufficient stocks held to allow distribution.
  - Indexed clean and healthy (pathogen free) material.
  - True to type material which is viable, with genetic integrity verified.
  - Documentation including passport data (as a minimum) and, if possible, also characterization and evaluation data.

13 Future trends in research

13.1 The rise of high throughput sequencing (HTS) technologies

The detection of plant viruses has been greatly affected by the development of High Throughput Sequencing (HTS or also called Next-Generation Sequencing - NGS) technologies and the bioinformatics analysis of the generated sequences (Massart et al., 2014). HTS technologies enable the simultaneous sequencing, and thus detection, of any organism present in a sample. These technologies are broad-spectrum and generic in nature and can potentially be used to replace a wide range of conventional methods (Olmos et al., 2018). Therefore, it is now conceptually possible to perform general viral and viroid indexing on a plant sample and to detect any virus or viral-like agents in any germplasm sample by high-throughput sequencing of the extracted nucleic acids and the identification of sequences from known and/or unknown viruses. The discovery of new viral species and distant variants of known species has been the most common HTS application in plant virus research, allowing the discovery of hundreds of new plant virus species, mainly from plants with symptoms of unknown aetiology. The discovery of a new virus by HTS technologies often immediately triggered the development of a molecular diagnostic protocol to detect the new virus, enlarging the number of targets for viral indexing of any plant.

In the short term, it is expected that HTS technologies will also be used commonly in routine diagnostics to: (a) understand the status of a pest in a region through surveillance programs, (b) certify nuclear stock and plant propagation material, (c) allow (post-entry) quarantine testing to prevent the introduction of pests into a country or area, and (d) monitor imported commodities for new potential risks (Olmos et al., 2018).

A first interesting application of HTS technologies in diagnostics laboratories with *Musa* germplasm is aimed to resolve the contradiction observed between the presence of viral particles or visible symptom expression and negative molecular tests. For example, HTS technologies have now been applied on several of the 22 accessions with contradictory results between indexing tests. The goal was to determine if the virions detected corresponded to a new viral species infecting banana or represented isolates presenting mismatches to the molecular detection primers.

HTS technologies have the potential to be used as molecular diagnostic tests for virus indexing of germplasm resources to guarantee the high health status of the accessions. The evaluation of HTS performance for pest detection in plant propagation material and nuclear stocks has been carried out for grapevine (Al Rwahnih et al., 2015) and fruit trees (Rott et al., 2017). An evaluation of HTS technologies for reliable virus indexing of *Musa* germplasm accessions is currently ongoing at the Germplasm Health Unit (GHU) of Bioversity International (Gembloux AgroBio Tech, Liege University, Belgium). This evaluation will allow in-depth comparison of diagnostic performance (sensitivity, specificity, repeatability, reproducibility) of HTS in virus indexing compared to methods used previously.

13.2 (RT)-PCR tests with improved inclusiveness

The application of HTS technologies to detect banana plant viruses generates large quantities of sequencing data, including the complete or near-complete genome of viral isolates. For example, several new whole genomes of BanMMV isolates have been generated (S. Massart, unpublished) adding significantly to the single publicly available genome at present. This deluge of genome sequences is particularly useful to evaluate and improve the inclusiveness of diagnostic primers, e.g. their ability to detect all isolates of a pest, as shown for little cherry virus 1 on *Prunus* (Katsiani et al., 2018) or barley yellow dwarf virus for cereals (M. Somera, personal communication). Similar large-scale analysis will soon be possible to be carried out with the banana viruses. The growing availability of viral sequences will reinforce the reliability of primer-based molecular detection protocols.

13.3 Phytoplasma screening

Many new phytoplasma species have been detected during the last decade, including a phytoplasma infecting banana plants and causing a wilt disease in Papua New Guinea (Davis et al., 2012) and the Solomon Islands (Davis et al., 2015). Phytoplasmas are obligate bacterial pathogens lacking a cell wall, living in the phloem tissue and are vectored by insects. The possible transmission of the banana phytoplasma through *in vitro* culture has not been documented but needs to be assessed. Currently, the detection of phytoplasma has been achieved using a nested PCR protocol described by Constable et al. (2003).

13.3.1 Banana seed testing

The virus indexing of seeds may be required to guarantee safe exchange of plants as the transmission of BBrMV through banana seeds has been confirmed recently (Selvarajan et al., 2020). The detection of BBrMV was carried out on seedlings, demonstrating true transmission, but it is not known whether the virus can be detected directly from seed. CMV (Gold, 1972) and possibly BSMYV (Daniells et al., 1995) have also been reported as seed-transmitted in banana. Future seed indexing might also require germination or embryo rescue combined with plant growth in a greenhouse to achieve reliable testing. Research in this area is still in its early stages.

14 Conclusion

For a sustainable future of banana production, which addresses the ongoing and increasing biotic and abiotic threats to banana production from pests, diseases, climate change and the loss of crop wild relatives, there is an urgent need to collect and safely maintain accessions of cultivated and wild germplasm. Environmental damage and forest destruction is eroding the natural distribution of wild, seeded *Musa,* and despite an estimated diversity of 500–1000 varieties, edible banana production is increasingly dominated by a single cultivar group – Cavendish. Preservation of the wild *Musa* genetic diversity is critical if we are to have available the range of genetic traits required for breeding programs into the future.

Financial support for these activities is crucial and is dependent on demonstrating the value of the activity. Thus, exchange of germplasm is of vital importance for security of the resource, and to maximise its evaluation and potential uses. First steps have been made as banana has recently been the subject of efforts to increase the representation of wild germplasm in collections (e.g. Sardos et al., 2018) and to better understand the genetic diversity of the crop and its wild relatives (Sardos et al., 2016, Christelová et al., 2017, Bawin et al., 2019).

In vitro conservation with slow growth storage or cryopreservation (Panis, 2009) has greatly facilitated storage of germplasm, but the risk of transfer of pests and pathogens, especially viruses, still hampers the exchange of germplasm. The high incidence of virus infection in germplasm accessions, unrecognised at the time of collection, highlights the need for careful and thorough indexing of accessions (De Clerck et al., 2017). A long-term solution for the problem of eBSV-containing B genome cultivars is needed. Some hope has been provided by the use of gene editing techniques (Tripathi et al., 2019) and also the development of improved breeding lines carrying only non-infectious alleles of eBSVs (Pichaut et al., 2013). High throughput sequencing is revolutionizing virus indexing, but together with the ability to rapidly and efficiently detect known and previously unrecognised viruses in germplasm comes the dilemma of creating quarantine issues when new viruses are found, without any supporting biological information. Nevertheless this valuable technology is destined to attain widespread acceptance (Massart et al., 2014).

Thus, there is an urgent need for biosecurity agencies to adopt a uniform and rational approach to this deluge of new viruses being discovered.

15 Where to look for further information

The Bioversity-CIAT Alliance (https://alliancebioversityciat.org) provides the scientific community and public with significant open-access resources relevant to the conservation and safe distribution of banana genetic resources. The ProMusa website (http://www.promusa.org/), managed by a secretariat provided by the Bioversity-CIAT Alliance, provides an online resource to foster scientific discussion on banana (*Musa*). It provides a wealth of information including resources on germplasm, production and plant protection, and supports extensive literature (http://www.musalit.org/) and photographic (http://www.musarama.org/) databases. The *Musa* Germplasm Information System (MGIS) is a database, managed by Bioversity-CIAT Alliance, of phenotypic and genomic information on accessions of bananas (both wild and cultivated) conserved in field and *in vitro* genebanks around the world. The database also provides access to the ITC accessions, many of which are available for distribution as *in vitro* plantlets or lyophilized leaves.

Many publications relevant to the conservation and safe movement of banana germplasm are available, including:

- MusaNet Technical Guidelines for the Safe Movement of *Musa* Germplasm. 3rd Edition:
  www.musalit.org/seeMore.php?id=15942.
- Cryopreservation of *Musa* germplasm. Technical Guidelines No. 9:
  www.musalit.org/seeMore.php?id=14476.
- Regeneration guidelines of banana:
  www.musalit.org/seeMore.php?id=14472.
- Global strategy for the conservation and use of *Musa* (Banana) genetic resources:
  www.musalit.org/seeMore.php?id=17120.
- Musalogue: A catalogue of Musa germplasm. Diversity in the genus *Musa*:
  www.musalit.org/seeMore.php?id=5935.

Several websites contain information on the international agreements covering the exchange of germplasm, including:

- The Convention on Biological Diversity (https://www.cbd.int/).
- The Nagoya Protocol (on Access to Genetic Resources and the Fair and Equitable Sharing of Benefits Arising from their Utilization to the Convention on Biological Diversity) (https://www.cbd.int/abs/about/).

- The Convention on Biological Diversity (http://www.fao.org/plant-treaty/overview/en/).

Research studies on the international dispersal of pests and pathogens of banana include:

- Blomme, G., Ploetz, R., Jones, D., De Langhe, E., Price, N., Gold, C., Geering, A., Viljoen, A., Karamura, D., Pillay, M., Tinzaara, W., Teycheney, P. Y., Lepoint, P., Karamura, E. and Buddenhagen, I. (2013). A historical overview of the appearance and spread of *Musa* pests and pathogens on the African continent: highlighting the importance of clean *Musa* planting materials and quarantine measures. *Annals of Applied Biology* 162(1), 4–26.
- Stainton, D., Martin, D., Muhire, B., Lolohea, S., Halafihi, Mi., Lepoint, P., Blomme, G,. Crew, K., Sharman, M., Kraberger, S., Dayaram, A., Walters, M., Collings, D., Mabvakure, B., Lemey, P., Harkins, G., Thomas, J. and Varsani, A. (2015) The global distribution of Banana bunchy top virus reveals little evidence for frequent recent, human-mediated long distance dispersal events. *Virus Evolution* 1: vev009.

16 References

Al Rwahnih, M., Daubert, S., Golino, D., Islas, C. and Rowhani, A. 2015. Comparison of next-generation sequencing versus biological indexing for the optimal detection of viral pathogens in grapevine. *Phytopathology* 105(6), 758–763.

Bawin, Y., Panis, B., Vanden Abeele, S., Li, Z., Sardos, J., Paofa, J., Ge, X., Mertens, A., Honnay, O. and Janssens, S. B. 2019. Genetic diversity and core subset selection in *ex situ* seed collections of the banana crop wild relative *Musa balbisiana*. *Plant Genetic Resources: Characterization and Utilization* 17(6), 536–544.

Benson, E., Harding, K., Debouck, D. G., Dumet, D., Escobar, R., Mafla, G., Panis, B., Panta, A., Tay, D. and Van den Houwe, I. 2011. Refinement and standardization of storage procedures for clonal crops. *Global Public Goods Phase 2. Part 2: Status of In Vitro Conservation Technologies for: Andean Root and Tuber Crops, Cassava, Musa, Potato, Sweetpotato and Yam*. Rome, Italy: The CGIAR System-wide Genetic Resources Programme (SGRP).

Bohra, P., Waman, A. A. and Jerard, B. A. 2020. Seed germination and storage studies in seed-fertile *Musa indandamanensis* and its conservation. *South African Journal of Botany* 128, 161–166.

Chabannes, M., Baurens, F. C., Duroy, P. O., Bocs, S., Vernerey, M. S., Rodier-Goud, M., Barbe, V., Gayral, P. and Iskra-Caruana, M. L. 2013. Three infectious viral species lying in wait in the banana genome. *Journal of Virology* 87(15), 8624–8637.

Christelová, P., De Langhe, E., Hřibová, E., Čížková, J., Sardos, J., Hušáková, M., Van den Houwe, I., Sutanto, A., Kepler, A. K., Swennen, R., Roux, N. and Doležel, J. 2017. Molecular and cytological characterization of the global *Musa* germplasm collection provides insights into the treasure of banana diversity. *Biodiversity and Conservation* 26(4), 801–824.

Constable, F. E., Gibb, K. S. and Symons, R. H. 2003. Seasonal distribution of phytoplasmas in Australian grapevines. *Plant Pathology* 52(3), 267–276.

Côte, F. X., Galzi, S., Folliot, M., Lamagnere, Y., Teycheney, P. Y. and Iskra-Caruana, M. L. 2010. Micropropagation by tissue culture triggers differential expression of infectious endogenous banana streak virus sequences (eBSV) present in the B genome of natural and synthetic interspecific banana plantains. *Molecular Plant Pathology* 11(1), 137–144.

Cronauer, S. S. and Krikorian, A. 1986. Regeneration in bananas and plantains. In: Vasil, I. (Ed.) *Plant Regeneration and Genetic Variability*. Orlando, FL, Academic Press.

Daniells, J., Thomas, J. E. and Smith, M. 1995. Seed transmission of banana streak virus confirmed. *Infomusa* 4, 1, 7.

Darnell-Smith, G. P. 1924. "Bunchy top" disease in banana. *Queensland Agricultural Journal* 21, 169–179.

Davis, R. I., Henderson, J., Jones, L. M., Mctaggart, A. R., O'Dwyer, C., Tsatsia, F., Fanai, C. and Rossel, J. B. 2015. First record of a wilt disease of banana plants associated with phytoplasmas in Solomon Islands. *Australasian Plant Disease Notes* 10(1), 14.

Davis, R. I., Kokoa, P., Jones, L. M., Mackie, J., Constable, F. E., Rodoni, B. C., Gunua, T. G. and Rossel, J. B. 2012. A new wilt disease of banana plants associated with phytoplasmas in Papua New Guinea (PNG). *Australasian Plant Disease Notes* 7(1), 91–97.

De Clerck, C., Crew, K., Van den Houwe, I., McMichael, L., Berhal, C., Lassois, L., Haissam Jijakli, M., Roux, N., Thomas, J. and Massart, S. 2017. Lessons learned from the virus indexing of *Musa* germplasm: insights from a multiyear collaboration. *Annals of Applied Biology* 171(1), 15–27.

Duroy, P. O., Perrier, X., Laboureau, N., Jacquemoud-Collet, J. P. and Iskra-Caruana, M. L. 2016. How endogenous plant pararetroviruses shed light on *Musa* evolution. *Annals of Botany* 117(4), 625–641.

Ellis, D., Escobar, R., Gueye, B., Panis, B., Vollmer, R. and Popova, E. 2018. Cryobanking clonal crops: when knowledge and practice are not the same. *Cryobiology* 85, 159.

Garming, H., Roux, N. and Van den Houwe, I. 2010. *The impact of the Musa International Transit Centre - Review of Its Services and Cost-Effectiveness and Recommendations for Rationalization of Its Operations*. Montpellier, France: Bioversity International.

Geering, A. D. W., Olszewski, N. E., Harper, G., Lockhart, B. E. L., Hull, R. and Thomas, J. E. 2005. Banana contains a diverse array of endogenous badnaviruses. *Journal of General Virology* 86(2), 511–520.

Gold, A. H. 1972. Seed transmission of banana viruses. *Phytopathology* 62, 760.

Helliot, B., Panis, B., Poumay, Y., Swennen, R., Lepoivre, P. and Frison, E. 2002. Cryopreservation for the elimination of cucumber mosaic and banana streak viruses from banana (*Musa* spp.). *Plant Cell Reports* 20(12), 1117–1122.

ICTV 2018. *Virus Taxonomy: 2018b Release* [Online]. Washington, DC: International Committee on Taxonomy of Viruses. Available at: https://talk.ictvonline.org/taxonomy/ [Accessed 5 March 2020].

IPPC 2020. *International Plant Protection Convention* [Online]. Rome, Italy: AGDI, Food and Agriculture Organization of the United Nations. Available at: https://www.ippc.int/en/.

Iskra-Caruana, M. L., Thomas, J. and Chabannes, M. 2019. Banana streak. In: Jones, D. R. (Ed.) *Handbook of Diseases of Banana, Abacá and Enset*. Wallingford, UK: CABI, 393–408.

Jones, D. R. 2019. *Handbook of Diseases of Banana, Abaca and Enset*. Wallingford, UK: CABI Publishing.

Katsiani, A., Maliogka, V. I., Katis, N., Svanella-Dumas, L., Olmos, A., Ruiz-García, A. B., Marais, A., Faure, C., Theil, S., Lotos, L. and Candresse, T. 2018. High-throughput sequencing reveals further diversity of little cherry virus 1 with implications for diagnostics. *Viruses* 10(7), 385.

Kumar, P. L., Selvarajan, R., Iskra-Caruana, M.-L., Chabannes, M. and Hanna, R. 2015. Biology, etiology, and control of virus diseases of banana and plantain. In: Gad, L. and Nikolaos, I. K. (Eds) *Advances in Virus Research*. Academic Press, Elsevier, Amsterdam, The Netherlands.

Laliberte, B. 2016. *Global Strategy for the Conservation and Use of Musa Genetic Resources*. Montpellier, France: Bioversity International.

Ma, S. and Shii, C.-T. 1972. *In vitro* formation of adventitious buds in banana shoot apex following decapitation. *Journal of the Chinese Society of Horticultural Science* 18, 135–142.

Massart, S., Olmos, A., Jijakli, H. and Candresse, T. 2014. Current impact and future directions of high throughput sequencing in plant virus diagnostics. *Virus Research* 188, 90–96.

Oliveira, R. P. D., Silva, S. D. O. E., Silva, K. M. D. and Silveira, D. G. 2000. *In vitro* conservation of diploid banana accessions. *Scientia Agricola* 57(2), 245–249.

Olmos, A., Boonham, N., Candresse, T., Gentit, P., Giovani, B., Kutnjak, D., Liefting, L., Maree, H. J., Minafra, A., Moreira, A., Nakhla, M. K., Petter, F., Ravnikar, M., Rodoni, B., Roenhorst, J. W., Rott, M., Ruiz-García, A. B., Santala, J., Stancanelli, G., Van Der Vlugt, R., Varveri, C., Westenberg, M., Wetzel, T., Ziebell, H. and Massart, S. 2018. High-throughput sequencing technologies for plant pest diagnosis: challenges and opportunities. *EPPO Bulletin* 48(2), 219–224.

Panis, B. 2009. *Cryopreservation of Musa Germplasm*. Montpellier, France: Bioversity International.

Panis, B., Piette, B. and Swennen, R. 2005. Droplet vitrification of apical meristems: a cryopreservation protocol applicable to all Musaceae. *Plant Science* 168(1), 45–55.

Perrier, X., De Langhe, E., Donohue, M., Lentfer, C., Vrydaghs, L., Bakry, F., Carreel, F., Hippolyte, I., Horry, J. P., Jenny, C., Lebot, V., Risterucci, A. M., Tomekpe, K., Doutrelepont, H., Ball, T., Manwaring, J., De Maret, P. and Denham, T. 2011. Multidisciplinary perspectives on banana (*Musa* spp.) domestication. *Proceedings of the National Academy of Sciences of the United States of America* 108(28), 11311-11318.

Pichaut, J.-P., Farinas, B., Umber, M., Chabannes, M., Laboureau, N., Duroy, P.-O., Bonheur, L., Salmon, F., Jenny, C., Iskra-Caruana, M.-L. and Teycheney, P.-Y. 2013. Taming infectious eBSV alleles for breeding new banana hybrid varieties: O06.005. In: You-Liang, P. and Zejian, G. (Eds) *International Congress of Plant Pathology*, 2013-08-25 / 2013-08-30, Beijing, China (restricted): Chinese Society for Plant Pathology, 99.

Pillay, M. S. R., Hartman, J., Vuylsteke, D., Talengera, D. and Tushemereirwe, W. 2004. Conventional breeding strategies to enhance the sustainability of *Musa* biodiversity conservation for endemic cultivars. *African Crop Science Journal* 12, 59–65.

Ploetz, R. 2019. Fungal diseases of the root, corm and pseudostem; Fusarium wilt. In: Jones, D. R. (Ed.) *Handbook of Diseases ofBanana, Abacá and Enset*. Wallingford, UK: CAB International.

Ramos, C. S. and Zamora, A. B. 1999. Isolation of banana plantlets cv. Señorita - Freed of banana bract mosaic virus. *The Philippine Agricultural Scientist* 82, 25–34.

Razdan, M. and Cocking, E. C. 2000. *Conservation of Plant Genetic Resources In Vitro: Applications and Limitations*. Enfield, NH: Science Publishing Inc.

Rott, M., Xiang, Y., Boyes, I., Belton, M., Saeed, H., Kesanakurti, P., Hayes, S., Lawrence, T., Birch, C., Bhagwat, B. and Rast, H. 2017. Application of next generation sequencing for diagnostic testing of tree fruit viruses and viroids. *Plant Disease* 101(8), 1489–1499.

Sardos, J., Christelová, P., Čížková, J., Paofa, J., Sachter-Smith, G. L., Janssens, S. B., Rauka, G., Ruas, M., Daniells, J. W., Doležel, J. and Roux, N. 2018. Collection of new diversity of wild and cultivated bananas (*Musa* spp.) in the Autonomous Region of Bougainville, Papua New Guinea. *Genetic Resources and Crop Evolution* 65(8), 2267–2286.

Sardos, J., Perrier, X., Doležel, J., Hřibová, E., Christelová, P., Van den Houwe, I., Kilian, A. and Roux, N. 2016. DArT whole genome profiling provides insights on the evolution and taxonomy of edible Banana (*Musa* spp.). *Annals of Botany* 118(7), 1269–1278.

Selvarajan, R., Balasubramanian, V., Priyanka, P., Jebakumar, R. M., Selvam, K. P. and Uma, S. 2020. Evidence of seed transmission of Banana bract mosaic virus in *Musa* synthetic diploid H-201, a possible threat to banana breeding. *European Journal of Plant Pathology* 156(4), 1181–1187.

Tarafdar, S., Gupta, K. S., Banerjee, S. and Shailja, D. 2017. Identification of SSR & ISSR markers associated to off-type somaclonal variants in micro-propagated banana. *Journal of Biotechnology and Biochemistry* 3(2), 18–22.

Thomas, J. 2019. Diseases caused by viruses. In: Jones, D. R. (Ed.) *Handbook of Diseases of Banana, Abacá and Enset*. Wallingford, UK, CABI Publishing.

Thomas, J., Iskra-Caruana, M. L., Kumar, P. L., Roux, N., Chabannes, M., Massart, S., Mathieu, Y., Chase, R. and Van den Houwe, I. 2015. *Position Paper on a Strategy to Distribute Banana (Musa) Germplasm with Endogenous Banana Streak Virus Genomes*. Montpellier, France: Bioversity International, MusaNet Conservation Thematic Group.

Thomas, J. E. 2015. *MusaNet Technical Guidelines for the Safe Movement of Musa Germplasm*. Rome, Italy: Bioversity International.

Thomas, J. E., Smith, M. K., Kessling, A. F. and Hamill, S. D. 1995. Inconsistent transmission of banana bunchy top virus in micropropagated bananas and its implication for germplasm screening. *Australian Journal of Agricultural Research* 46(3), 663–671.

Tripathi, J. N., Ntui, V. O., Ron, M., Muiruri, S. K., Britt, A. and Tripathi, L. 2019. CRISPR/Cas9 editing of endogenous banana streak virus in the B genome of *Musa* spp. overcomes a major challenge in banana breeding. *Communications Biology* 2, 46.

Uma, S., Lakshmi, S., Saraswathi, M. S., Akbar, A. and Mustaffa, M. M. 2011. Embryo rescue and plant regeneration in banana (*Musa* spp.). *Plant Cell, Tissue and Organ Culture (PCTOC)* 105(1), 105–111.

Van den Houwe, I., De Smet, K., Du Montcel, H. T. and Swennen, R. 1995. Variability in storage potential of banana shoot cultures under medium term storage conditions. *Plant Cell, Tissue and Organ Culture* 42(3), 269–274.

Van den Houwe, I. and Swennen, R. 2000. Characterization and control of bacterial contaminants in *in vitro* cultivars of banana (*Musa* spp.). *Acta Horticulturae* 530, 69.

Wardlaw, C. W. (Ed.) 1961. The virus diseases: bunchy top. In: *Banana Diseases, Including Plantains and Abaca*. London: Longmans.

Part 3

Genetic improvement through breeding

Chapter 10

Making banana breeding more effective

F. Bakry, J. P. Horry and C. Jenny, CIRAD, UMR AGAP and AGAP, Université de Montpellier, CIRAD, INRAE, Institut Agro, France

1 Introduction

Bananas are of high economic importance, either as a staple food in tropical and subtropical countries and as an export commodity for some countries. Around 85% of the global production, mostly produced in non-intensive cropping systems (backyard gardens, small plots, small farms and agroforestry settings), are used for self-consumption, or traded at local and regional markets. The remaining 15% produced in highly intensive monoculture plantations by agribusinesses are exported from tropical countries to international markets in developed countries.

Global banana production has steadily increased since the 1960s, to reach today about 145 million tons per year. Despite the increases in global banana production, the production systems are concentrated in certain areas, lack genetic diversity and are vulnerable to outbreaks of diseases. Genetic vulnerability is extreme with nearly 95% of the world production originating from just a few genetic combinations, involving a very limited number of natural cross events (from 7 to 14 at the most). One group of cultivars 'Cavendish' originating from one single recombining event represents half of the total production, export and local markets taken together. Such lack of genetic diversity in a cultivated crop of major importance used as a staple food for millions of people is a critical problem. There is no other equivalent among

http://dx.doi.org/10.19103/AS.2020.0070.12
Published by Burleigh Dodds Science Publishing Limited, 2021.

our cultivated crop plants with such a low level of genetic diversity, even in other vegetatively propagated crops like sugarcane, potatoes or strawberries. These crops are also polyploids but display much higher levels of genetic diversity compared to cultivated bananas. Bananas are different as during their domestication period, spanning thousands of years (Perrier et al., 2011), they have been selected for sterility. Consequently, in the absence of genetic recombination, the currently cultivated banana cultivars are nearly completely frozen at the polyploid level, with no gene flow and without any genetic evolution, except the occurrence of somatic mutations or bud-sporting due to strict vegetative propagation (i.e. cloning) over long periods. Within, for example, the 'frozen genotype' Cavendish, selection by growers for somaclonal variants has given rise to several cultivars (e.g. Williams, Grand Naine, Valery) which show limited variation for agronomic characters such as height and bunch characteristics.

2 The need for sustainable genetic solutions

Over the last 50 years the environment in which bananas are grown has changed dramatically. This change is mainly due to the global spread of fungal diseases. Numerous studies have described in detail how the banana black leaf streak fungus (*Pseudocercospora fijiensis;* previously aka *Mycosphaerella fijiensis*) has spread globally under the combined effect of human-mediated pathogen spread (on an intercontinental scale) and wind-driven aerial dispersal in cultivated areas (Rieux et al., 2013). Another pathogen currently experiencing global spread is *Fusarium oxysporum* spp *cubense* TR4, a soilborne fungus, for which human activities are the only cause of long-distance spread of the disease (García-Bastidas et al., 2019; Maymon et al., 2018; Zheng et al., 2018; Ploetz, 2015).

Although highly susceptible to black leaf streak the production of Cavendish bananas has continued through regular aerial application of a range of different fungicides at a very high ecological and economic cost. Although the disease can be controlled through regular application of fungicides, these production methods are coming under increased scrutiny. On the other hand, for Fusarium with TR4 there is no 'agro-pesticide' control solution or other effective field management solutions such that a genetic solution is the only option. Hence, in the long term, the only solution is through breeding or genetic engineering new varieties resistant to these diseases. Therefore, the development of new varieties, resistant both to black leaf streak and Fusarium wilt diseases, is a topical issue.

Banana production systems are also under threat from changes in the global climate. Water resources are becoming scarce and in some cases, dry seasons in the tropics become increasingly chaotic and accompanied by even higher temperatures. Thus in addition to disease resistance tolerance to drought and high temperatures are also important selection criteria in a banana breeding program.

This combination of biotic and abiotic stress factors forces us to review our breeding objectives. We no longer can limit ourselves in the genetic improvement of bananas just in terms of productive ideotypes close to current varieties to which we would have added an additional beneficial trait (such as resistance to a disease). Instead we need to focus on developing completely new products bringing new advantages, new characteristics and which are well-adapted to constantly changing constraints. This demands an urgent need to broaden the genetic base of cultivated bananas and release of new cultivars, resistant to pest and diseases, which take advantage of the richness of the still underexploited banana genetic diversity, including the crop's wild relatives.

There is a need for more diverse bananas with far more diverse genetic backgrounds which meet specific objectives such as fresh fruit, cooking bananas or varieties dedicated to processing. The aim of this chapter is to show how diversification objectives can be achieved taking into account sexual reproductive barriers. The objective is to produce large populations of hybrids to facilitate selection of genotypes with combinations of the right characteristics.

3 Evolution and domestication resulting in gamete sterility of contemporary varieties and consequences for breeding

3.1 Wild Musa species

Banana originates from Southeast Asia and the Pacific region. These bananas growing in the wild are diploids and produce non-edible fruits full of seeds. Over 50 species of seed-producing bananas are thought to exist, but two species, *Musa acuminata* and *Musa balbisiana*, have largely contributed to the occurrence and subsequent domestication of most edible bananas (Simmonds and Shepherd, 1955; Perrier et al., 2009). Many plants of both species and subspecies are resistant or immune to several diseases and pests such as *Fusarium* wilt and black leaf streak disease and provide excellent sources of natural resistance.

The natural range of *M. acuminata* extends from West to East, from eastern India to New Guinea, including the Philippines and Indonesia (Fig. 1). Along that axis, *M. acuminata* displays substantial morphological variability, which gave rise to classification into several subspecies (Cheesman, 1947b). Within these subspecies there is a considerable amount of genetic diversity and they vary in height and are often slender, and display great variations in bunch and fruit shapes and lengths. At least nine subspecies have been defined depending on the authors (De Langhe and Devreux, 1960; Cheesman, 1947b; Häkkinen and Väre, 2008; Simmonds, 1962). It is generally accepted that the origin of edibility

Published by Burleigh Dodds Science Publishing Limited, 2021.

Figure 1 Natural range of the extent of *M. acuminata* subspecies in Asia.

due to parthenocarpy genes originates from the complex of the *M. acuminata* subspecies.

Musa balbisiana is also a seed-producing species, which exhibits less genetic variability than *M. acuminata*. It also originates from Southeast Asia, but from regions much further north in South Asia than the *M. acuminata* subspecies. It is probably the most common and most widespread of all the species of the genus *Musa* (Cheesman, 1947a). Plants are very vigorous and tall, displaying strong vegetative vigor, strong suckering and very good rooting in the soil. In addition, varieties containing B genome show increased tolerance to abiotic stress. In fact, some indirect evidences suggest that elements of the *Musa balbisiana* (B) genome confer greater tolerance to drought than those of the *Musa acuminata* (A) genome (Kissel et al., 2016). *Musa balbisiana* is also reputed for its resistance to cold. In a trial conducted by CIRAD in 2006 under controlled conditions on 45-day-old in vitro plants, *M. basjoo* and *M. balbisiana* in vitro plants were found to be more cold tolerant than *M. acuminata* spp. *siamea* and *Cavendish* (AAA) genotypes (personal observations).

For their part, seed-producing bananas have kept both reproductive systems, firstly sexual as they are perfectly fertile and readily produce seeds, and secondly vegetative, through suckers like edible varieties (Fig. 2). Under hand pollination, the number of seeds obtained per bunch proved to vary considerably from one accession to another: from 4000 to about 15000 seeds/bunch for *M. acuminata* genotypes but over 40000 seeds/bunch for

Figure 2 Seminiferous banana reproduces also by seeds. (a) Bunch of *M. acuminata* seminiferous banana. (b) Seedy fruit of *M. balbisiana* (scale in cm). (c) Seeds of *M. acuminata* (scale in cm). (d) Open seed of *M. balbisina* (Emb: embryo; End: endosperm; Int: seed integument).

M. balbisiana accessions. In the case of hand pollination, seeds are mostly perfectly formed and germinate readily when tested in vitro.

3.2 Edible varieties

The situation is radically different for edible seedless varieties. Edible bananas are thought to have been domesticated first in Southeast Asia more than 7000 years ago (Perrier et al., 2011). About 2500 years ago bananas and plantains were brought to Africa by early seafarers resulting in wide spread cultivation and making East Africa and central Africa secondary centers of diversity. A distinct group of cultivars, Fe'i bananas, developed independently in the Pacific region. Today, nearly 1000 different cultivated banana varieties exist. Most domesticated bananas are triploids, with three sets of chromosomes, and their parthenocarpic fruits do not contain seeds. Apart from a few exceptions, triploid varieties are typically bigger than diploid varieties with correspondingly bulkier fruits. This triploidy in banana was the start of the domestication of the banana as we know it today (Perrier et al., 2009). Humans were not involved in making the triploids. They found the seedless triploids which are natural hybrids and started cultivating them and selecting further in these lines in a process that we call domestication.

Over the course of evolution, these triploid varieties arose naturally from irregular meiosis of edible AAcvs and ABcvs, sometimes producing unreduced gametes (Simmonds, 1962). A direct consequence of fusion of diploid gametes with haploid gametes was the emergence of triploid genotypes AAA and AAB or ABB by interspecific hybridizations (Fig. 3), leading to the diversity of the current cultivated triploids, maintained under human selection via vegetative propagation.

A group of diploid bananas called Mchare, which currently are only still known to occur in parts in East Africa where they were brought by early seafarers about 2500 years ago, is an important link to our current banana cultivars. The cultivars Cavendish and Gros Michel can be considered as three-way hybrids resulting from a cross between a Mchare 2× clone (which can be interpreted as a natural hybrid F1 AA *banksii* × *zebrina/microcarpa*) donor of the 2n

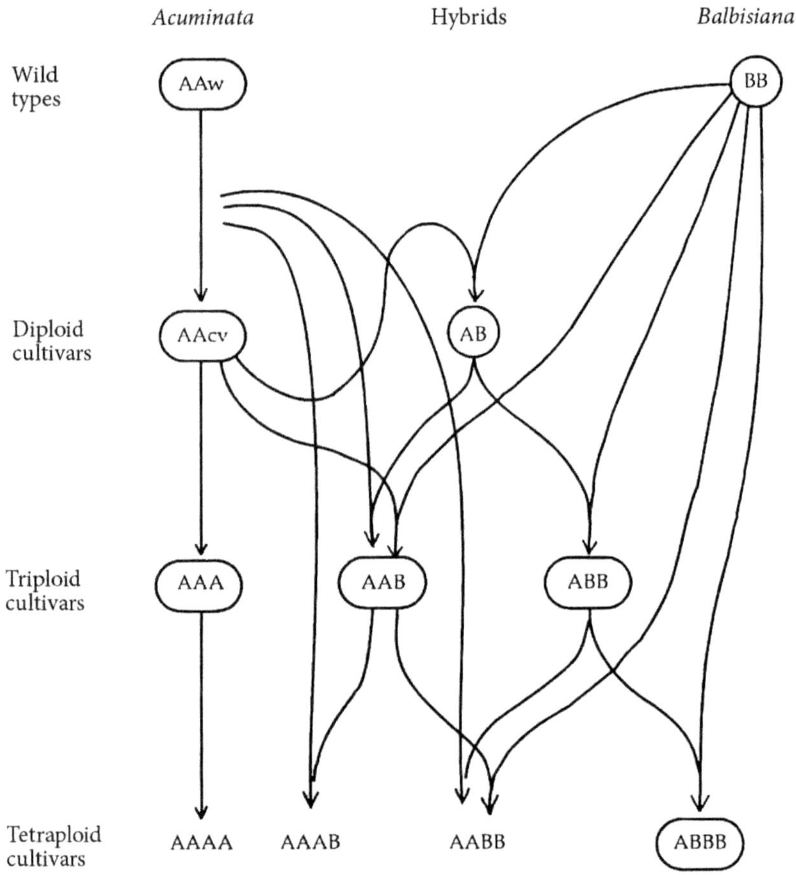

Figure 3 Evolution of bananas (from Simmonds and Shepherd, 1955). From wild types to present triploid varieties (w-wild; cv-cultivated).

Published by Burleigh Dodds Science Publishing Limited, 2021.

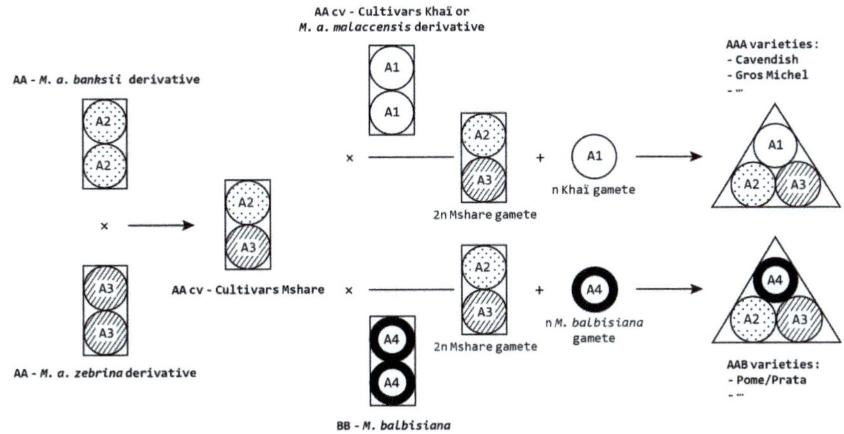

Figure 4 Evolution of edible bananas (from Perrier *et al.*, 2009). Detailed origin of some AAA and AAB varieties (Cavendish, Gros Michel, Pome).

gamete and a *malaccensis* genome from the Malaysian peninsula originating somewhere between New Guinea and Indonesia. By their specific genomic constitution (Fig. 4) these cultivars present the highest level of heterozygosity that can be found in triploid AAA bananas. Heterozygosity is even higher in Prata/Pome (AAB) clones of interspecific constitution, which are three-way hybrids also having a Mchare *acuminata* background shared with Cavendish and Gros Michel but in which the *M. acuminata* ssp. *malaccensis* background is replaced by *M. balbisiana* (Fig. 4). The same is true for plantains or other AAB and ABB which are also interspecific hybrids combining in their genome two species that separated more than 1 million years ago (Baurens et al., 2019).

Triploidization events remained rare in the evolution of domesticated bananas. It occurred independently in various contact areas likely between edible diploids as donor of 2n gametes and seminiferous diploids, cultiwilds or protocultivars but still gamete fertile (n gamete donor). It started probably early after the AAcv emergence and is certainly still ongoing, as demonstrated by the recent origin inferred for the genome of several AAAcvs in New Guinea (Perrier et al., 2009). The natural scarcity of these triploidization events combined with vegetative propagation and the extraordinary ability of triploid genotypes to adapt to a wide variety of environments can explain the very narrow genetic basis of the varieties cultivated today throughout the world.

3.3 Gamete sterility and low seed-set handicap breeding in edible bananas

The low fertility of cultivated bananas is a handicap for breeders. Triploid varieties may produce a few seeds when pollinated. In triploids, seed set is overall higher in ABB in comparison to AAA and AAB. Stover and Simmonds (1987) reported that Pisang Awak (ABB) in backyards in Malaysia produced so many seeds that it was almost inedible. In controlled conditions, seed set is often lower and may vary from no seed to a maximum of 200 seeds/bunch in some Popoulu AAB clones when pollinated with male fertile diploid genotype (Bakry and Horry, 1992). In Cavendish, Aguilar Morán (2013) reported that pollination of 20000 bunches, approximately 2 million of female flowers, yielded 200 seeds of which 40 viable embryos were rescued which eventually resulted in 20 plants. On average, one germinated hybrid was obtained for every 100000 pollinated flowers. These results illustrate how much labor is required to obtain a few hybrids using this cross strategy.

In triploid varieties, the number of fruits per bunch varies from 100 to 250 and the number of ovules ranks from 300 to 600 per fruit. In the absence of gamete sterility, the seed potential could be theoretically estimated between 30 000 and 150 000 seeds per bunch. Consequently, it may be estimated that in triploid bananas, reproductive barriers induce a reduction of over 99% of female fertility.

Abnormalities in the meiosis in diploid and triploid edible varieties are very frequent showing chromosome mispairing in place of regular bivalents (Shepherd, 1999). At triploid level, mispairing is reinforced in presence of three sets of haploid chromosomes, occurrence of aneuploid and abortive gametes is frequently reported and sometimes, in addition, along with the formation of 2n to 4n gametes. In interspecific diploids, globally, AB clones are fairly completely sterile because of the partial homeology between the *M. acuminata* and *M. balbisiana* chromosomes (Jeridi et al., 2011). In general AA varieties show a wide range of male and female fertility. Nevertheless, the overall fertility of AA varieties is higher than those of AAA or AAB triploids. Structural heterozygosity which is present in most diploid AA varieties contributes to formation of frequent abortive and/or aneuploid gametes with 12–16 chromosomes instead of 11 (Bakry, personal observation). These irregularities lead to the formation of gamete distributions with strong shifts for recombination in addition to the random distribution of alleles (Hippolyte et al., 2012; Martin et al., 2017). However, structural heterozygosity cannot always be considered as the main source of gamete sterility in *Musa*. For example, some seedy types like *M. acuminata ssp. microcarpa* are completely gamete fertile despite being heterozygote for no less than three structural events.

But the sterility of edible varieties is not only limited to meiosis anomalies which are almost equivalent for micro- and macro-sporogenesis (Dodds, 1943). For in fact, the edible varieties have much lower female fertility than that observed on the male side and female sterility has been intensified as a consequence of human selection for parthenocarpy. Often, no penetration of the pollen tubes into the embryonic sacs is observed. As a result, many varieties carrying viable embryonic sacs produce no seeds in hand pollination. When fertilization occurs, the auxinic metabolism of the parthenocarpic fruits, genomic interspecificity and unbalanced chromosomal ratios between embryos and maternal tissues are likely to favor embryo and endosperm abortions. For instance, absence of endosperm is very frequent in interspecific *M. acuminata* × *M. balbisiana* crosses (Fig. 5a). Moreover, seeds containing no endosperm usually show embryo malformations compared to wild clones (Fig. 5a). Therefore, the germination rate of seeds coming from pollinated edible varieties rarely exceeds 20% under greenhouse conditions. This explains why in vitro embryo rescue is systematically used in banana breeding (Fig. 5b,c,d), allowing an 85% increase germination rate (Bakry, 2008).

Finally, the environmental conditions in which the bananas are cultivated (soil quality, nutrition, heat, light and humidity) are also known to have an influence on gamete fertility and seed set of banana clones. For example, it has been observed in Cameroon that edible varieties grown at 500 m altitude produce more fertile pollen than at sea level (Bakry personal observations).

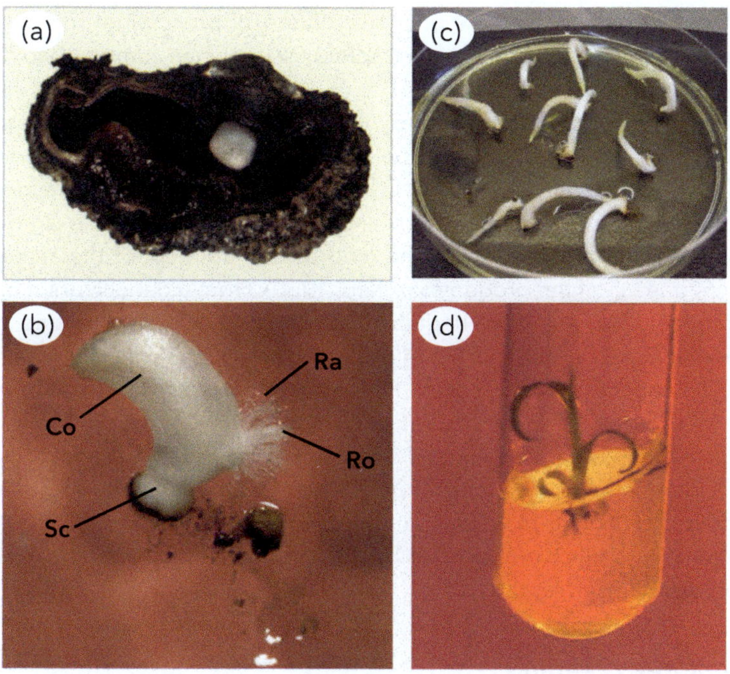

Figure 5 In vitro embryo rescue in banana. (a) Banana seed containing viable malformed embryo without endosperm. (b) A 10-day-old *M. balbisina* germinating embryo - detailed observations show coleoptile development, and sometimes, emergence of radicles and one primary rootlet. Co: coleoptyl; Sc: scutellum; Ro: primary root; Ra: radicles. (c) One-month-old *M. balbisana* germinated embryos. (d) A 45-day-old embryo rooted on growth medium.

Published by Burleigh Dodds Science Publishing Limited, 2021.

4 Strategies in banana breeding

Triploid hybrids are thought to be the best objective for banana breeding, as triploidy gives a selective advantage over other ploidy levels. Diploids are not suitable as final breeding products as they usually are less productive and less vigorous than triploids, with smaller bunches and sometimes contain seeds in their fruits. However, some commercial trades of significant importance rely on diploids like the 'Kunnan' varieties (ABcv) in India (20% of the national production) or 'Pisang Mas'/'Sucrier' (AAcv) in Colombia, favored at niche markets but not fit for mass consumption at export markets. Tetraploidy does not seem appropriate too as these types are extremely rare among traditional banana landraces and never had a significant impact on local agriculture and trade. The reason remains unexplored but it has been noticed that tetraploids used to show droopy leaves that may no longer act as an effective rain collector as do erect banana trees (Cattan et al., 2009) and are, furthermore, prone to wind breakage. In addition, despite bunches and fruit size of synthetic tetraploid hybrids may be equivalent to triploids, the postharvest qualities never meet the requirements of regional and international trades as fresh fruits. In addition, tetraploid fruits are poor substitutes for processing. This might be due to the higher water content in tetraploid tissues when compared to diploid and triploid cultivars.

The improvement of bananas by cross-breeding is essentially based on two different selection strategies. The first, that we call 'Pedigree Breeding', is based on the selection of individuals or their descendants for the superiority of certain traits such as vigor, bunch weight or other favorable agronomic features. It derives its name from the records that are kept of the ancestry or pedigree of each of the progenies (Allard, 1960). From a genetic point of view, this approach implies a high heritability of easily selectable traits and choice of progenitors to be crossed being mostly oriented on their own phenotypic value. The second strategy, called 'Reconstructive Breeding' is based on the search of the best combining abilities between two diploids of which one is the donor of diplo-gametes. Based on complementarities, this approach aims to associate the favorable traits brought by both parents and to maximize the heterozygosity in the triploid progenies (Sanford, 1983). In this case, the choice of progenitors is much more driven by the value of the triploid progenies than the parents themselves.

4.1 Pedigree breeding

4.1.1 Primary tetraploids

The first breeding strategies implemented by British banana breeders as early as 1922 were based on the selection of tetraploid hybrids by pollination of the cultivated triploid variety producing a 2n gamete by diploid male parents producing n gametes. The aim is to exploit triploid varieties showing residual fertility when pollinated with a diploid clone. Indeed, despite their high degree of sterility, some triploid varieties produce few seeds once hand pollinated. Among them, some arise from a few unreduced triploid eggs cells (n = 3x = 33) that could be fertilized with normal haploid pollen to give tetraploid embryos (Fig. 6.) (Bakry and Horry, 1992). Because of the very small amount of seeds produced in these crosses and by selecting only tetraploid hybrids in the progenies, the main genetic objective supporting this approach was to avoid as far as possible genetic recombination of the female triploid parents producing 2n gametes (coming from apomeisosis or meiotic first-division restitution (FDR) or second-division restitution (SDR)) resulting in the tetraploid hybrids. Indeed, most of the natural triploid landraces are very heterozygous (Raboin et al., 2005). They must be seen as the result of natural crossing events resulting in rare and unique genomic structures that have been preserved by vegetative propagation during human domestication over the millennia. Under these circumstances, it is clear that recombination events associated with regular meiosis will break down these particular genetic structures with very

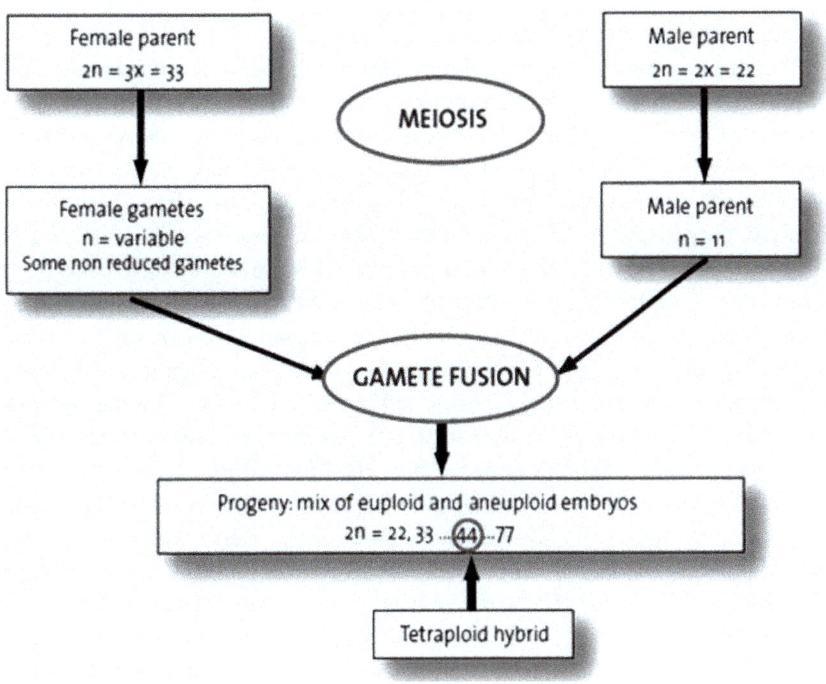

Figure 6 Pedigree Breeding scheme for creating tetraploid hybrids from triploid and diploid parents.

little chance of selecting improved genetic structures given the small amount of viable seeds, either at the level of gametes or progeny. The genetic basis of this cross-pathway has been clearly explained by Menendez and Shepherd (1975) in a very pedagogical scheme showing how primary tetraploid hybrids can be seen as the result of an addition of triploid and haploid genomes (rather than coming from recombination) as could be done, for example, by protoplast fusion (Fig. 7).

Pedigree breeding of bananas was developed back in 1922 to confer *Fusarium* wilt resistance to cv Gros Michel and taken up again with other varieties of dessert and cooking bananas to confer resistance to the black leaf streak disease and nematodes. The low fertility of the rare parental triploid cultivars means that only small populations of tetraploid hybrids can be generated, and a considerable number of crosses have to increase the probability of useful results. The dessert tetraploid hybrids developed at the Fundación Hondureña de Investigación Agrícola (FHIA) arose from numerous pollinations of dwarf mutants of 'Gros Michel' and 'Prata' with improved diploids resistant to the *Fusarium* wilt, the Sigatoka complex and the nematode *Radopholus similis.*

The choice of the diploid male parent is obviously crucial in pedigree breeding. Since the triploid egg cells of the maternal parent are rather homogenous, all breeding efforts should be based on the improvement and selection of the diploid male parent.

The great genetic diversity of the diploids collected in Southeast Asia and India has led to the selection of fertile diploids and some parthenocarpic clones as male parent. But, wild seedy diploids, such as the seminiferous 'Calcutta 4' (*M. acuminata ssp. burmanicoïdes*) accession, which expresses resistance to *Fusarium* wilt and Sigatoka leaf spot disease, have very few of the outward appearances of commercial bananas. Their fingers are short and seedy, and their bunches with the exception of 'Calcutta 4' are rarely pendulous. Parthenocarpic edible diploids often show a slight male fertility, have a better bunch shape and larger fruits, but all are not resistant to diseases. Therefore, intensive breeding at the diploid level has been conducted to combine favorable traits. Notable among these is M53 (resistant to the Sigatoka complex and to *Fusarium* wilt race 1), bred in the 1950s by the breeding program of the former Jamaican Banana Board. Later, FHIA focused on crossings among diploids to create hybrids with multiple resistance to diseases, combined with outstanding agronomical characteristics: dwarf types with enormous bunches and resistance to black leaf streak, *Fusarium* wilt race 1 and nematodes (Rowe and Rosales, 1992). It should be mentioned that the creation and selection for favorable characteristics among these improved diploids can be very complex (Fig. 8). Since that time, numerous monospecific AA and interspecific AB diploid hybrids have been created in particular by the National Research Center for Banana (NRCB) and the Tamil Nadu Agricultural University (TNAU) in India.

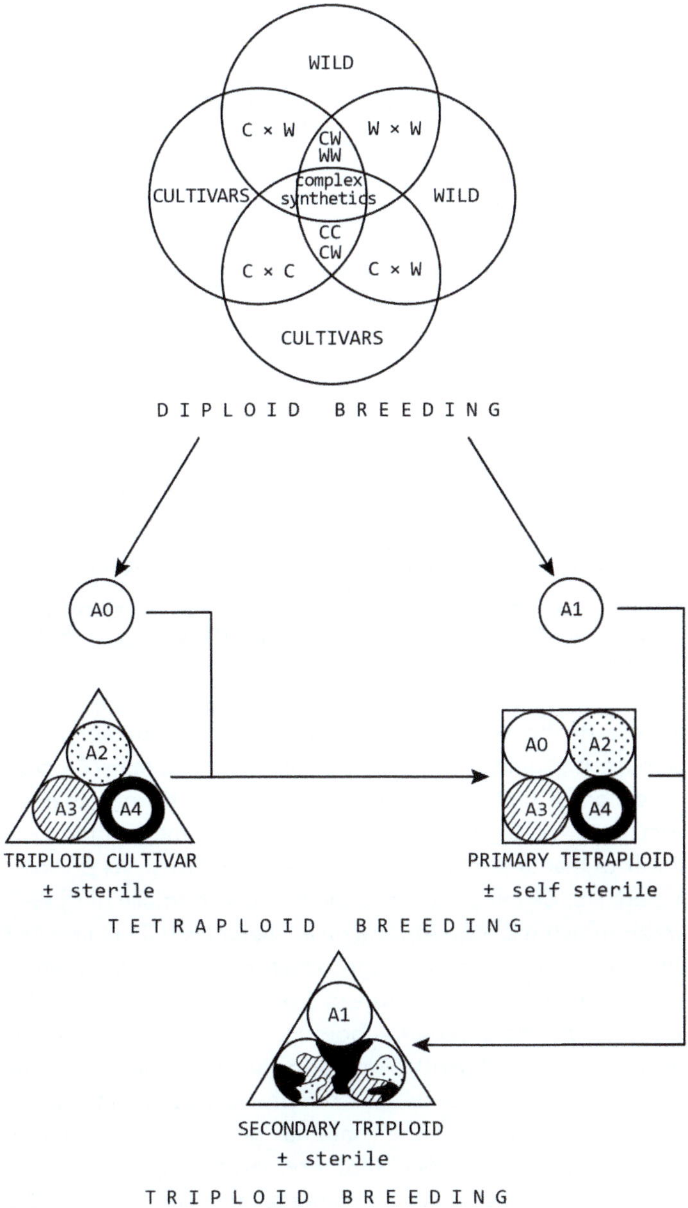

Figure 7 Pedigree breeding: diagram of breeding based on *Musa acuminata* (from Menendez and Shepherd, 1975). A0 ... A4= individual sets *M. acuminata* genes (n=11). This image is an exact reproduction of the figure of the two authors. Figures 4 and 10 in this chapter were created according to the same concept and the same graphic design in homage to the memory of these two authors.

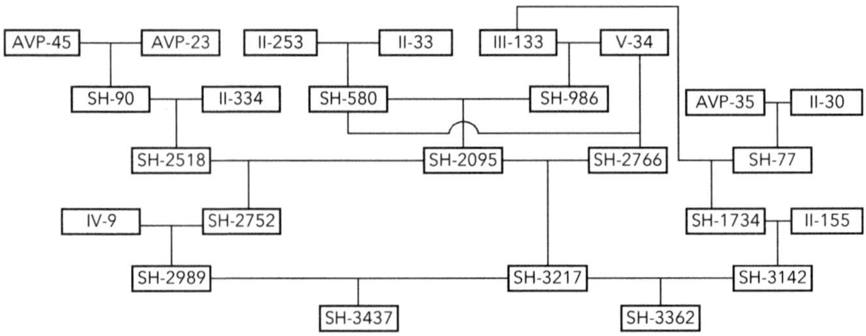

AVP-23: 'Congo', a Cavendish clone collected in Cuba (the only triploid in this pedigree chart);
AVP-35: 'Pisang Lidi', a parthenocarpic *Musa acuminata* subsp. *malaccensis* derivative from Sumatra;
AVP-45: a wild *M. a.* subsp. *zebrina* accession from Java:
II-30: a wild *M. a.* subsp. *errans* accession from the Philippines:
II-33: 'Guyod', a parthenocarpic *M. a.* subsp. *banksii* derivative from the Philippines:
II-155: 'Pisang Jari Buaya', a parthenocarpic *M. a.* subsp. *microcarpa* derivative from Sabah;
II-263: 'Pisang Surong', a wild *M. a.* subsp. *malaccensis* accession from Malaya;
II-334: 'Rangis', a wild *M. a.* subsp. *siamea* accession from Malaya;
III-133: 'Sinwobogi', a parthenocarpic *M. a.* subsp. *banksii* derivative from Irian Jaya;
IV-9: a wild *M. a.* subsp. *burmannica* accession from Burma;
V-34: 'Tjau Lagada', a parthenocarpic *M. a.* subsp. *microcarpa* derivative from Jaya;

Figure 8 Pedigree breeding: chart showing the diverse genetic backgrounds of SH-2905, SH-3142, SH-3217, SH-3362 and SH-3437 used in diploid breeding by FHIA (from Rowe and Rosales, 1989).

Regardless of the success of specific strategies for the production of hybrid diploids, it should be also noticed that the Plantains/2× crosses also enables selection of numerous types of AA diploid hybrids that are pollen fertile, resistant to the Sigatoka complex and, for some of them, with a fruit quality similar to that of plantains. They were used directly as pollinators or as starting material for developing elite diploids. This approach has been used by banana breeders at the Centre Africain de Recherche sur Bananiers et Plantains (CARBAP, Cameroon), the National Agricultural Research Organisation (NARO, Uganda) and the International Institute of Tropical Agriculture (IITA, Nigeria) to develop diploids specific to certain subgroups such as Plantains (AAB) and East African Highland bananas (EAHB) (AAA) also called as Mutika bananas.

In practice, tetraploid hybrids are often phenotypically close to the female parent, suggesting transfer of the maternal genome via apospory or FDR gametes. Dodds and Simmonds (1946) reported that apospory may occur in *Musa* making the genome of the 2n gametes identical to that of the mother plant, but not only that. Indeed, phenotypic segregation has been observed in tetraploid offsprings of Plantains (Ortiz and Vuylsteke, 1998b) or of Popoulu (Bakry and Horry, 1992) which are likely to be related, when meiosis occurs, to the mode of restitution of 2n gametes (Sanford, 1983; Veilleux, 1985). The fact that heptaploid hybrids in progenies can be obtained from some triploid varieties (Cheesman and Dodds, 1942) clearly suggests the formation of

4n gametes in *Musa* arising from both FDR and SDR of the mother parent. Some tetraploid hybrids, however, clearly show a dramatic decrease in vigor compared to the triploid mother plant (pers. obs.) indicating that 2n gametes were probably derived from the SDR. But these cases are rare as in general, the synthetic tetraploids, like the triploid parent, are vigorous and very heterozygous containing four distant sub-genomes in their genome.

As for dessert types, a significant number of disease-resistant tetraploid hybrids has been selected. Unfortunately, despite numerous efforts and investments, no tetraploid hybrid coming from these breeding programs has been developed until today by the banana industry for export markets partly due to the poor postharvest qualities of the fruits (rapid maturing and softening of the fruit).

For local and regional purposes, several research institutes have released through pedigree selection plantain-type hybrids as cooking bananas. The best known are 'FHIA-21' (Dzomeku et al., 2006, 2007; Hauser, 2010) from the FHIA breeding program, 'PITA 14' (Ortiz and Vuylsteke, 1998a) from the IITA breeding program and 'CRBP39' (Cohan et al., 2003) from the collaborative CIRAD-CARBAP plantain breeding program. New plantain-like hybrids sometimes performed poorly because they mainly focus on tolerance/resistance to black leaf streak *(Pseudocercospora fijiensis)* (Jenny, Auboiron and Beveraggi, 1994), which is not the major driver for plantain production in smallholder plantain-based cropping systems. Second, all were interspecific tetraploid (AAAB) hybrids containing the endogenous banana streak virus (eBSV) sequences that are integrated into the B genome, which are prone to activation under in-field abiotic stress or rapid in-vitro propagation (Côte et al., 2010). Moreover, tetraploid hybrids, AAAB as well as AAAA and AABB are often much more fertile than the triploid mother plant. Consequently, they are able to produce fruits containing some seeds if they are pollinated, which obviously reduces their acceptability in the market place. Nonetheless, as success story, FHIA21 has been developed for local consumption to replace plantain landraces in some locations of Latin America and the Caribbean.

4.1.2 Secondary triploids

In view of the problems associated with the 3×/2× strategy, the fertility of tetraploids and concerns of activation of e-BSV sequences in AAAB hybrids thus obtained, the latter were soon regarded as intermediate products for the synthesis of secondary triploids when crossed with diploids. This material has been exploited on a large scale using fertile tetraploids and diploid hybrid parents, which are capable of generating larger F2 hybrid progenies that are in fact the grandchildren of popular triploid cultivars. However, one should bear in mind that the genetic gain obtained by nuclear restitution of the preceding stage (i.e. the transmission of heterosis from the triploid variety to the tetraploid hybrid) will be lost by the redistribution and the recombination of the alleles at each locus that will occur during meiosis of the tetraploid hybrid (Fig. 7).

The synthesis of secondary triploids has been applied with some success with cooking bananas. Because of the high homozygosity of the *M. acuminata* genome in cooking bananas, these results may be explained by a low genotypic variance of the diploid gamete population and by the additive gene effects involved in the inheritance of major agronomic traits. Indeed, this strategy has allowed IITA, CARBAP and FHIA to obtain progenies of sufficient size to make some selections in secondary triploid hybrids of cooking bananas. Conversely, this ultimate approach is much more questionable for dessert bananas which show a higher level of heterozygosity of the *M. acuminata* genome. As for the F1 cross, the choice of diploid parents is also very important for the F2 cross and will depend on the quality or the reconstitution of the secondary triploid hybrids. CARBAP used this approach to generate some dwarf-type triploid hybrids obtained from primary tetraploid hybrids that themselves arose from triploid cultivars of dwarf Plantain. The main advantage of this two-stage strategy is to obtain, via these primary tetraploid hybrids, several hundred second-generation descendants from very sterile triploid cultivars.

NARO in Uganda applied this strategy with some success on AAA Mutika in East Africa (Tumuhimbise et al., 2018). But for B-containing genomes, to our knowledge, all plantain-derived hybrids (either primary AAAB tetraploids or secondary AAB triploids) produced and known to date by international breeding programs contain integrated sequences of the banana streak virus which hampers the development of a new variety. One exception has been recently presented (Dépigny et al., 2018) of a new cooking secondary triploid hybrid called K74 that was selected in partnership between CARBAP and CIRAD. 'CARBAP K74' is a triploid plantain-like hybrid resulting from a cross between an improved plantain-like tetraploid (CARBAP 832 – AAAB) parental line and an improved cooking-type diploid (AA). Molecular studies showed that this hybrid is free of any endogenous banana streak virus (eBSV) sequences. This lack of integrated sequences is probably due to the recombination between genomes A and B on chromosomes 1 and 2 during meiosis of the maternal parent AAAB. As shown by Noumbissié et al. (2016), this situation is quite common and occurs in about 20% of individuals of this type of progeny.

An agronomic evaluation of K74 was performed over two crop cycles to quantify its productivity and evaluate its reaction to pests and diseases. In trial plots (Fig. 9), damage assessments highlighted a lower susceptibility to major pests and diseases of plantains. The agronomic potential of 'CARBAP K74' allows the production of acceptable yields with few inputs over several cycles without replanting, underscoring its potential for smallholders as well as for

Published by Burleigh Dodds Science Publishing Limited, 2021.

Figure 9 Pedigree breeding: K74 selection - secondary triploid, AAB cooking type. (a) Field plot in third ratoon at CARBAP, Cameroon. (b) Bunch of K74 (approximatively 35 kg). (c) Fruit at harvest (about 25 cm long). (d) Fruit maturity point in use as plantain traditional landraces.

larger plantations. Furthermore, postharvest traits and consumer testing of the fruit showed that 'CARBAP K74' is a promising candidate for consumption and processing. Thus, 'CARBAP K74' is an innovative promising hybrid, which could help farmers to achieve agronomic performance, limit environmental impacts of banana production and alleviate food security issues.

4.2 Reconstructive breeding

As Stebbins stated (1947), 'there is no doubt that the more we know of how polyploidy has operated in the past to produce new species and races of plants, the better we shall be able to use experimental polyploidy as a means of improving our useful cultivated plants'.

Several years ago CIRAD came up with an alternative banana-breeding strategy called 'Reconstructive Breeding' based on the development of triploids from spontaneous or improved diploid plant material (Bakry et al., 1990). In this strategy, triploid hybrids are thought to be the best objective for banana breeding, as triploidy conferred, at world scale, a selective advantage over other ploidy levels in the cultivation of landraces of banana. In addition, from a banana breeder's perspective, rather than an obstacle, triploidy should be thought as a fantastic opportunity to combine three or more distinct genetic pools in a single genome to maximize heterozygosity and to the benefit from all the favorable consequences arising thereof (Gallais, 2009).

4.2.1 The principle behind reconstructive breeding

In contrast to 'Pedigree Breeding', the reconstructive breeding strategy is based on optimization of combining ability between two diploids of which one is the donor of diplo-gametes. The aim is to combine the favorable traits brought by both parents and to maximize the heterozygosity in the triploid progenies (Sanford, 1983) whether of mono- or inter-specific constitution. In this case, the choice of the couples of progenitors is primarily governed by the value of the triploid progenies and less by the value of the parents themselves.

However, as production of 2n gametes is uncontrolled and rare in diploid clones (Dodds, 1943), regular production of diplo-gametes has been achieved through the chromosome doubling of diploid varieties (Bakry and Horry, 1994; Bakry et al., 2007). These double-diploids are then crossed with other diploids to produce triploid individuals (Fig. 10).

This reconstructive breeding strategy has several advantages: (i) genotypes selected at the diploid level can be largely transferred to the triploid descendants by the 2n gametes; (ii) the hybrids display the desired sterility because of their triploid nature; (iii) utilization of very fertile seminiferous parents sometimes allows production of large triploid progenies which enable effective selection; (iv) this strategy enables breeders to tap the wide diversity in diploid *Musa* germplasm (Fig. 11). New selection criteria can also be included at different stages – and through the use of new diploid parents – reconstructive breeding can respond quickly either to the appearance of new races of pathogenic fungi or to meet other selection objectives.

Recent advances in understanding the evolutionary pathways of the banana complex through genomic studies have reinforced the utility of this breeding strategy. By identifying the most likely diploid ancestors of current triploid cultivars (Raboin et al., 2005), these studies have made it possible to select the most relevant parental combinations from the wide array of *Musa* germplasm available. It is in such a context of knowledge that banana breeders will find much more benefit in the use of polyploidy as a tool, combined in various ways with hybridization and selection, than in the creation of tetraploids which might be expected to have immediate value. Viewed in this light, the importance of artificial tetraploidization in banana breeding is by no means diminished, but it must be considered as one intermediary step, which is integrated with many others in producing various types of new triploid varieties.

4.2.2 The process of reconstructive breeding

The process of reconstructive breeding starts with selection of genotypes intended for colchicine treatment according to the type of banana to be developed (cooking or dessert), their agronomic characteristics, their resistance

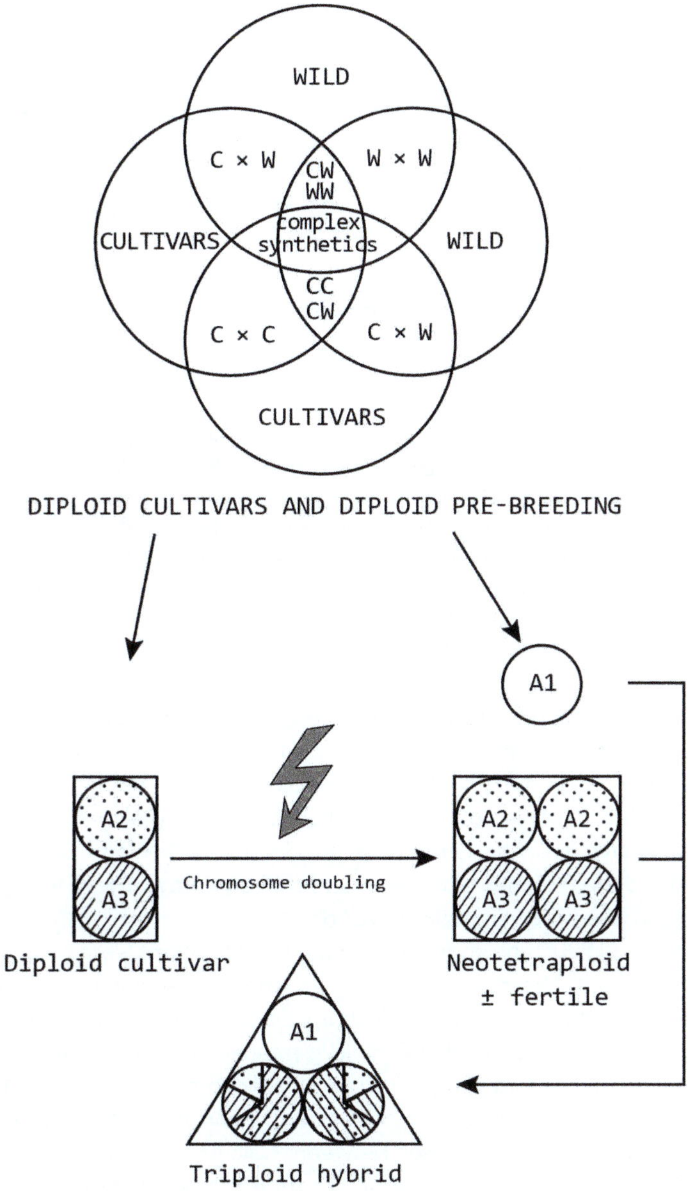

Figure 10 Reconstructive breeding: diagram based on *Musa acuminata*. A1 ... A4= individual sets *M. acuminata* genomes (n=11).

to diseases and pests as well as their paternal and/or maternal fertility. Improved diploids obtained by hybridization and expressing high resistance to various diseases are also included in this treatment.

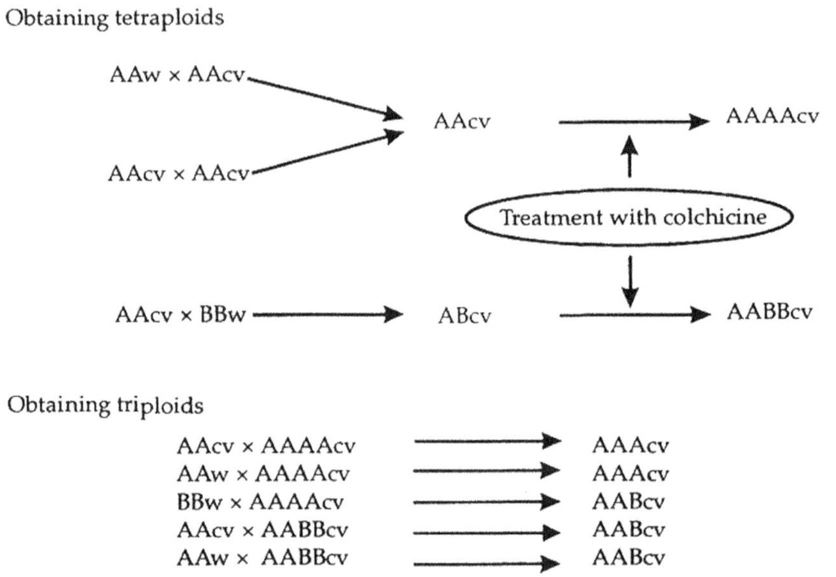

Figure 11 Reconstructive breeding: scheme for the synthesis of new triploid varieties of different genomic constitutions from diploid germplasm (w – wild; cv – cultivated).

Since the beginning of this program in the nineties, application of colchicine treatments on in vitro shoot proliferations of edible diploids has resulted in more than 25 non-chimeric AAAA autotetraploids and AABB allotetraploids as selected by chromosome counts and/or flow cytometry (Bakry et al., 2007). All these plants have flourished in tropical conditions in Guadeloupe. Reciprocal crosses with diploid clones have been undertaken to investigate the synthesis of mono- and inter-specific triploid hybrids.

In the early 1990s, chromosome number investigations were carried out on plants issued from intraspecific crosses between the seminiferous 'Calcutta 4' clone, used as tester, and four doubled-diploid clones (Guyod, Galéo, IDN110 and Tjau Lagada) as donor of polyploid gametes. The selfing of 'Calcutta 4' resulted in strictly diploid progenies (22 chromosomes), proving that all the male and female gametes of 'Calcutta 4' have 11 chromosomes. Almost all the 347 hybrids resulting from crosses between 'Calcutta 4'' and the four distinct double-diploid varieties turned out to be triploid (over 98% of the hybrids) with a number of chromosomes varying from 32 to 38 according to the pair of parents (unpublished data). Moreover, no difference was observed between reciprocal crosses of the same parents. Another study about the progeny of an interspecific cross [BB ♀ (seminiferous type: Pisang Klutuk Wulung) × AAAA ♂ (variety: doubled IDN110)] showed, by flow cytometry, that all the 227 hybrids were triploid.

These first results which formed the base of the CIRAD-breeding activities, strongly indicated that the studied double-diploid clones are almost exclusively producing efficient polyploid gametes containing, for a major part, exactly 22 chromosomes but also, in a smaller proportion, gametes ranging from 20 to 27 chromosomes. These results were to be confirmed later on a larger number of doubled genotypes. The evidence suggests that whatever the direction of the crosses, crossing double-diploid varieties with diploid clones did not lead to lineages of very variable chromosomal constitution as one may have supposed. On the contrary, the progeny is essentially triploid, which is the expected objective. These first outcomes obtained through 2×/4× crosses at CIRAD did validate the original pathways of banana improvement through reconstructive breeding.

The gamete fertility of the doubled AA diploids is rather unpredictable: some clones are slightly fertile at both the diploid and tetraploid level, while others are completely sterile at the tetraploid level. In practice, efforts in getting seeds from crosses between *M. acuminata* accessions are equivalent to those implemented for pedigree breeding. However, in some circumstances, the barrier of sterility has been effectively circumvented and allowed us to obtain quality seeds in large quantities. In that respect, some AB interspecific clones are sterile at the diploid level, but showed to be fertile as male and female parents in the AABB allotetraploid configuration (Bakry, pers. obs.), indicating that gamete sterility in AB clones is probably due to incomplete chromosome pairing at meiosis (Jeridi et al., 2012). This result i particularly favorable for interspecific breeding and has been implemented for synthesis of many valuable AAB and ABB hybrids (Jenny et al., 2013).

Another example of reconstructive breeding is based on pollination of seedy accessions by double-diploid parthenocarpic clones as male parents. In such crosses, sterility factors are no longer a constraint because the female parent is fully fertile and the tetraploid male parent always provides enough fertile pollen, even if the proportion of sterile pollen is high. In those conditions, it is common to get anywhere from 200 to 3000 seeds per pollinated bunch with no major difficulties. If parents are accurately selected, a very high proportion of triploid descendants are parthenocarpic with no seed production similar to those of natural triploid varieties (Bakry and Horry, 1994) in such combinations.

One of the main advantages of reconstructive breeding compared to pedigree breeding is the ability to transfer most of the human and field resources away from pollination and seed collecting and deploy these resources to selection among hybrid progeny, which is the core activity of any breeder in an effort to achieve genetic gain (Table 1).

Table 1 Comparison of progenitor fertility, hybrid production and selection ratios (% success) for the two banana crossing strategies

	Development of tetrapoids 3x × 2x → 4x ♀ parent: cultivar		Development of triploids 2x × 4x → 3x ♀ parent: wild	
	Quantity	Ratio (%)	Quantity	Ratio (%)
Hybrid production				
Pollinated bunches	1000		1	
Bunches nearing maturity	860		1	
Bunches containing seeds	258	26	1	100
Seeds per bunch	3		3000	
Seeds (total)	774		3000	
Embryos cultured in vitro	774		500	
Embryos germinated in vitro	193	25	250	50
Individuals in the field (total)	174	22	250	50
Hybrid selection				
For hybrids of plantains				
First selection of probable 4x hybrids	8	5		
Final selection of 4x hybrids	2	1		
For Silk banana types hybrids				
First selection of 3x hybrids			50	20
Final selection of 3x hybrids			15	6

4.2.3 Maximizing heterozygosity and trait complementarity in triploid progenies

Maximizing heterozygosity in triploid hybrids depends on three factors; (i) on the genetic distance between the two parents; (ii) on the level of heterozygosity of the 2n gamete donor-diploid parent (after chromosome doubling) and; (iii) on the proportion of this parental heterozygosity transmitted to triploid progeny via the 2n gametes.

In AABB neo-allotetraploids, derived from the natural AB landraces by chromosome doubling, there are two sets of chromosomes, one from each species. Each chromosome is supposed to pair only with its homologous form during meiosis and only bivalents are formed (Stebbins, 1947). This pairing behavior results in disomic inheritance with 100% of the diploid interspecific heterozygosity being transmitted by each 2n gamete (Fig. 12). From a practical point of view, one should expect the genomic and phenotypic value of the original AB variety to be fully transmitted, and without recombination, to the triploid progeny by the AB gametes.

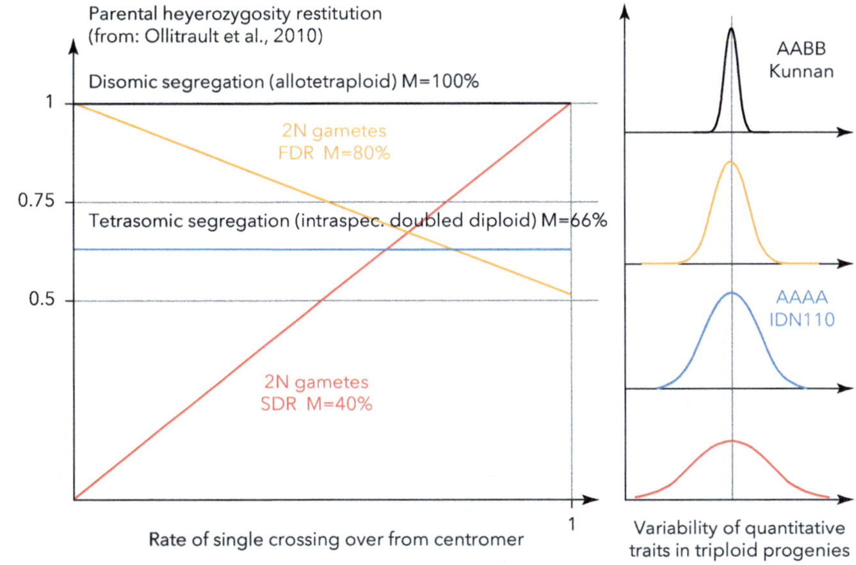

Figure 12 Synthesis of triploids from diploid germplasm in banana: theoretical proportion of heterozygosity of diploid parental lines transmitted to triploid progeny by 2n gametes (from Ollitrault et al., 2008 - Jeridi et al., 2012). FDR or SDR 2n gametes: rare and not controlled in banana diploid germplasm - In vitro chromosome doubling: convenient and easy in banana by colchicin treatment → non-chimerical doubled-diploids (AAAA, AABB, BBBB).

However, a case study on Kunnan (Indian AB landrace) and its tetraploid derivative highlighted that A and B chromosomes are partially homologous and can form from 3 to 8 pairs of chromosomes. A mixed inheritance pattern was proposed for 'Kunnan 4×', with three sets of chromosomes segregating in tetrasomic inheritance, three sets in disomic inheritance and five sets in an intermediary pattern. 'Kunnan 4×' is considered here as being a '3/5/3' segmental allotetraploid, with chromosome associations resulting in various configurations, including cross-structures, rings and chains between homoeologous and homologous chromosomes (Sybenga, 1975, 1996). The genetic consequences of these results are that transmission of heterozygosity from the 'Kunnan' AB landrace to triploid progenies is probably lower and the genetic variability of the 2n gametes probably higher than expected in a strict amphidiploidy configuration. Nevertheless, segmental allo-tetraploidy should allow the efficient transfer of complex traits and multi-traits phenotypes of the parental diploid 'Kunnan' to triploid progenies.

In AAAA neo-autotetraploids, derived from chromosome doubling of AA accessions, the presence of four homologous chromosomes, instead of two, results in equal opportunities to pair at meiosis leading to multivalent

Table 2 Percentage of parental heterozygosity transmitted through gametes arising from different mechanisms (from Sanford, 1983). The starting material in each case is assumed to be diploid.

Origin of gamete	Parental heterozygosity (%)
Apospory	100
Asynapsis plus first division restitution	100
First division restitution:	75–83
Somatic doubling followed by regular meiosis	67
Second division restitution	33–50
Post-meiotic doubling	0
Normal reductional meiosis	0

formation and tetrasomic inheritance. For doubled-diploids, it hypothetically leads to 66% restitution of the diploid parental heterozygosity to triploid progenies (Table 2 and Fig. 12). However, as indicated by Stebbins (1947), large translocations or inversions will undoubtedly affect metaphase pairing. Therefore, many autopolyploids derived from diploid structural hybrids should also be considered as partial segmental polyploids with chromosomes that preferentially pair in disomic inheritance for those involved in structural events and in tetrasomic inheritance for the rest of the chromosome set.

In *Musa*, all AAcvs. incorporated in our breeding activities are at least bearing one chromosomic structural event (Fauré, Bakry and Gonzalez de Leon, 1992; Martin et al., 2017). And in most cases, two to three chromosomes are implied in translocated, inverted or duplicated chromosomic segments. As a concrete case, we have shown, using microsatellite markers, that the IDN110 accession, heterozygous for at least three structural events, transmits about 73% of its parental heterozygosity through 2n gametes to the triploid progeny, that is, a little more than the theoretical model (Bakry and Jenny, pers. data).

4.2.4 Examples of reconstructive breeding

CIRAD has engaged since 1993 in an intensive program of banana breeding involving hybridization among various parents, leading to the production of numerous 'dessert' and 'cooking' triploid hybrids. During this process a better understanding was gained of the respective contributions of each different species in the phenotypic expression of the hybrid. It became evident that the cooking versus sweet traits of the fruits were provided by the *M. acuminata* parents, whereas *M. balbisiana* conferred a strong vegetative strength to the hybrids while resistance to black leaf streak disease could be brought by any of the two species. Moreover, observations revealed that, in comparison with diploids, AAA and AAB triploid hybrids show more vigor and bunches of a better average size and weight. Interfamily variability of progeny depends mainly on the parents selected, but it is always higher than intrafamily variability. Lastly, while the parthenocarpy trait may segregate in triploid progeny when *M. balbisiana* or *M. acuminata* seminiferous types are used as a female parent, all triploid hybrids are parthenocarpic when both parents are edible varieties.

4.2.4.1 Breeding for sweet-acid bananas

Sweet-acid bananas (AAB Silk and Pome) are highly susceptible to *Fusarium* wilt race 1 and tropical race 4 and Sigatoka diseases. Breeding programs aim at the development of new disease-resistant sweet-acid banana cultivars that retain the organoleptic and productivity of the traditional landraces. Common pedigree breeding strategies for these groups consisted so far in the selection of AAAB tetraploid hybrids, directly issued from crosses between AAB sweet-acid cultivars pollinated with AA clones carrying disease-resistance genes. However, this crossing pathway is hampered by low gamete fertility and the rare occurrence of desired 2n gametes on the AAB female side. In addition, AAAB hybrids contain infectious e-BSV alleles (Duroy et al., 2016) which can be activated to cause infection of plants with BSV particles. At CIRAD, using a 'reconstructive approach', we have tested the synthesis of triploid hybrids directly from the 'Kunnan' landrace (ABcv), chosen as elite diploid because of its high consumer value in India. Its genomic constitution suggests that it would come from a cross between a *M. acuminata* spp. *malaccensis*-derived edible diploid and a *M. balbisiana* seminiferous relative.

Natural AB clones are often sterile but their AABB neo-allotetraploid counterparts obtained through colchicine treatment are fertile. Therefore, both as male and female parent, the gamete fertility of the tetraploid 'Kunnan' was made exploited in crosses with AA and BB accessions to generate directly triploid hybrids (Jenny et al., 2013). In crosses with seminiferous and edible disease-resistant *M. acuminata* spp. *malaccensis* derivatives, we obtained several AAB hybrids ranging from Silk-like to Mysore-like (another AAB important sweet-acid banana) cultivars, exhibiting resistance to Fusarium wilt (race 1 and TR4) and Sigatoka leaf diseases. On the other hand, in crosses with *M. balbisiana* accessions, several ABB hybrids were selected with a phenotype very similar to 'Pisang Awak' clones.

CIRAD reconstructive breeding program has also investigated other breeding pathways for sweet-acid bananas based on interspecific crosses in which seminiferous *M. balbisiana* accessions are directly pollinated as female parent. Crossing the 'Pisang Klutuk Wulung' BB accession with the double-diploid line IDN110 AAAA as male parent yielded 227 triploid hybrids of BAA genomic composition in a single bunch pollination. Preliminary results showed promise with the selection of some new exceptional interspecific hybrids, resistant to diseases and agronomic and consumer traits close to natural varieties (Bakry and Horry, 1994). However, these triploid hybrids displayed unfortunately infections by BSV particles likely resulting from stress-related activation of eBSV carried by the *balbisiana* parent (Lheureux et al., 2003). Therefore, due to these disappointing and unexpected results, these interspecific crossing activities were stopped, and CIRAD developed a pre-breeding program in search of a *M. balbisiana* free from infectious eBSV. Molecular markers such as immunocapture-PCR and Southern-blot were developed and routinely used to support the pre-breeding strategy (Umber et al., 2011, 2016). Marker-assisted selection eventually resulted in new seminiferous lines of BB free of any e-BSV sequences that can be activated, which have been used in crosses with double-diploid *M. acuminata* varieties (Fig. 13). A new triploid progeny has recently been obtained from the same double-diploid IDN110 male genitor, free of e-BSV sequences, in which most hybrids surpass natural AAB varieties not only in terms of disease resistance but also in agronomic performance while having excellent fruit quality (Fig. 14).

Figure 13 Reconstructive breeding: in 2017, creation of the first interspecific hybrid population (BAA) without risk of activation of infectious eBSVs (from F. Salmon, G. Pressat, J. P. Horry, P. Y. Teycheney & F. Bakry - 2018). Building new banana varieties resistant to diseases (TR4 and others): constraints and solutions. https://apsnet.confex.com/apsnet/ICPP2018/meetingapp.cgi/Session/1994.

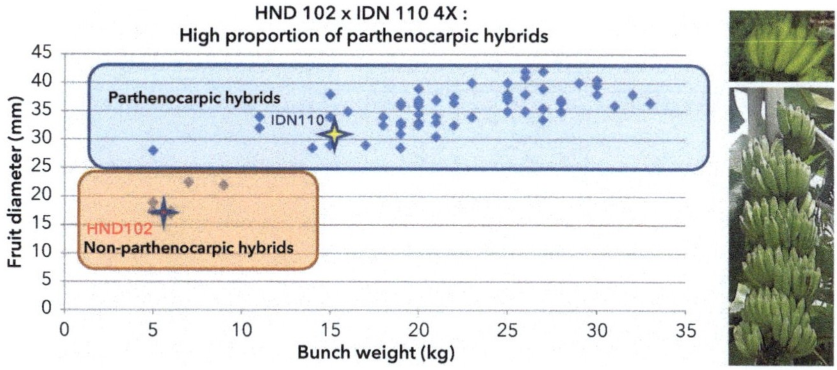

Figure 14 Reconstructive breeding: performances of the BAA hybrid population. Trait distribution for bunch weight and fruit diameter. Blue box: parthenocarpic hybrids (92%) - maroon box: non parthenocarpic hybrids (8%).

4.2.4.2 Breeding for AAA hybrid bananas

There exists a whole range of natural diploid clones, cultivated all around Southeast Asia as landraces, and displaying a much-diversified landscape of tastes and diseases resistances, along with highly varied phenotypes. These varieties evolved from combinations of the numerous *M. acuminata* subspecies disseminated in the region, thus bearing various levels of heterozygosity in what could seem to be at first glance a rather continuous gradient without much structure. Molecular studies (Carreel et al., 1994; Grapin et al., 1998; Perrier et al., 2011) have shown that a grouping of these landraces could be achieved in rather homogenous clusters at the genetic level. CIRAD has thus extensively evaluated this material in order to select promising clones among this diversity, with sufficient residual fertility to be used in creation of new varieties through reconstructive breeding. Around fifteen of these clones, mainly belonging to the 'Khai' and 'Figue' clusters, were doubled to produce intermediary autotetraploids. An intensive crossing schedule was set up to combine these autotetraploid plants, used as 2n gamete donors with other diploid landraces from these same clusters. Maximizing the genetic distance between the parents was one of the choice criteria, along with self-value evaluation of each parent, fertility, genetic combining ability and phylogenetic data (Fauré et al., 1994; Hippolyte et al., 2012; Carreel et al., 2002). Depending on the crosses and thanks to in vitro embryo rescue, it was possible to obtain large progenies and large populations in the field of triploid hybrids, in which a real selection program could be implemented. Larger numbers of progeny also enable the evaluation of the phenotypic combining ability of these parental lines.

In the progeny, a very high degree of phenotypic variation was observed. Every important character was evaluated: yield range and factors, vegetative development, resistances to diseases and pests as well as fruit quality. It is interesting to note that true families were still easily recognizable, with an

		Nr obs.		Height	Weight	Hands	NrF	LF
Diploid parents				cm	kg			mm
Female	Pisang Madu	18	Mean	415	28	14	351	150
Male	IDN 110	5	Mean	253	16	11	170	175
Combination	**Pisang Madu × IDN 110 4x**	535	Mean	374	22	11	217	149
			CV (%)	20	45	22	30	19
			Min	172	2	5	12	13
			Max	620	64	18	420	240

→ CIRAD 918, CIRAD 925

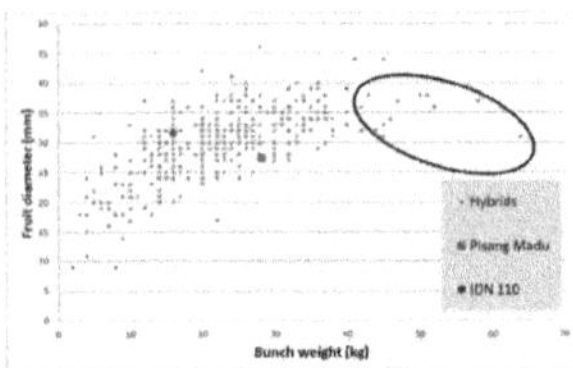

		Nr obs.		Height	Weight	Hands	NrF	LF
Diploid parents				cm	kg			mm
Female	Manang	10	Mean	379	21	12	209	168
Male	Pisang Lilin	10	Mean	249	5	7	106	131
Combination	**Manang × Pisang Lilin 4x**	99	Mean	376	23	9	154	184
			CV (%)	23	43	21	30	23
			Min	192	3	6	13	20
			Max	569	48	13	256	298

→CIRAD 916

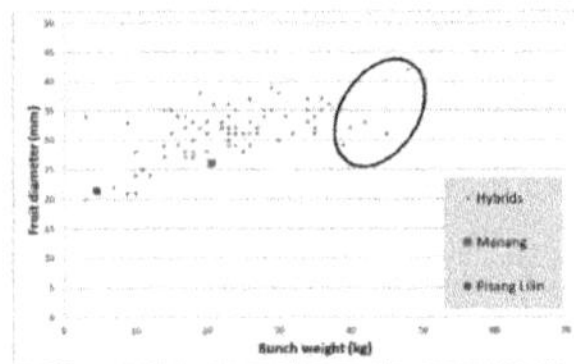

		Nr obs.		Height	Weight	Hands	NrF	LF
Diploid parents				cm	kg			mm
Female	IDN 110	5	Mean	253	16	11	170	175
Male	Khai Nai On	9	Mean	432	32	10	206	0
Combination	**IDN 110 4x × Khai Nai On**	77	Mean	347	23	10	168	154
			CV (%)	23	47	25	32	20
			Min	205	6	6	73	100
			Max	540	52	16	298	230

→ CIRAD 920

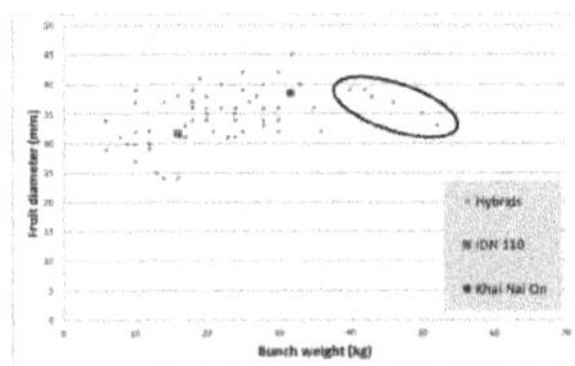

0 value means no data.

Figure 15 Reconstructive breeding: expression of heterosis in three AAA families. Trait distribution for bunch weight and fruit diameter. In O: high-performance hybrids.

important influence of the tetraploid parent. Some promising hybrids have been selected following this program, including CIRAD 916, CIRAD 918, CIRAD 920 and CIRAD 925 (Fig. 15).

All of these hybrids proved to display at least a strong partial resistance to BLSD *Mycosphaerella fiijiensis*, which was at that time the first goal of the banana breeding program. The global annual yield of these hybrids was close to what is observed in dessert commercial plantations with Cavendish clones. But the diversity was coming from the different components making up this yield (numerous smaller bunches vs. one and a half large bunch for instance), the sizes of the fruits and the number of hands, and finally the taste which ranged from common sweet to sweet-acid along with various scents (Fig. 16). One particular selection, CIRAD 925, has been intensively evaluated by CIRAD and banana growers in Guadeloupe and Martinique, and is now being developed for the organic market. Since 2020, it is produced on a larger scale and sold in French supermarkets under the 'Pointe d'Or' brand.

This scheme has also been tested using a non-parthenocarpic seminiferous *M. acuminata* as diploid female parent. It was very interesting to see that a significant part of the progeny was bearing fruits with a diameter equal or larger than the parthenocarpic parent, with bunch weights exceeding those of both parents, hence showing a clear heterosis effect (Jenny et al., 2013).

Figure 16 Reconstructive breeding: example of six advanced AAA CIRAD hybrids selected for further field evaluation at a larger scale. Dessert varieties selected in tropical conditions (Guadeloupe and Martinique, F.W.I.) - All resistant to yellow Sigatoka and black leaf streak - All resistant to FOC Race 1 - Various patterns of resistance to FOC TR4 (evaluated in early screening at WUR and/or in field conditions in Northern Territories, Australia) - Different patterns of resistance to nematods - Productivity, fruit quality: variable. R: resistant - M: moderately susceptible - S: susceptible.

4.2.4.3 Focus on Cavendish-like banana hybrids

Due to the crisis in Cavendish production linked to the high susceptibility to BLSD and TR4, breeding activities have focused on the use of 'Mchare' AAcv accessions introduced from the Comoros islands (Perrier et al., 2018). A set of these landraces were selected based on agronomic performance and, after chromosome doubling, they were tested in crosses with AA accessions of the 'Khai' cluster (Perrier et al., 2011). Despite the low fertility of most of the double-diploids (Goigoux, Salmon and Bakry, 2013), some triploid progenies were obtained, showing highly variable performances. Among them, one hybrid was looking very similar to the variety 'Lacatan', which is a giant variety within the Cavendish subgroup, both in terms of plant stature and in fruit quality, thus demonstrating the proof of concept of this novel breeding approach (Fig. 17). The susceptibility to Sigatoka complex was expected as no surprise, as both Mchare and Khai parents used in this cross are susceptible.

Natural diploid AA cv. progenitors belonging to the 'Khai' cluster (*M. acuminata* spp. *malaccensis* background) and combining gamete fertility, resistance to diseases and favorable agronomic features are rare. Therefore, pre-breeding is required within the 'Khai' cluster to develop fertile improved progenitors containing various sources of disease-resistant genes derived from crosses between edible diploid clones and seminiferous relatives. It is likely that the exploitation of *Musa* genetic maps, the development of high-density molecular markers, and the capacity in the near future to identify various allelic forms of numerous genes will help in the implementation of these new pre-breeding activities, akin to a 'Breeding by design' approach.

Figure 17 Reconstructive breeding: 'Cavendish like' AAA hybrid coming from Pisang Madu (AAcv; ♀) X 'tetraploid Nzumoheli' (AAAAcv; ♂) cross.

4.2.4.4 Proposals for the future to improve cooking bananas by reconstructive breeding

The high potential of the neo-allotetraploids presented above for sweet-acid bananas, offers a wide range of possibilities in breeding for cooking bananas too. In the present situation, there is a lack of diploid AB cooking types among landraces. We therefore suggest to first create new AB diploid parthenocarpic hybrids by pollinating eBSV-free BB lines with AA diploid 'cooking type' cultivars from *M. acuminata spp. banksii* origin, to be found in Papua New Guinea derivatives (Perrier et al., 2009). The direction of the crosses must be oriented to increase the quantity of seeds, using the seminiferous accession as female parent. This first step is fundamental and must be carried out with great skill and care. Indeed, the selection of new AB hybrids at the diploid level in a single step implies the concomitant performance of two improvement steps which are often staggered in time in cross-breeding:

- first, the selection of a haploid set of genes on the female *balbisiana* parent on one side and a haploid set on the male *acuminata* parent on the other side;
- secondly, the selection for good specific combinations of these two haploid sets of genes in the selected AB hybrid.

The result of this pre-breeding approach has resulted in interesting diploid AB hybrids (Fig. 18).

Figure 18 Reconstructive breeding: de novo synthesis of BA parthenocarpic hybrids. Bunch of a cooking parthenocarpic-diploid BA hybrid coming from a cross between Pisang Klutuk Wulung (BB seminiferous) as female parent and Tomolo (AAcv - Cooking diploid) as male parent.

It is likely that most of these diploid interspecific hybrids selected for multi-trait purposes will show more or less strong gamete sterility. However, according to our results on sweet-acid bananas, chromosome doubling should first restore gamete fertility, then allow back-crossing of these neo-allotetraploids with other interesting AA 'cooking genotypes'.

Restoration of fertility in parental tetraploids and better handling of heterozygosity transmitted to triploid progenies is much easier than pedigree breeding pathways. By its capacity to gain advantage from any triparental combination (including unimproved wild relatives and/or any combined source of resistance), reconstructive breeding offers unlimited potential to combine cooking features, maximizing heterozygosity, incorporation of new sources of resistance and non-infectious e-BSV patterns in new triploid-cooking AAB hybrids or even ABB hybrids for better drought tolerance.

5 Conclusion

As written by Heslop-Harrison (2011): 'Superdomestication underpinned by genomic research with collection and assessment of the biodiversity in the genus are likely to ensure the future of the Musa crop.' That sentence is now more valid than ever before.

In recent decades, several genomic tools have been successfully used to decipher the complexity of the *Musa* genepool. Molecular markers have not only proven to be effective to distinguish between genotypes, but have also provided a clear understanding of the history of the evolution and domestication of the crop from the ancestor seminiferous species to the present time landraces and cultivars. The acquired knowledge of the probable diploid ancestors of most triploid cultivars now allows us to build breeding schemes that can mimic the sequence of crossings and selections that have occurred over several millennia. Nevertheless, for the future, a wider knowledge of diploids may also open the way to novel genetic combinations. This is particularly the case for the identification of new sources of resistance to Fusarium wilt TR4 that could be incorporated into improved hybrids. On the other hand, the challenges for banana improvement are enormous. Contemporary food security issues, the uncertainties linked to the ongoing climate change, the impact of man on biodiversity conservation, the lack of resources to collect the remaining wild *Musa* germplasm are all constraints that must be included in improvement strategies and objectives. The exploitation of this potential neo-diversity in crossbreeding is certainly one of the major assets we have at our disposal to address these issues. Broadening the genetic base of banana through plant breeding will support the diversification of agrosystems, extend potential production areas, improve fruit processing and reach new consumers. For example, the inclusion of *Musa yunnanensis*, identified by Häkkinen and Hong (2007) in the mountains of southern China, in the improvement strategies could pave the way for the creation of cold-tolerant varieties.

Recent advances in *Musa* genomics, from the first complete sequence of the *Musa* genome published in 2012 (D'Hont et al., 2012) to the on-going re-sequencing of hundreds of different *Musa* genotypes, will no doubt increase the efficiency of banana breeding. The *Musa* genome can now be deeply explored for the characterization of desirable genes, a prelude for their use in marker-assisted selection. The very fine structure of the genome, the arrangement of the chromosomes and genes of interest can now be studied with unprecedented precision, adding to the characterization of the genetic complex and the understanding of the evolutionary pathways. For banana improvement, this approach is particularly important in a biological context of gamete sterility (and difficulties to get large quantities of seeds) for which recombination and implementation of introgression strategies remain the greatest challenge.

Conventional breeding through hybridization has been the way by which almost all cultivated plants have historically been improved. As stated by Rowe and Rosales (2000), banana was recalcitrant in this regard, but it has now joined the long list of crops that are improved by cross-pollinations and selection. However what should emerge from this chapter and various others in this book is that conventional breeding methods should be viewed as just one part - albeit an important part - of a more general genetic improvement strategy. Unconventional techniques need to be used to broaden the arsenal of tools available to genetically improve bananas (Novak, 1992; Sagi et al., 1995; Jenny et al., 2002; Dale et al., 2017). By increasing phenotypic diversity, induced mutation can result, for example, in hybrids with reduction in plant size and improved bunch and fruit shapes. On the other hand, conferring resistance to banana bunchy top virus through a GMO approach (Mware, 2016) to natural or improved plant material is regarded as a central issue for developing countries in Africa. It is therefore essential that conventional *Musa* breeders and biotechnologists work together in a complementary fashion to accelerate improvement of this important food crop.

6 Where to look for further information

Website:

Global strategy for the conservation and use of Musa genetic resources: https://www.bioversityinternational.org/e-library/publications/detail/global-strategy-for-the-conservation-and-use-of-musa-genetic-resources/ - Chapter 12: Genetic Improvement.

Books:

Bakry F. et al. 2009. *Genetic Improvement of Banana in Breeding Plantation Tree Crops: Tropical Species*. Jain, S. Mohan and Priyadarshan, P.M. (Eds.).

Simmonds, N.W. 1962. *The Evolution of the Bananas. Tropical Science Series*. Longmans, London (GBR). 170p.

Simmonds, N.W. 1966. *Bananas*. 2nd ed., *Tropical Agricultural Series*. Longman, New York. 512p.

7 Acknowledgements

Special thanks to Frédéric Salmon (Banana breeder, CIRAD, Guadeloupe), who provided illustrations and data collected in Guadeloupe for the HND102 × IDN110 T cross. We also would like to thank Xavier Perrier (CIRAD) for the fruitful discussions which accompanied this work. Finally, the authors also acknowledge Laurence Rodriguez (CIRAD) for her assistance with figure production.

8 References

Aguilar Morán, J. F. (2013). Improvement of cavendish banana cultivars through conventional breeding. *Acta Horticulturae* 986, 205–208.

Allard, R. W. (1960). *Principles of Plant Breeding*. Google-books-ID: W0xGcAAACAAJ.Wiley.

Bakry, F. (2008). Zygotic embryo rescue in bananas. *Fruits* 63(2), 111–115.

Bakry, F. and Horry, J. P. (1992). Tetraploid hybrids from interploid 3x/2x crosses in cooking bananas. *Fruits* 47(6), 641–655.

Bakry, F. and Horry, J. P. (1994). *Musa* breeding at CIRAD-FLHOR. In: Jones, D. R. (Ed.). First Global Conference of the IMTP held at FHIA. The Improvement and Testing of Musa: A Global Partnership. INIBAP, La Lima (HND). pp. 169–175.

Bakry, F., Horry, J. P., Teisson, C. and Tézenas du Montcel, H. (1990). L'amélioration génétique des bananiers à l'IRFA/CIRAD. *Fruits*. Spécial bananes-Réunion annuelle 'bananiers' 1989, 25–40.

Bakry, F., Paulo De La Reberdiere, N. P., Pichot, S. and Jenny, C. (2007). In liquid medium colchicine treatment induces non chimerical doubled-diploids in a wide range of mono- and interspecific diploid banana clones. *Fruits* 62(1), 3–12.

Baurens, F. C., Martin, G., Hervouet, C., Salmon, F., Yohomé, D., Ricci, S., Rouard, M., Habas, R., Lemainque, A., Yahiaoui, N. and D'Hont, A. (2019). Recombination and large structural variations shape interspecific edible bananas genomes. *Molecular Biology and Evolution* 36(1), 97–111.

Carreel, F., Fauré, S., de León, D. G., Lagoda, P., Perrier, X., Bakry, F., du Montcel, H. T., Lanaud, C. and Horry, J. (1994). Evaluation de la diversité génétique chez les bananiers diploïdes (*Musa* sp.). *Genetics Selection Evolution* 26(Suppl 1), 125–136.

Carreel, F., Gonzalez de Leon, D., Lagoda, P. J. L., Lanaud, C., Jenny, C., Horry, J. P. and Tézenas du Montcel, H. (2002). Ascertaining maternal and paternal lineage within *Musa* by chloroplast and mitochondrial DNA RFLP analyses. *Genome* 45(4), 679–692.

Cattan, P., Ruy, S. M., Cabidoche, Y.-M., Findeling, A., Desbois, P. and Charlier, J. B. (2009). Effect on runoff of rainfall redistribution by the impluvium-shaped canopy of banana cultivated on an andosol with a high infiltration rate. *Journal of Hydrology* 368(1–4), 251–261.

Cheesman, E. E. (1947a). Classification of the bananas – II. The genus *Musa* L. *Kew Bulletin* 2, 106–117.

Cheesman, E. E. (1947b). Classification of the bananas. III. Critical notes on species. b. *M. acuminata*. *Kew Bulletin* 2, 17–30.

Cheesman, E. E. and Dodds, K. S. (1942). Genetical and cytological studies of *Musa*. IV. Certain triploid clones. *Journal of Genetics* 43(3), 337–357.

Cohan, J., Abadie, C., Tomekpe, K. and Tchango Tchango, J. (2003). Agronomic performance and resistance to black leaf streak of the hybrid 'CRBP-39'. *InfoMusa* 12(1), 29–32.

Côte, F. X., Galzi, S., Folliot, M., Lamagnère, Y., Teycheney, P. Y. and Iskra-Caruana, M. L. (2010). Micropropagation by tissue culture triggers differential expression of infectious endogenous *banana streak virus* sequences (eBSV) present in the B genome of natural and synthetic interspecific banana plantains. *Molecular Plant Pathology* 11(1), 137–144.

Dale, J., James, A., Paul, J. Y., Khanna, H., Smith, M., Peraza-Echeverria, S., Garcia-Bastidas, F., Kema, G., Waterhouse, P., Mengersen, K. and Harding, R. (2017). Transgenic Cavendish bananas with resistance to Fusarium wilt tropical race 4. *Nature Communications* 8(1), 1496..

De Langhe, E. and Devreux, M. (1960). Une sous-espèce nouvelle de *Musa acuminata* Colla. *Bulletin du Jardin botanique de l'état, Bruxelles* 30(3), 375–388.

Dépigny, S., Noupadja, P., Tomekpe, K., Bonneviot, P., Lescot, T. and Bakry, F. (2018). 'CARBAP K74': a triploid plantain-like hybrid designed to promote sustainable plantain-based cropping systems. *Acta Horticulturae* 1196, 63–70.

D'Hont, A., Denoeud, F., Aury, J.-M., Baurens, F.-C., Carreel, F., Garsmeur, O., Noel, B., Bocs, S., Droc, G., Rouard, M., Silva, C. D., Jabbari, K., Cardi, C., Poulain, J., Souquet, M., Labadie, K., Jourda, C., Lengellé, J., Rodier-Goud, M., Alberti, A., Bernard, M., Correa, M., Ayyampalayam, S., Mckain, M. R., Leebens-Mack, J., Burgess, D., Freeling, M., Mbéguié-A-Mbéguié, D., Chabannes, M., Wicker, T., Panaud, O., Barbosa, J., Hribova, E., Heslop-Harrison, P., Habas, R., Rivallan, R., Francois, P., Poiron, C., Kilian, A., Burthia, D., Jenny, C., Bakry, F., Brown, S., Guignon, V., Kema, G., Dita, M., Waalwijk, C., Joseph, S., Dievart, A., Jaillon, O., Leclercq, J., Argout, X., Lyons, E., Almeida, A., Jeridi, M., Dolezel, J., Roux, N., Risterucci, A.-M., Jean Weissenbach, Ruiz, M., Glaszmann, J.-C., Quétier, F., Yahiaoui, N. and Wincker, P. (2012). The banana *Musa acuminata* genome and the evolution of monocotyledonous plants. *Nature* 488, 213–217.

Dodds, K. S. (1943). Genetical and cytological studies of *Musa*. V. Certain edible diploids. *Journal of Genetics* 45(2), 113–138.

Dodds, K. S. and Simmonds, N. W. (1946). Genetical and cytological studies of *Musa*. VIII. The formation of polyploid spores. *Journal of Genetics* 47(3), 223–241.

Duroy, P. O., Perrier, X., Laboureau, N., Jacquemoud-Collet, J. P. and Iskra-Caruana, M. L. (2016). How endogeneous pararetro viruses could enlighten *Musa* evolution? *Annals of Botany* 117(4), 625–641.

Dzomeku, B. M., Armo-Annor, F., Adjei-Gwen, K., Nkakwa, A., Akyeampong, E. and Baning, I. S. (2006). Evaluation of four Musa hybrids in Ghana. *Tropical Science* 46(3), 176–179.

Dzomeku, B. M., Quain, M. D., Lamptey, J. N. L., Anno-Nyako, F. O., Aubyn, A. and Darkey, S. K. (2007). Agronomic and sensory evaluation of some IITA hybrids in Ghana. *International Journal of Agricultural Research* 2(3), 307–311.

Fauré, S., Bakry, F. and Gonzalez de Leon, D. (1992). Cytogenetic studies of diploid bananas. In: Ganry, J. (Ed.). *Breeding Bananas and Plantains.*: Proceedings of the an International Symposium on Genetic Improvement of Bananas for Their Resistance to Diseases and Pests. CIRAD-FLHOR, Montpellier, France. pp. 77-92.

Fauré, S., Noyer, J. L., Carreel, F., Horry, J. P., Bakry, F. and Lanaud, C. (1994). Maternal inheritance of chloroplast genome ant paternal inheritance of mitochondrial genome in bananas (*Musa acuminata*). *Current Genetics* 25, 265–269.

Gallais, A. (2009). *Hétérosis et variétés hybrides en amélioration des plantes*. Google-books-ID: M7iQF2sDmjoC. Editions Quae.

García-Bastidas, F. A., Quintero-Vargas, J. C., Ayala-Vasquez, M., Schermer, T., Seidl, M. F., Santos-Paiva, M., Noguera, A. M., Aguilera-Galvez, C., Wittenberg, A., Hofstede, R., Sørensen, A. and Kema, G. H. J. (2019). First Report of Fusarium Wilt Tropical Race 4 in Cavendish Bananas Caused by Fusarium odoratissimum in Colombia. *Plant Disease* 104(3), 994.

Goigoux, S., Salmon, F. and Bakry, F. (2013). Evaluation of pollen fertility of diploid and doubled-diploid clones of MLALI and their potential use for banana breeding. *Acta Horticulturae* 986, 195–204.

Grapin, A., Noyer, J. L., Carreel, F., Dambier, D., Baurens, F. C., Lanaud, C. and Lagoda, P. J. L. (1998). Diploid Musa acuminata genetic diversity assayed with sequence-tagged microsatellite sites. *Electrophoresis* 19(8–9), 1374–1380.

Häkkinen, M. and Hong, W. (2007). New species and variety of *Musa* (Musaceae) from Yunnan, China. *Novon: A Journal for Botanical Nomenclature* 17(4), 440–446.

Häkkinen, M. and Väre, H. (2008). Typification and check-list of *Musa* L. names (Musaceae) with nomenclatural notes. *Adansonia* 30(1), 63–112.

Hauser, S. (2010). Growth and yield response of the plantain (*Musa* spp.) hybrid 'FHIA 21' to shading and rooting by Inga edulis on a Southern Cameroonian ultisol. *Acta Horticulturae* 879, 487–494.

Heslop-Harrison, J. S. (2011). Genomics, banana breeding and superdomestication. *Acta Horticulturae* 897, 55–62.

Hippolyte, I., Jenny, C., Gardes, L., Bakry, F., Rivallan, R., Pomies, V., Cubry, P., Tomekpe, K., Risterucci, A. M., Roux, N., Rouard, M., Arnaud, E., Kolesnikova-Allen, M. and Perrier, X. (2012). Foundation characteristics of edible *Musa* triploids revealed from allelic distribution of SSR markers. *Annals of Botany* 109(5), 937–951.

Jenny, C., Auboiron, E. and Beveraggi, A. (1994). Breeding plantain-type hybrids at CRBP. In: Jones, D. R. (Ed.). *The Improvement and Testing of Musa: A Global Partnership*. Proceedings of the First Global Conference of the IMTP. FHIA, La Lima (Honduras) (vol. 1994). pp. 176–187.

Jenny, C., Holtz, Y., Horry, J.-P. and Bakry, F. (2013). Synthesis of new interspecific triploid hybrids from natural AB germplasm in banana (*Musa* sp.). In: ISHS-ProMusa International Symposium. Bahia Othon Palace Hotel, Salvador, Bahia, Brazil.

Jenny, C., Tomekpé, K., Bakry, F. and Escalant, J. V. (2002). Conventional breeding of bananas. In: Jacome, L., Lepoivre, P., Marin, D., Ortiz, R., Romero, R. and Escalant,

J. V. (Eds). *Mycosphaerella Leaf Spot Diseases of Bananas: Present Status and Outlook*.INIBAP, San José, Costa Rica; Montpellier, France. pp. 199–208.

Jeridi, M., Bakry, F., Escoute, J., Fondi, E., Carreel, F., Ferchichi, A., D'Hont, A. and Rodier-Goud, M. (2011). Homoeologous chromosome pairing between the A and B genomes of *Musa* spp. revealed by genomic in situ hybridization. *Annals of Botany* 108(5), 975–981.

Jeridi, M., Perrier, X., Rodier-Goud, M., Ferchichi, A., D'Hont, A. and Bakry, F. (2012). Cytogenetic evidence of mixed disomic and polysomic inheritance in an allotetraploid (AABB) *Musa* genotype. *Annals of Botany* 110(8), 1593–1606.

Kissel, E., Vanhove, A. C., Garcia, S., Panis, B., Rouard, M., Cenci, A., Roux, N., Zorrilla, J., Swennen, R. and Carpentier, S. C. (2016). Abiotic stress research in crops using -omics approaches: drought stress and banana in the spotlight. *Acta Horticulturae* 1114, 81–90.

Lheureux, F., Carreel, F., Jenny, C., Lockhart, B. E. L. and Iskra Caruana, M. L. (2003). Identification of genetic markers linked to banana streak disease expression in inter-specific *Musa* hybrids. *Theoretical and Applied Genetics*. 106, 594–598.

Martin, G., Carreel, F., Coriton, O., Hervouet, C., Cardi, C., Derouault, P., Roques, D., Salmon, F., Rouard, M., Sardos, J., Labadie, K., Baurens, F. C. and D'Hont, A. (2017). Evolution of the banana genome (*Musa acuminata*) is impacted by large chromosomal translocations. *Molecular Biology and Evolution* 34(9), 2140–2152.

Maymon, M., Shpatz, U., Harel, Y. M., Levy, E., Elkind, G., Teverovsky, E., Gofman, R., Haberman, A., Zemorski, R., Ezra, N., Levi, Y., Or, G., Galpaz, N., Israeli, Y. and Freeman, S. (2018). First report of *Fusarium oxysporum* f. sp. cubense tropical race 4 causing Fusarium Wilt of cavendish bananas in Israel. *Plant Disease* 102(12), 2655.

Menendez, T. and Shepherd, K. (1975). Breeding new bananas. *World Crops* May/June 1975, 104–112.

Mware, B. O. (2016). Development of banana bunchy top virus resistance in bananas: RNAi approach. Queensland University of Technology. Available at: https://eprints.qut.edu.au/95736/ [Accessed 23 April 2020].

Noumbissié, G. B., Chabannes, M., Bakry, F., Ricci, S., Cardi, C., Njembele, J. -C., Yohoume, D., Tomekpe, K., Iskra-Caruana, M.-L., D'Hont, A. and Baurens, F. -C. (2016). Chromosome segregation in an allotetraploid banana hybrid (AAAB) suggests a translocation between the A and B genomes and results in eBSV-free offsprings. *Molecular Breeding* 36(4). Available at: http://link.springer.com/10.1007/s11032-016-0459-xdoi:10.1007/s11032-016-0459-x [Accessed 30 March 2016].

Novak, F. J. (1992). *Musa* (bananas and plantains). In: Hammerschlag, F. A. and Litz, R. E. (Eds). *Biotechnology of Perennial Fruit Crops*. CAB International, Wallingford. pp. 449–487.

Ollitrault, P., Dambier, D., Luro, F., Froelicher, Y. (2008). Ploidy manipulation for breeding seedless triploid citrus. In: Jules, J. (Ed.). *Plant Breeding Reviews, Volume 30*. Hoboken: J. Wiley [Etats-Unis], pp. 323–352. (Plant Breeding Reviews, 30).

Ortiz, R. and Vuylsteke, D. (1998a). 'Pita-14': a black Sigatoka-resistant tetraploid plantain hybrid with virus tolerance. *HorstScience* 33(2), 360–361.

Ortiz, R. and Vuylsteke, D. (1998b). Segregation of bunch orientation in plantain and banana hybrids. *Euphytica* 101(1), 79–82.

Perrier, X., Bakry, F., Carreel, F., Jenny, C., Horry, J.-P., Lebot, V. and Hippolyte, I. (2009). Combining biological approaches to shed light on the evolution of edible bananas. *Ethnobotany Research and Applications* 7, 199–216.

Perrier, X., Jenny, C., Bakry, F., Karamura, D., Kitavi, M., Dubois, C., Hervouet, C., Philippson, G. and De Langhe, E. (2018). East African diploid and triploid bananas: a genetic complex transported from South-East Asia. *Annals of Botany* 123(1), 19–36. Available at: https://academic.oup.com/aob/advance-article/doi/10.1093/aob/mcy156/5104470doi:10.1093/aob/mcy156 [Accessed 25 September 2018].

Perrier, X., de Langhe, E., Donohue, M., Lentfer, C. J., Vrydaghs, L., Bakry, F., Carreel, F., Hippolyte, I., Horry, J. P., Jenny, C., Lebot, V., Risterucci, A. M., Tomekpe, K., Doutrelepont, H., Ball, T., Manwaring, J., de Maret, P. and Denham, T. (2011). Multidisciplinary perspectives on banana (*Musa* spp.) domestication. *Proceedings of the National Academy of Sciences of the United States of America*. 108(28), 11311–11318.

Ploetz, R. C. (2015). Fusarium wilt of banana. *Phytopathology®* 105(12), 1512–1521.

Raboin, L. M., Carreel, F., Noyer, J.-L., Baurens, F.-C., Horry, J. P., Bakry, F., Tézenas du Montcel, H. T. D., Ganry, J., Lanaud, C. and Lagoda, P. J. L. (2005). Diploid ancestors of triploid export banana cultivars: molecular identification of 2n restitution gamete donors and n gamete donors. *Molecular Breeding* 16, 333–341.

Rieux, A., Bellaire, L. D. L. D., Zapater, M. F., Ravigne, V. and Carlier, J. (2013). Recent range expansion and agricultural landscape heterogeneity have only minimal effect on the spatial genetic structure of the plant pathogenic fungus Mycosphaerella fijiensis. *Heredity* 110(1), 29–38.

Rowe, P. and Rosales, F. (1989). banana and plantain improvement program. Genetic improvement. Informe Annual 1989. Honduran Foundation for Agricultural Research. pp. 15–23.

Rowe, P. and Rosales, F. (7 September 1992). Genetic improvement of bananas, plantains and cooking bananas in FHIA, Honduras. In: Ganry, J. (Ed.). *Breeding Bananas and Plantains*. Proceedings of the an International Symposium on Genetic Improvement of Bananas for Their Resistance to Diseases and Pests. CIRAD-FLHOR, Montpellier, France. pp. 243–266.

Rowe, P. and Rosales, F. (2000). A triploid parental line for breeding disease-resistant cooking bananas for East Africa. *Acta Horticulturae* 540, 169–175.

Sagi, L., Remy, S., Verelst, B., Swennen, R. and Panis, B. (1995). Genetic transformation in *Musa* species (banana). In: Bajaj, Y. P. S. (Ed.). *Biotechnology in Agriculture and Forestry*. Springer-Verlag Berl. pp, 214–227.

Sanford, J. C. (1983) Ploidy manipulations, chap. 7. In: Moore, J. N. and Janick, J. (Eds). *Methods in Fruit Breeding*. Purdue Univ. Press, West Lafayette, IN.

Shepherd, K. (1999). *Cytogenetics of the Genus Musa*. IPGRI, Montpellier.

Simmonds, N. W. (1962). *The Evolution of the Wild Bananas Tropical Science Series*. Longmans, Green & Co, London (GBR).

Simmonds, N. W. and Shepherd, K. (1955). The taxonomy and origins of the cultivated bananas. *Journal of the Linnean Society of London, Botany* 55, 302–312.

Stebbins, G. L. (1947). Types of polyploids: their classification and significance. In: Demerec, M. (Ed). *Advances in Genetics*. Academic Press. pp. 403–429. Available at: http://www.sciencedirect.com/science/article/pii/S0065266008604903doi:10.1016/S0065-2660(08)60490-3 [Accessed 23 April 2020].

Stover, R. H. and Simmonds, N. W. (1987). *Bananas* (3rd edn). New York: Longman Scientific & Technical.

Sybenga, J. (1975). *Meiotic Cconfigurations: A source of information for estimating genetic Parameters. Monographs on Theoretical and Applied Genetics*. Berlin Heidelberg: Springer-Verlag. Available at: https://link.springer.com/book/10.1007/978-3-642-8 0960-6 [Accessed 24 April 2020].

Sybenga, J. (1996). Chromosome pairing affinity and quadrivalent formation in polyploids: do segmental allopolyploids exist? *Genome* 39(6), 1176–1184.

Tumuhimbise, R., Barekye, A., Kubiriba, J., Akankwasa, K., Arinaitwe, I. K., Karamura, D. and Tushemereirwe, W. K. (2018). New high-yield cooking banana cultivars with multiple resistances to pests and diseases ('NAROBan1', 'NAROBan2', 'NAROBan3', and 'NAROBan4') released in Uganda. *HortScience* 53(9), 1387–1389.

Umber, M., Bonheur, L., Jenny, C. and Teycheney, P.-Y. (2011). Towards the elimination of infectious endogenous *Banana streak* virussequences from *Musa balbisiana*. In: 13èmes Rencontres de virologie végétale (RVV 2011), Aussois, France, 16–20 janvier 2011. SFP. s.l.: s.n., Résumé, 1 p. Rencontres de virologie végétale. 13, Aussois, France, 16 Janvier 2011/20 Janvier 2011. Available at: http://agritrop.cirad.fr /559618/.

Umber, M., Pichaut, J.-P., Farinas, B., Laboureau, N., Janzac, B., Plaisir-Pineau, K., Pressat, G., Baurens, F.-C., Chabannes, M., Duroy, P.-O., Guiougou, C., Delos, J.-M., Jenny, C., Iskra-Caruana, M.-L., Salmon, F. and Teycheney, P. (2016). Marker-assisted breeding of *Musa balbisiana* genitors devoid of infectious endogenous banana streak virus sequences. *Molecular Breeding* 36(6). Available at: http://link.springer.com/10.10 07/s11032-016-0493-8doi:10.1007/s11032-016-0493-8 [Accessed 7 June 2016].

Veilleux, R. (1985). Diploid and polyploid gametes in crop plants: mechanisms of formation and utilization in plant breeding. In: *Plant Breeding Reviews*. John Wiley & Sons, Ltd. pp. 253–288. Available at: https://onlinelibrary.wiley.com/doi/abs/10 .1002/9781118061008.ch6doi:10.1002/9781118061008.ch6 [Accessed 23 April 2020].

Zheng, S. J., García-Bastidas, F. A., Li, X., Zeng, L., Bai, T., Xu, S., Yin, K., Li, H., Fu, G., Yu, Y., Yang, L., Nguyen, H. C., Douangboupha, B., Khaing, A. A., Drenth, A., Seidl, M. F., Meijer, H. J. G. and Kema, G. H. J. (2018). New geographical insights of the latest expansion of *Fusarium oxysporum* f.sp. cubense tropical race 4 into the greater Mekong subregion. *Frontiers in Plant Science* 9, 457. Available at: https://www.fro ntiersin.org/articles/10.3389/fpls.2018.00457/fulldoi:10.3389/fpls.2018.00457 [Accessed 26 June 2020].

Published by Burleigh Dodds Science Publishing Limited, 2021.

Chapter 11

Overcoming the fertility crisis in bananas (*Musa* spp.)

Delphine Amah, International Institute of Tropical Agriculture (IITA), Nigeria; David W. Turner, The University of Western Australia, Australia; D. Jane Gibbs, Consultant, Australia; Allan Waniale, Makerere University and National Agricultural Research Laboratories, Uganda; Gil Gram, International Institute of Tropical Agriculture (IITA), Uganda and Katholieke University of Leuven (KUL), Belgium; and Rony Swennen, International Institute of Tropical Agriculture (IITA), Tanzania and Katholieke University of Leuven (KUL), Belgium

1 Introduction

Significant opportunities exist for improving edible bananas to tackle emerging productivity challenges due to the continuous spread of diseases and to improve nutritional/compositional qualities of the fruit through crop breeding. Hybridization, a means of generating genetic diversity for crop improvement is central to most crop-breeding schemes. However, the possibilities for hybridizations in bananas are hindered by several reproductive barriers. Edible bananas evolved from the seed-fertile diploid progenitors *Musa acuminata* and *Musa balbisiana,* but most cultivated varieties are triploids and (to a smaller extent) diploids, which are parthenocarpic and seedless (Simmonds, 1962; Brown et al., 2017). While seed production is important for breeding to ensure genetic recombination and provide viable progeny for recurrent selection, parthenocarpy and female sterility are paramount to ensure fruit edibility in

http://dx.doi.org/10.19103/AS.2020.0070.13

banana cultivars. Consequently, inter- or intra-ploidy crosses involving cultivated varieties, aiming towards their improvement, result in too few or no viable seeds leading to limited progeny hence constituting a 'fertility crisis'. This fertility crisis is unquestionably the biggest challenge for genetic improvement of banana.

Hybridizations in banana were first attempted in Trinidad in the 1920s aiming to introduce resistance genes from diploids to commercial triploids to create a Fusarium wilt resistant 'Gros Michel' (AAA) export dessert banana (Simmonds, 1966; Shepherd, 1974). From that time onwards banana breeding programs have mostly relied on conventional sexual hybridization involving crossing of triploid cultivars which, at most, have residual fertility with wild or cultivated fertile diploids to generate primary tetraploids. Subsequently fertile tetraploid plants are crossed with diploid hybrids to generate secondary triploids which are mostly sterile (Tenkouano and Swennen, 2004; Bakry et al., 2009). However, only a limited proportion of the wide diversity of *Musa* spp. has yielded viable hybrids in breeding programs while much of this diversity is still inaccessible for breeding. The majority of banana cultivars and hybrids have at most residual male and female fertility. Limited hybrid production also complicates the use of routine conventional breeding techniques and modern molecular techniques applicable to most seed producing crops. Banana breeders therefore tend to devote vastly more effort towards producing progeny rather than evaluating them (Tenkouano et al., 2011).

Several key reproductive steps in the life cycle of a flowering plant are responsible for adequate seed production. Pollination and subsequent fertilization of the ovules in the fruit are crucial processes in producing viable seeds. Knowledge of reproductive biology and breeding systems of bananas has progressed considerably although in-depth studies on flower development are limited to only a few banana cultivars (Fortescue and Turner, 2011) or *Musa* species (Kirchoff, 2017). Understanding the specific morphological, physiological and environmental factors surrounding the processes responsible for effective pollination, fertilization and viable seed production as well as identifying and overcoming possible barriers to these processes are essential for successful banana breeding.

This chapter discusses the key processes surrounding effective fertilization and viable seed production in relation to our current knowledge on banana-reproductive biology and the gaps to further our knowledge. Throughout the chapter we focus on possible limiting processes and provide insights on ways to overcome the fertility crisis to expand the possibilities for gene recombination through intra- and interspecific crosses.

2 Reproductive biology of banana

The reproductive biology of banana can be summarised as a series of processes (Fig. 1). The production of fertile seed can be derailed at numerous locations throughout the series due to genetic and environmental challenges. For a single gamete, failure at any 'decision' point (Fig. 1), in either the female or male series limits viable seed production. Some gametes will succeed and so the fusion of two fertile gametes may result in the production of a seed. The eventual seed set in an ovary will reflect to some degree the sequential success (or failure) of processes in the series. Furthermore, breeding schemes experience additional challenges at the level of seed and embryo germination which also needs to be addressed.

Historically the explanations for sterility emphasized meiotic errors arising from polyploidy, changes in chromosome structure, or inter (sub)species hybridity all leading to gametic sterility. However, research has revealed that there are more processes involved in sterility. For example, morphological errors occurring either before (multiple archesporia) or after gametogenesis (failure of embryo sac development) have been reported (Fortescue and Turner, 2005a). Subsequent errors such as pollen tube growth inhibition in the style or in the ovary, or failure of embryo development post fertilization have also been reported to contribute to infertility (Simmonds, 1962). Specific female sterility genes may also be involved (Sardos et al., 2016).

2.1 The inflorescence

Bananas can be reproduced vegetatively through suckers and sexually through botanical seeds. The banana inflorescence is a thyrse with each node of flowers best described as a cincinnus (Turner and Gibbs, 2018). Most bananas are monoecious (female and male flowers produced on the same plant) and dichogamous (pistils [carpels] and stamens mature at different times), resulting in temporal and spatial separation of the two sexes within an inflorescence. The hands of female flowers are formed before the hands of male flowers, with each hand on the same floral axis and each subtended by a bract. Anthesis for each hand of flowers commences with their exposure upon lift of the subtending bract. Flowering time depends on cultivar and environmental conditions.

In the early stage of inflorescence development, the female and male flowers are morphologically almost indistinguishable, with gynoecial (pistils) and androecial (stamens) parts of the flower developing in both, although with subtle differences in, the pattern and the size of organ development (Kirchoff, 2017). Ultimately, however, whether the mature flower is female or male is not due to small differences in the pattern of organogenesis but rather is concerned with the size and fertility of the gynoecium and androecium in the flowers (Kirchoff, 2017). By the time the developing inflorescence is about halfway up the pseudostem, the ovaries of female flowers have become longer than those of the male allowing female flowers to be distinguished from male

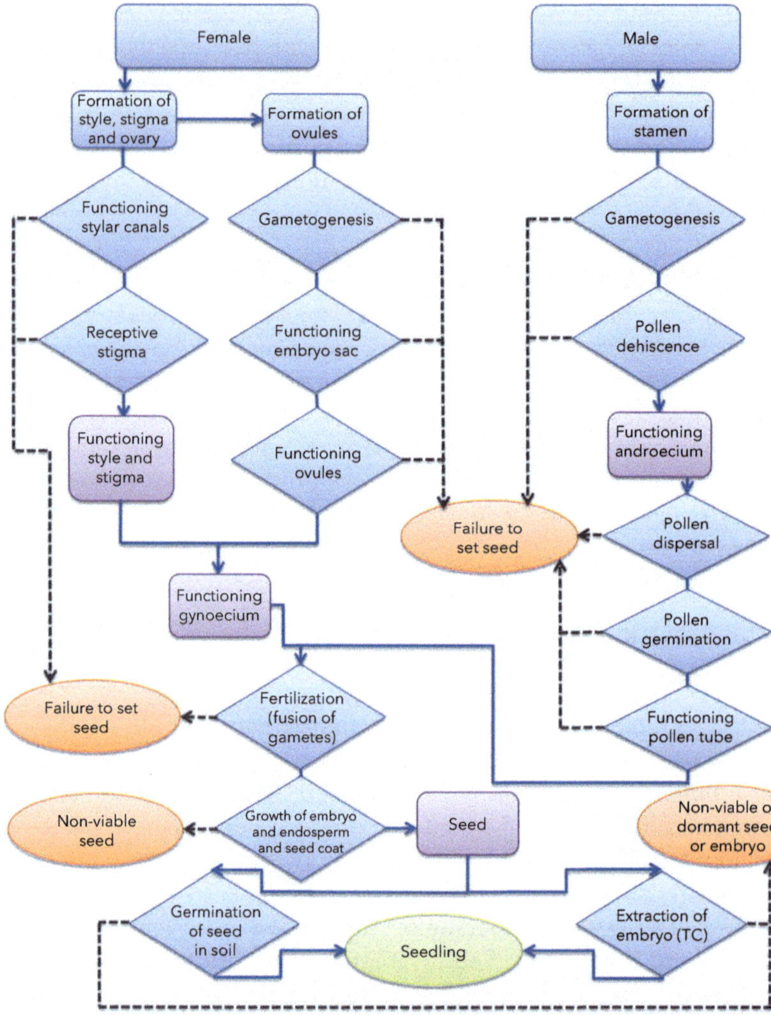

Figure 1 Key pre- and post-fertilization processes in reproductive biology, leading to success or failure in producing a seedling. This diagram is conceptual, focusing on the processes around the successful fusion of a female gamete with a male gamete in a banana ovary. It does not represent the sequential development of female and then male flowers that occurs on a single inflorescence. The sequence flows from flower formation to seedling. It starts with the flower type and moves from formation of organs (blue rectangles) by the solid lines to the functional components (purple rectangles) needed for seedling production (green oval). The diamonds in each part of the sequence represent 'decision' points between each stage where events may fail (broken lines) and lead to unproductive end-points indicated by orange-coloured ovals. The outcome of these 'decision' points for an individual gamete or for a population of gametes is discussed in the accompanying text (Sections 2-6). Development of the inflorescence is described in Section 2.1. A key step in female reproductive biology is meiosis, which occurs about 5-7 days before the stigma becomes receptive; for the oldest female

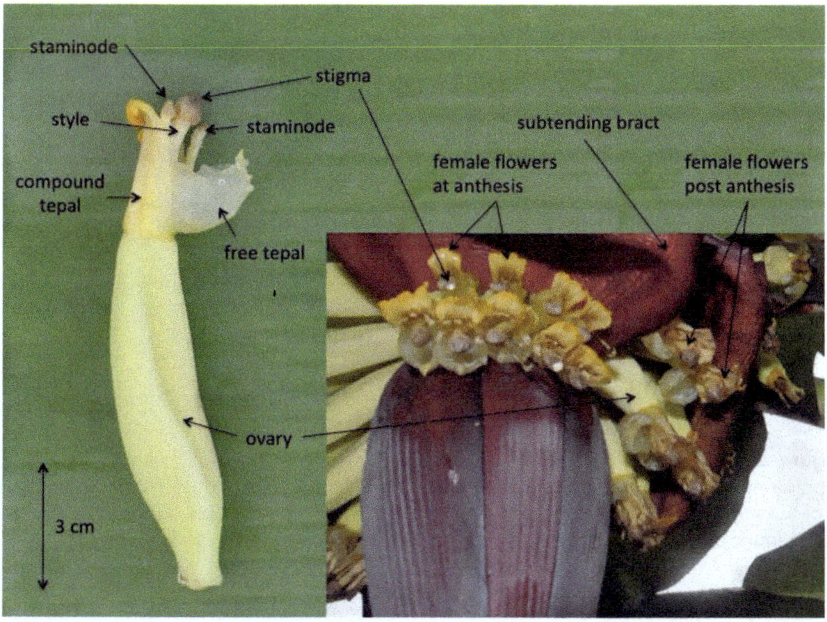

Figure 2 A female flower of *Musa* (AAB) plantain 'Bobby Tannap', at anthesis showing its position within a hand beneath a subtending bract (inset) and parts of the individual flower. The ovary forms the fruit, the style and stigma are massive and the androecium is reduced to staminodes. The compound and free tepals enclose the style, stigma and staminodes. The 3 cm scale applies to the single fruit.

flowers (Stover and Simmonds, 1987). Functionally-female flowers, that is, fruit-forming flowers, typically have a non-functioning androecium, with less-well to poorly developed or absent staminodes and no pollen (Turner and Gibbs, 2018) (Figs 2 and 3). In male flowers, the ovary is much smaller and non-fruit forming, the style and stigma slender and the androecium morphologically well developed (Figs 4 and 5).

In some *Musa* species, hermaphrodite fruit-forming flowers with pollen-producing stamens capable of self-fertilization are found (Simmonds, 1966;

Figure 1 (*Continued*)
flowers, this is about the time the bunch emerges from the pseudostem (Simmonds, 1962). Gametogenesis, including meiosis, is discussed in Section 2.2. Anthesis of female flowers occurs as or soon after the bract sub-tending the flower lifts (Section 3), at which time the stigma becomes receptive to pollen (Section 2.5.1). Following pollination and germination of pollen on the receptive stigma, it takes about 12 hours for the pollen tube to grow down the style; fertilization then occurs after which the zygote remains single-celled for about 4 weeks, although the endosperm begins development (Simmonds, 1962). For the style (2.3.1), pollen tube growth and pollen guidance, see Section 2.6. For pollen viability, see Section 2.5.2. For fertilization and seed set see Section 4 and for seed and embryo dormancy see Section 5.

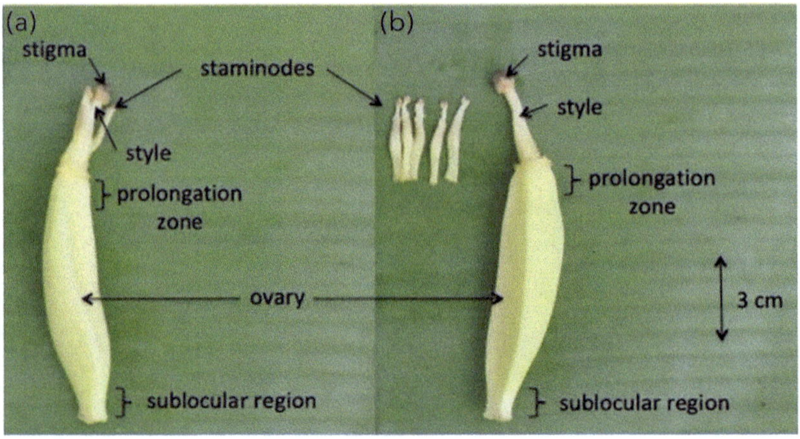

Figure 3 A female flower of *Musa* (AAB) plantain 'Bobby Tannap', at anthesis with the tepals removed. (a) The gynoecium is intact with ovary, basal sublocular region, distal prolongation zone and massive style and stigma. Staminodes intact. (b) As in (a), but the five staminodes removed to reveal style and stigma. Scale applies to (a) and (b).

Argent, 1976). There are no reports of the occurrence of such flowers in edible cultivars. Moreover, there is no evidence of hermaphrodite fruit-forming flowers in *M. balbisiana* (BB), one of the foundation species for some of the edible bananas (Argent, 1976). This opens *M. balbisiana* to out-crossing, in contrast to hermaphrodite flowers which can self-fertilise. Functional stamens in fruit-forming flowers of wild species might increase selfing and limit species hybridisation (Shepherd, 1999).

2.2 Gametogenesis

Gametogenesis is the process by which gametes required for sexual reproduction are produced. A central part of the process is meiosis, a cell division mechanism whereby genetic material is exchanged in cross-overs between arms of paired homologous chromosomes, which then separate to form gametes (Harrison et al., 2010). Megaspores and pollen mother cells (2n) undergo meiosis to produce haploid cells (n). In flowering plants the cells formed by meiosis are the egg cell in the embryo sac in the case of female gametes and pollen in the case of male gametes. Meiosis involves two cell divisions and has a number of phases between each division - prophase, metaphase, anaphase and telophase.

2.2.1 Why is gametogenesis an issue in banana?

In banana hybrids, whether or not viable gametes are produced during gametogenesis depends in part on ploidy, and the interaction between the

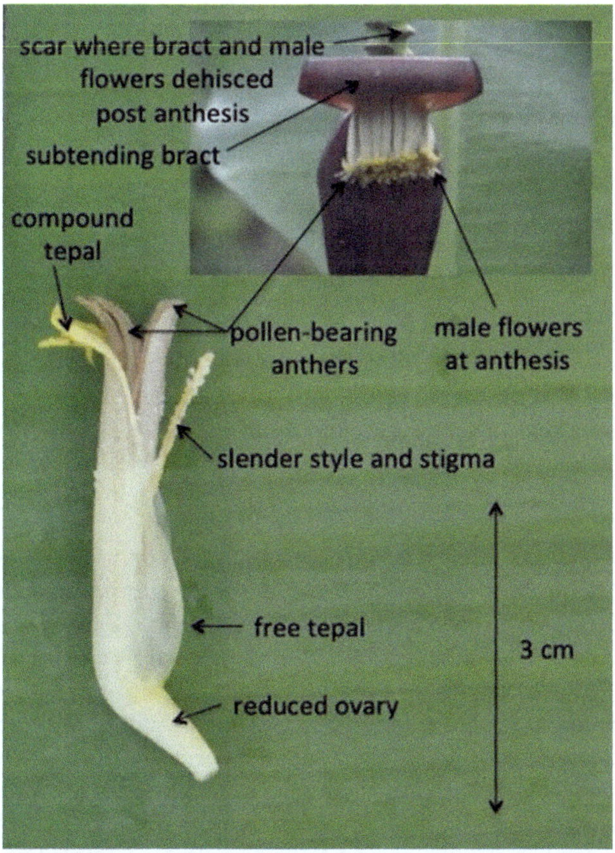

Figure 4 A male flower of *Musa acuminata* ssp. burmannica 'Calcutta 4' (AA), at anthesis showing its position within a hand (inset), subtending bract and scars on the male peduncle from which preceding hands of male flowers have abscised post-anthesis. The gynoecium is greatly reduced, especially the ovary (compare with Figure 2), with a slender style and stigma. The androecium is well developed with large stamens and pollen-producing anthers. The 3 cm scale applies to the single male flower.

genomes, especially those arising from different species, such as *M. acuminata* and *M. balbisiana*. In addition, chromosome structural changes can contribute non-viable gametes resulting in seedless edible diploids (Dodds, 1943). These influences are expressed at the level of meiosis in ways such that Dodds and Simmonds (1946) concluded that failure can occur at any point during meiosis. Here we follow the association between fertility and unpaired chromosomes (univalents) for illustration.

Musa species follow the normal pattern of meiosis, but many edible bananas (triploids and diploids) have been selected by people over time for female sterility and parthenocarpic development of pulp. For triploids the

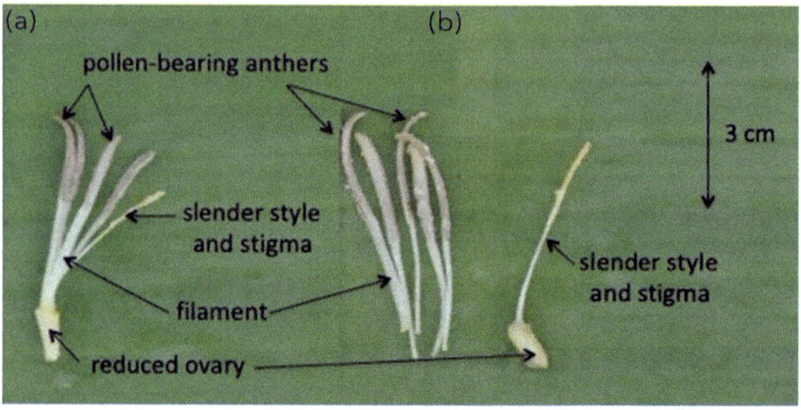

Figure 5 The male flower of *Musa acuminata* ssp. burmannica 'Calcutta 4' (AA), with tepals removed to show (a) pollen bearing anthers and (b) the slender style and stigma separated from the androecium.

matching of pairs of chromosomes is incomplete or unbalanced, leading to infertility (Griffiths et al., 2015). For pairing of chromosomes in triploids in meiosis there are two possibilities: formation of trivalents or, bivalents + univalents. Bivalents are paired homologous chromosomes and univalents are unpaired chromosomes. In banana breeding, when a triploid female and diploid male are crossed, the segregation of the progeny into different ploidy levels allows inferences to be made about the successful fusion of female and male gametes (Wilson, 1946a). Experiments in Jamaica, in which the same diploid male parent was used, produced progeny of different ploidy from three different cultivars within the AAA genome (Wilson, 1946a). The progeny of 'Gros Michel' were mainly tetraploids, for 'Red' triploids dominated and for 'Cavendish' no progeny were produced. 'Bluggoe', an ABB cultivar tended to produce diploid progeny. The complexity of female and male fertility and their interaction in *Musa* has been long known (Cheesman and Dodds, 1942). An additional dimension is the species to species interactions between *acuminata* (A) and *balbisiana* (B) genomes in triploids and diploids (Jeridi et al., 2011).

In banana, studies of chromosome behaviour during meiosis have a long history (Cheesman, 1932) continuing to this day (e.g. Shepherd, 1999; Adeleke et al., 2004), and recently in the seminal work of Jeridi et al. (2011) and that of Šimoníková et al. (2019). Lack of pairing (asynapsis) or subsequent separation of pairs (desynapsis) may contribute to sterility. Asynapsis produces univalents that can be counted (Wilson, 1946b; Adeleke et al., 2004). Simmonds and Dodds (1949) found failure of pairing to be rare in seeded diploid *Musa* spp. In some genotypes many chromosomes show asynapsis but in others only one chromosome pair may be affected (Wilson, 1946b). The broader negative

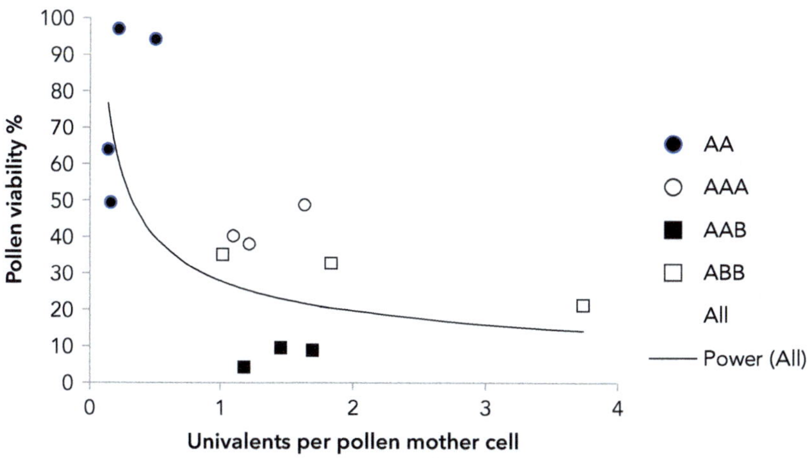

Figure 6 The relationship between the number of univalents per pollen mother cell and pollen viability in a range of diploid and triploid banana clones. Data from Adeleke et al. (2004).

association between the number of univalents per pollen mother cell, for example, and pollen viability (determined by staining techniques) (Fig. 6) had some distinct components. Adeleke et al. (2004) counted univalents in 4 diploids and 9 triploids. The triploids had 1-2 univalents per pollen mother cell with pollen viability of 1-50%, compared with diploids with less than 0.5 univalents and pollen viability ranging from 60% to 97% (Fig. 6). Variation within these broad groupings of pollen viability showed that it was associated with factors other than just the number of univalents.

Under tropical conditions usually one or two hands of male flowers appear each day. For a seeded diploid meiosis in the pollen mother cells (prophase to young tetrads) is completed within the equivalent time taken for one or two male hands to open, that is one to two days (Dodds and Simmonds, 1946). From tetrads to pollen grains takes another 13 hands, that is, 13 days. In contrast, for a diploid hybrid with double restitution, tetraploid pollen was rapidly produced over 4 days because the two normal cell divisions in meiosis did not occur. In this case the stages of meiosis from first metaphase to tetrads were not detected. Shepherd (1999) described this as 'meiosis avoidance'. The tetraploid pollen grains once formed, quickly degenerated and were not detected when the hand of male flowers opened some two weeks later.

Another process giving rise to seedless fruit may occur through meiotic restitution which is the production of 2n gametes (unreduced gametes) by incomplete cell division in meiosis. Male meiotic restitution can be induced through low temperatures as has been shown for *Arabidopsis* (Liu et al., 2018).

On the other hand, a possible role for temperature in the production of 2n female gametes in banana is associated with high temperatures, high solar radiation and low relative humidity (Ortiz and Vuylsteke, 1995a). The formation of unreduced 2n gametes under high temperatures was not considered a disadvantage but rather, a useful strategy for breeding tetraploid progeny from 2n gametes, avoiding some of the complexities of triploids (Ortiz and Vuylsteke, 1995a). Cool weather between archesporia (cells forming spores) and embryo sac development, the period during which meiosis occurs, was associated with ovule malformation (Fortescue and Turner, 2005a) suggesting that environmental conditions may be critical at this stage.

The proportion of mature embryo sacs found by Simmonds (1960) in three edible triploids varied from 2% to 36%. This low proportion arises from a combination of morphological abnormality, in the form of multiple archesporia, and from irregular meiosis during the development of female gametes. While we know that univalent and other meiotic errors are involved in sterility, predicting which ones might be important in any particular cross remains perilous.

2.3 The gynoecium

The gynoecium (carpel) consists of the ovary, style and stigma (Fig. 2). The female flower of banana is normally made up of three carpels that fuse to form the ovary (Fahn et al., 1961). In *Musa* tri-carpellary ovaries are normal and bi-carpellary ovaries are formed infrequently according to Ram et al. (1962) who examined inflorescences grown in the tropics. Fahn et al. (1961) observed bi-carpellary ovaries, which they regarded as abnormal, and were associated with bunches flowering during May (late spring) in the northern hemisphere Mediterranean environment of Israel.

Each carpel has, among other flower parts, a terminal style and these fuse to form what then appears as a single style. The points of fusion of the styles form grooves along the exterior length of the composite style. Near its distal section each style bifurcates and this creates the 6 lobed stigma (Ram et al., 1962) consisting of stigmatic papillae. The lobes on the bi-furcated styles may appear as single lobes leading to the view of a three lobed stigma for the flower (Fig. 3).

We were unable to find in the published literature systematic studies of the anatomy of the banana style using transverse sections from the stigma through to the ovary. There are however numerous studies that focus on the ovary of female flowers and their ovules (eg. White, 1928; Dodds, 1945; Fortescue and Turner, 2005b,c), presumably because the ovary develops parthenocarpically to provide the edible fruit (e.g. White, 1928; Ram et al., 1962). Moreover, the formation of embryo sacs in ovules within the ovaries is an issue for fertility in edible banana cultivars because the frequency of normal embryo sacs is low

(Dodds, 1945; Shepherd, 1954). Even so, there are hundreds of ovules in an ovary and even with low frequency of normal embryo sacs, more seeds should be set than is currently the case.

2.3.1 The style

The morphology and anatomy of the style have been deduced from three sources. First, from the diploid species, *Musa velutina* and *M. ornata* and a triploid edible banana *Musa* (AAB) Silk 'Go Sai Yung' (Kirchoff, 1992). Secondly, from closely related genera, such as *Ensete* (Inta et al., 2015) and thirdly from members of the Zingiberales order, to which the Musaceae family belongs. The selected genera were *Strelitzia* (Kronestedt and Walles, 1986; Kronestedt et al., 1986) and *Canna* (Glinos and Cocucci, 2011).

Flowers made up of three fused carpels are a feature of all eight families within the Zingiberales (Specht et al., 2012). While evolutionary development of other floral parts such as the stigma, anthers, petals and sepals (Figs 2 and 3) differ across families according to interaction with pollinators, the architecture of the ovary and style (Figs 2 and 3) has remained largely unchanged (Specht et al., 2012). The three-carpelled flowers have a composite style formed by the fusion of the three styles. These surround an inner tube that forms the stylar canal or transmitting tract. The opening, or mouth, of the canal near the top of the composite style is clearly seen in *Ensete* (Inta et al., 2015). The transmitting tissue may have a three- or six-armed 'star' shape in transverse section, such as within the style of *Strelitzia reginae* (Kronestedt et al., 1986) or an elongated channel as in *Canna* (Glinos and Cocucci, 2011). The cells of the styles adjacent to the transmitting tissue are secretory and provide the nutrients and directional compounds to guide pollen tubes towards the ovary. Near the base of the style the transmitting tissue divides (from the view point of a pollen tube travelling towards the ovary) into three, one channel going to each locule of the ovary (*Stretlizia reginae* - Kronestedt and Walles, 1986; *Musa velutina* - Kirchoff, 1992). This partitioning is more obvious in the prolongation region of the ovary.

In *Strelitzia* the shape of the transverse section of transmitting tissue differs depending on the presence of pollen tubes. Empty stylar canals have a star shape but the tissue is able to expand to accommodate the pollen tubes and then has a more rounded appearance. Not all pollen tubes that enter the transmitting tissue make it to the base of the style and into the locules, as also observed in *Musa* (Dodds, 1945; Uma and Arun, 2016). The transmitting tissue may therefore act as a point of selection for pollen tubes. However, since many pollen tubes arrive within the locules (Fig. 7) but few ovules are fertilised, the pathway and guidance (Section 2.6.1) of pollen tubes from the stylar-conducting tissues to the ovules is a focus for future investigations.

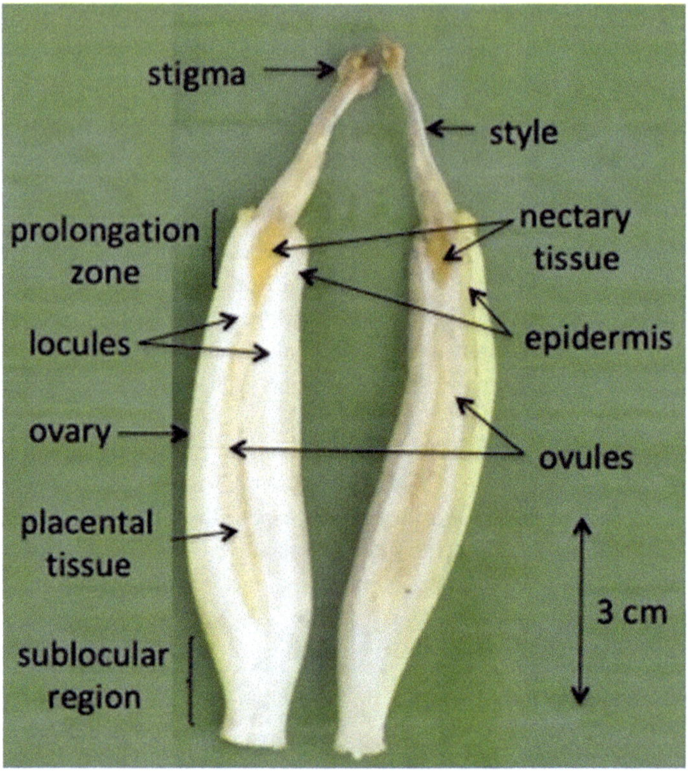

Figure 7 A longitudinal section of a female flower of *Musa* (AAB) plantain 'Bobby Tannap', showing distribution of ovules located centrally within locules and attached to the placental tissues. The placenta lies axially within the ovary between the sublocular region and the prolongation zone. The innermost section of the prolongation zone contains nectary tissue. The stylar canal is centred within the style and separates into three channels distal to the nectary tissue. Within the prolongation zone the channels are placed between the nectary tissue and the epidermis. Each channel then enters a locule.

2.4 The androecium

Musaceae are grouped taxonomically within the Zingiberales by their androecium, which is described as non-petaloid with 5–6 stamens (Simmonds, 1962). Typically, the male flower of *Musa* species has five polliniferous stamens (Kirchoff, 1992), three in an outer stamen whorl and two in the inner whorl around the gynoecium, with the posterior stamen of the inner whorl absent (Kirchoff, 2017). The filaments are free and stiff, and the anthers are bilobed and tetrasporangiate. They dehisce via vertical slits and release their pollen toward the centre of the flower. This occurs just after the bract subtending each flower cluster rises (Stover and Simmonds, 1987) (Figs 4 and 5). It is assumed this androecial structure and behaviour applies to edible bananas, except that in the male flowers of most edible bananas, the stamens bear little or no pollen.

2.4.1 Pollen production

For a banana genotype to be used as a male parent in breeding, it must produce ample, viable pollen that can germinate and fertilize the ovule. If it fails to meet these criteria, it cannot be utilized even if it is a source for desirable traits to address breeding objectives (Fortescue and Turner, 2011). The young anther of a male fertile, diploid, *M. acuminata* (AA) can contain up to 2200 pollen mother cells, with up to 9000 microspores at maturity (Dodds, 1943). In contrast, the anthers of many, but not all edible diploids lack pollen as a consequence of meiotic errors arising from chromosomal structural changes leading to gametic sterility (Simmonds, 1962). Microspore numbers in mature anthers of four edible diploids was one hundredth of the numbers of pollen mother cells in the young anthers, with many mature anthers completely empty (Dodds, 1943). Triploid bananas also have little or no pollen, as a consequence of uneven chromosome pairing during meiosis (Stover and Simmonds, 1987).

Scarce pollen production at anthesis was observed in 28 out of 77 *Musa* landrace accessions (Dumpe and Ortiz, 1996), with about 60% of these landraces having less than moderate pollen production. Of these, all the AAA, AAB, ABB and AAAB genotypes and 40% of the AA genotypes exhibited less than moderate pollen production, confirming that a lack of pollen production is not confined to triploids. On the other hand, all five BB genotypes had moderate to abundant pollen production. The quantity of pollen produced at anthesis was positively associated with pollen viability ($r = 0.61$, $P \leq 0.01$), measured as % pollen stainability using acetocarmine (Dumpe and Ortiz, 1996). Not only the quality, but the quantity of pollen produced at anthesis by the male parent was critical for seed production, at least in some *Musa* accessions. For example, the genotype, *Musa* (AAA or AA Dochez et al., 2006) 'Marau', which produced a moderate amount of pollen with about 25% viability was sufficiently male fertile in crossing with female-fertile *Musa* (AAB) plantain 'Mbi Egome' for successful seed set, while genotype *Musa* (AA) 'Tuu Gia', which produced scarce pollen but with 75% viability, was not despite crossing with several female-fertile partners (Dumpe and Ortiz, 1996). Whether this was due merely to the number of viable pollen grains available for pollination is unclear. In any event, in seeking male parents for breeding, it is clear that identifying those accessions that are sufficiently male fertile is key. Using the criteria of pollen production and pollen stainability applied by Dumpe and Ortiz (1996), 30 of the 77 landrace accessions tested might be considered sufficiently fertile for use as male parents.

2.5 Anthesis

At anthesis a flower is open and functional. For pollination to be successful the stigma must be receptive and pollen must be viable.

2.5.1 Stigma receptivity

A stigma is said to be receptive when it provides favourable conditions to support pollen germination. Banana stigmas are six-lobed and when receptive, they are constantly wet with a sticky papillate surface. The sticky substance is thought to come from extracellular secretions that contain carbohydrates, enzymes, phenolic compounds and amino acids. The extracellular secretions interact with pollen for species recognition before pollen can germinate. Unfortunately, this area has not received a lot of attention in *Musa* to further understand the specific constituents and underlying biochemical reactions. The timing and duration of stigma receptivity determines the effective pollination period that ensures efficient fertilization and seed production from compatible genotypes.

Studies to determine stigma receptivity require assessing morphological changes on the surface of the stigma, carrying out pollen germination on the stigma and staining or testing for enzymatic activity (He et al., 2017). A more direct assessment of pollen germination, fertilization and seed set following controlled pollinations would be more precise. Uma and Arun (2016) studied seed set on 218 accessions from various banana groups and noted a correlation between the sticky secretion that accompanies stigma receptivity and seed set. Genotypes of the AAA-Cavendish group which had the lowest seed set were associated with lowest stickiness while all other banana groups which set more seeds had moderate to high stickiness. Similarly, they noted that production of stigmatic secretions lasted from 1 day (in AAB-plantains) to 4 days (in ABB-cooking bananas) implying that pollinations can be carried out over a period of 1–3 days for successful seed set. Furthermore, they observed a higher level of enzyme (peroxidase and catalase) activity in seed fertile cultivars than sterile cultivars with implications for stigma function.

Other research revealed four stages of stigma receptivity in banana and plantains (IITA, 1993; Ssebuliba et al., 2009), distinguished on the basis of shape, colour, and amount of mucilaginous mass on the stigma surface. Stage I stigmas have a smooth surface characterized by a conspicuous ridge and yellow- to bone-white coloured mucilaginous mass. Stage II stigmas develop a somewhat rounded shape with a cream coloured corrugated surface as a result of abundant mucilage. Stage III stigmas have a dark brown-coloured surface with black patches and less mucilage while stage IV stigmas have completely black surfaces with a few brown patches with no visible mucilage. In *Musa* (AAB) plantain, stages II and III were reported to be the most receptive in cultivars 'Bobby Tannap' and 'Obino l'Ewai' (IITA, 1993). On the other hand, stage III was reported to be the most receptive in *Musa* (AAA) East African Highland Banana

(EAHB) Matooke cultivars 'Bitambi,' 'Enzirabahima,' 'Kazirakwe,' 'Nakayonga,' and 'Nfuuka' (Ssebuliba et al., 2009). In addition to different stages of stigma receptivity, genetic differences and compatibility between parents are thought to also play a role in seed set in bananas.

Exploring the application of pollen germination and tube growth stimulation substances on the stigma before pollination may provide an opportunity to enhance seed set in bananas.

2.5.2 Pollen viability

Depending on the pollen source used, the timing and method of application, seed yield from banana crosses varies between good to very poor seed with the latter having no embryos (Simmonds, 1952). Variation in seed set rates for various cross combinations calls for the need to evaluate pollen viability and genetic compatibility to identify potential fertile combinations of cultivars and hybrids for use in breeding. Pollen viability is crucial for the pollen-pistil interaction necessary to ensure fertilization and seed formation in flowering plants. To a limited extent, banana researchers have centred on pollen structure, quantification, viability, germination and growth of pollen tubes as components of breeding programs.

Pollen viability can be assessed using various methods such as staining with vital and nuclear dyes, *in vitro* and *in vivo* germination tests and direct assessment of final seed set (Heslop-Harrison et al., 1984). Ortiz et al. (1998) documented that pollen viability measured by staining did not correlate with seed set in triploid-diploid crosses implying that seed set in this case was affected by other factors in addition to pollen viability. Pollen viability assessment in banana has mainly involved staining with acetocarmine (Dumpe and Ortiz, 1996; Ssebuliba et al., 2008; Goigoux et al., 2013; Oselebe et al., 2014), Alexander (Fortescue and Turner, 2004) and 2,3,5-triphenyl tetrazolium chloride (Soares et al., 2008, 2015). *In vitro* germination tests using a sucrose solution and sucrose-based media has also been reported for bananas (Nyine and Pillay, 2007; Soares et al., 2015). Staining generally reported relatively higher viability values (possibly overestimating fertility) than germination tests. This discrepancy may be linked to inefficiencies of the pollen germination media to simulate stigma and pistil conditions, necessitating further optimization of germination media. This is clearly demonstrated by Nyine and Pillay (2007) who recorded a two-fold increase in pollen germination rates on diluted banana nectar compared with a sucrose solution.

In a broader context, pollen viability is affected by several factors including genotype, flowering time, developmental stage of the flower, temperature and relative humidity. Understanding the conditions that affect pollen viability will influence the practical aspects of optimizing pollen quantity and quality, collection and use in breeding schemes. However, most studies on pollen viability have focused on the effect of genotype (Dumpe and Ortiz, 1996; Fortescue and Turner, 2004; Ssebuliba et al., 2008; Soares et al., 2008). Collectively, these studies indicate that banana cultivars and hybrids differ widely in their capacities to produce viable pollen either within or among genome groups. *Musa* species typically produce abundant viable pollen while cultivated diploids as well as diploid hybrids produce relatively more viable pollen than tetraploids and triploids. Low pollen production in triploids usually precludes their use as male parents.

Studies of environmental effects on pollen viability in bananas are scarce. Ortiz et al. (1998) observed a seasonal variation in pollen stainability for, *M. acuminata* ssp. *burmannica* 'Calcutta 4', *Musa* AA 'Galeo' and 'Pisang lilin' and a plantain-banana hybrid TMP2x1297-3 examined for a year in Nigeria. High pollen stainability was positively associated with high solar radiation, high temperature and high evaporation and negatively associated with high rainfall and high relative humidity. Soares et al. (2015) assessed pollen viability at different times of the day and two seasons of the year (summer and winter) in Brazil. They observed that *in vitro* germination rates and % viability from staining tests were highest at 8 am and lowest at 4 pm, and that pollen viability was higher in the summer than in the winter. Uma and Arun (2016) evaluated pollen viability following exposure to a range of temperatures (25–45°C for 1 hour) and noted a reduced germination rate (43–72%) for pollen exposed to temperatures over 35°C.

Nevertheless, pollen viability assessment is only indicative of fertility. Hence further studies should be targeted towards a functional analysis of *in vivo* pollen germination, pollen tube development and fertilization as a true reflection of pollen fertility for desirable genotypes.

2.6 Pistil: pollen interaction

To reach an ovule within an ovary of banana a pollen grain and the tube arising from it has a journey consisting of several steps. The pollen needs to arrive at the stigma, germinate and produce a pollen tube. About an hour is needed for pollen to germinate and penetrate the stigma (Shepherd et al., 1987). The pollen tube then travels across/through the stigmatic tissue, into the transmitting tract of the style. After travelling down the style and through the prolongation section it enters the upper section of one of the locules of the ovary. Once in a locule it needs to locate an ovule and enter its micropyle. The stylar part of this journey (30–50 mm) takes about 12 h (Dodds, 1945). The whole journey may cover a distance from 50 mm to 100 mm or more and take about 18 h. These values represent rates of elongation of the pollen tube of 2.5–5.6 mm/h which are similar to the range of rates recorded by Shepherd et al. (1987) in Brazil. The sequential steps in this journey mean that stimulation or inhibition of the pollen or pollen tube can be exercised at a number of points (Fig. 7).

The style is essential for successful conventional breeding, but opinions about its significance in influencing female fertility vary. Dodds (1945) working in Trinidad, examined style function and found no inhibition of pollen tube growth in the styles of edible bananas. More pollen tubes arrived in the locules than were needed to fertilise the number of embryo sacs present. Even so, Dodds (1945) noted that most embryo-sacs do not receive pollen tubes. Shepherd et al. (1987) commented on good pollen tube growth in the styles of Cavendish (AAA) cultivars in Brazil, although these cultivars did not produce seeds when pollinated. Further experiments by Shepherd et al. (1987) suggested some difference between cultivars of the Pome group (AAB) in the rate of movement of pollen tubes along the style and in the number of pollen tubes within the style.

Uma and Arun (2016) working in Southern India, observed a lack of germination of pollen from *M. acuminata* ssp. *burmannica* 'Calcutta 4' and *Musa* (AA) 'Pisang lilin' on the stigmatic surface of *Musa* (AAA) Cavendish 'Grand Naine' but ample germination on the stigma of *Musa* (AAB) Silk 'Rasthali'. These observations on members of the Cavendish subgroup differ from those of Shepherd et al. (1987) who found no limitations to pollen germination on stigma of Cavendish bananas or movement through the style. Uma and Arun (2016) observed thinning of the pollen tube population within the style, thus reducing the number of pollen tubes likely to enter the ovary - a different observation from that of Dodds (1945). Soares et al. (2014) working in Brazil, observed necrosis in the prolongation region of the ovary in triploid cultivars (AAA and AAB), whether pollinated or not. In diploid cultivars no necrosis was observed. They thought necrosis at this location, between the style and the ovules, may interfere with pollen tube transfer at the point of entry into the locules of the ovary and consequently reduce fertilisation and seed fertility. Ssebuliba et al. (2005) working with *Musa* (AAA) EAHB found a negative association between seed set and style length. However, the residuals were very high and they concluded that style length, among other features of pistil anatomy were unlikely to be large contributors to seed fertility.

At present, observations differ about the importance of the stigmatic and stylar tissues in controlling the number of pollen tubes likely to arrive within locules of the ovary and so it is not possible to make general conclusions about points of control. Knowledge of the anatomy of the pathway and identification of the points of control, if any, are needed for a broader picture of pollen/pistil interactions in *Musa*.

2.6.1 Pollen guidance

To form a seed following pollination, a pollen tube must move through the stigma and style, enter the locule, locate an ovule, enter the micropyle and fertilise the embryo sac. An ovule needs only one pollen tube for fertilisation and so a mechanism for rejection of other pollen tubes is required once pollination has occurred. Mizuta and Higashiyama (2018) provide a very useful recent summary of the known mechanisms in flowering plants with emphasis on *Arabidopsis*. They refer to pre-ovular guidance from the stigma to the ovary and to ovular guidance that allows a single pollen tube to locate a single ovule within the ovary.

As far as is known, no studies of pollen guidance *per se* have been made in *Musa*, although failure of the guidance system could well contribute to poor seed set. There are a number of observations in the pollination of *Musa* flowers that seem consistent with a non-functional or partly functional pollen guidance system. For example, Shepherd (1960) observed two types of failure of pollen tubes moving towards the ovary. The first was failure to cross the stigma and was related to the perceived maturity of the style. The second was the failure of pollen tubes, once in the style, to reach the ovary. There were differences between female genotypes in the expression of these failures in pollen tube growth.

2.6.2 Distribution of seeds within a fruit

Shepherd (1954) attributed the high concentration of seeds at the stylar end of the ovary of *Musa* (AAA) 'Gros Michel' to pollen tube 'growth factors'. In that case, 86% of the seeds in the fruit formed in the quarter nearest the stylar end. Shepherd observed considerable variation in the rate of pollen tube growth and concluded that when compared with the growth of pollen tubes in *Musa* species, pollen tube growth in styles of 'Gros Michel' was 'erratic'. However, more than half of the pollen tubes within the style grew sufficiently to reach the ovary.

In *Musa* species with a pollen guidance system intact the distribution of seeds is expected to be uniform along the length of the ovary, even in genotypes with long fruit such as *M. acuminata* ssp. *banksii* (Simmonds, 1953). Ovaries may be 9–15 cm in length in edible *Musa* (AAA) EAHB (Ssebuliba et al., 2005). The concentration of seeds at the stylar end in this sub-group may have a number of causes, including a non-functional pollen guidance system. In the absence of a guidance system, the growth of pollen tubes, all of which enter the locules from the stylar end, is likely to approximate a distribution where, with lack of direction, pollen tubes concentrate at the stylar end and few reach the base of the ovary. This is likely to be reflected in an exponential decline in seed number from the stylar end of the ovary to the base.

2.6.3 Quantifying seed distribution within fruit

Shepherd (1960) presented data on the proportional distribution of seeds through four quarters of the fruit for a number of crosses of female diploid, triploid and tetraploid clones with pollen from either *M. acuminata* ssp. *malaccensis* or ssp. *burmannica*. Seed set tended to be greatest at the stylar end of the ovary (Fig. 8). Here the cumulative distribution of seed from the stylar to the basal end of the ovary is compared against a theoretical uniform distribution of seed to estimate the functioning of a pollen tube guidance system. We assume that competent ovules are uniformly distributed along the ovary and that an abundance of pollen tubes enter the locule.

The function used to compare distributions is:

$$P = 1 - \exp(-kQ) \tag{1}$$

Where P is the cumulative proportion of seeds from the stylar end, Q is the quarter numbered from the stylar end and k is a coefficient of seed distribution within the ovary. A high value of k indicated a strong bias of seed distribution

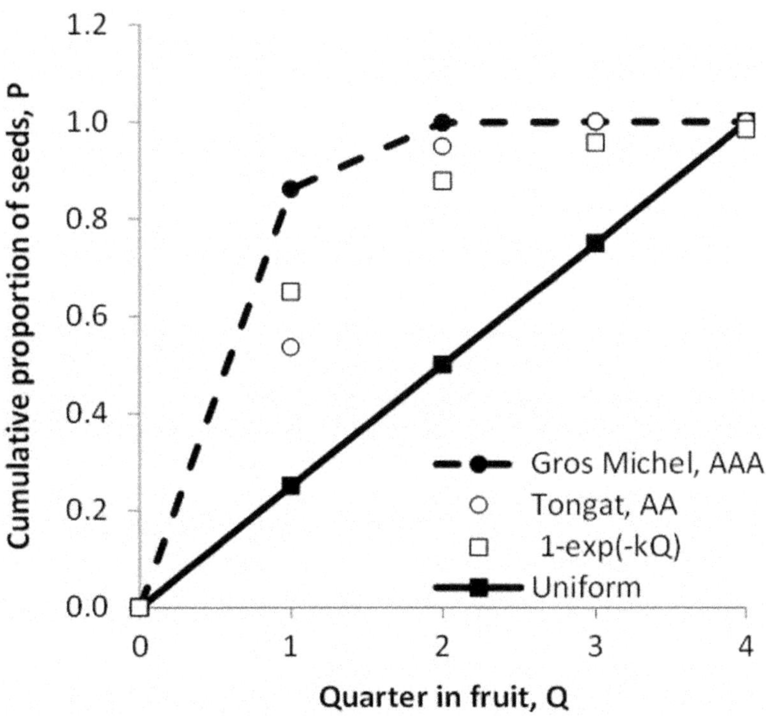

Figure 8 The cumulative proportional distribution of seed set in quarters from the stylar (1) to the basal end (4) of ovaries of 'Gros Michel' AAA (dashed line) and 'Tongat' AA clones. In the legend, Uniform (solid line) is a theoretical uniform distribution of seed throughout the quarters of the ovary. Points above the solid line show bias towards the stylar end of the fruit and below the solid line bias towards the basal end. The open squares are the fit for the equation 1-exp(-kQ) for 'Tongat' with k = 1.05.

towards the stylar end as shown for 'Gros Michel' (AAA) where k = 1.9 (Fig. 8). For 'Tongat' (AA), the concentration at the stylar end is reduced compared with 'Gros Michel' and k = 1.05.

3 Bract opening in banana

The opening of bracts that subtend hands of flowers on an inflorescence of bananas allows pollination of female flowers in *Musa* species and access for hand pollination in edible bananas. Matching pollination with maximum receptivity of the stigma increases the likelihood of success in generating seed for banana breeding, especially in edible cultivars that have low fertility (Simmonds, 1960). Time of stigma receptivity is short, less than a day and possibly only a few hours (IITA, 1993; Ssebuliba et al., 2006). In the practice of banana breeding, pollen is collected at dawn and female flowers pollinated during early to mid-morning (Shepherd, 1954; Rowe and Richardson, 1975; Swennen and Vuylsteke, 1993; Vuylsteke et al., 1993a,b). If flowers become available for pollination during the night, they may well have reduced receptivity by the time daylight arrives. Knowledge of the diurnal behaviour of bract lifting and curling and subsequent appearance of female and male flowers could lead to improved breeding efficiency.

3.1 The process of bract opening

In cultivars of *Musa* AA and AAA, bracts that subtend nodes of female flowers undergo two main events that expose flowers to pollinators at anthesis. The bract lifts away from the floral bud with the base of the bract as the fulcrum. Then, sometimes concurrently, there is an involute curling of the bract, starting from the tip, thereby exposing the flowers to pollinators. Once 'hands' of female flowers are opened the bracts dehisce. Unlike the bracts of female flowers, which always dehisce, for bracts subtending the transitional and male hands dehiscence depends on the cultivar, with bracts of male flowers in some cultivars drying on the peduncle. Bracts usually lift, curl and dehisce in an acropetal sequence, first exposing the hands of female flowers and then the hands of male flowers on the inflorescence.

3.2 Diurnal behaviour of bract opening

In Kunming, China (21°N), during October and November (autumn/early winter) most bracts of hands of female flowers of *Musa itinerans* open at night with a second major bract opening late in the afternoon (Liu et al., 2002) (Fig. 9).

Flowers of *M. itinerans* produced nectar in the 2–5 h after each set of bract openings and again 12 h later. At these times flowers were visited by either bats

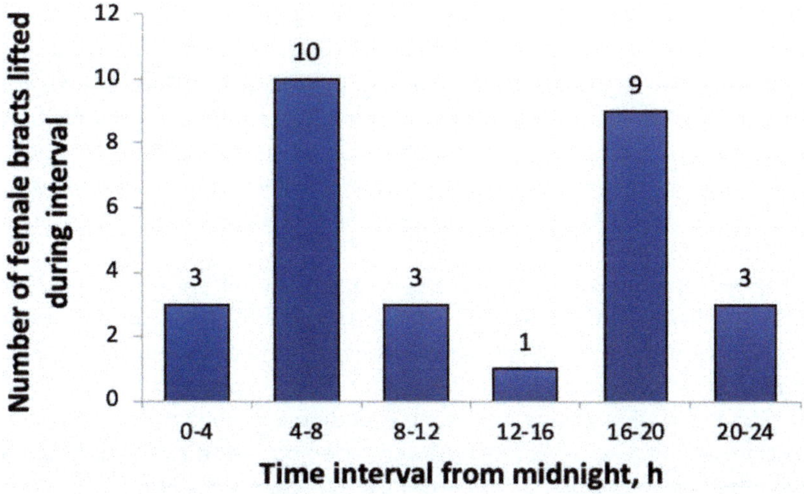

Figure 9 The number of female hands opening in each interval from midnight in *Musa itinerans* at Kunming, China. Data are for 18 inflorescences from Liu et al. (2002). Sunrise was at 7.15 am and day-length was 11.2 h.

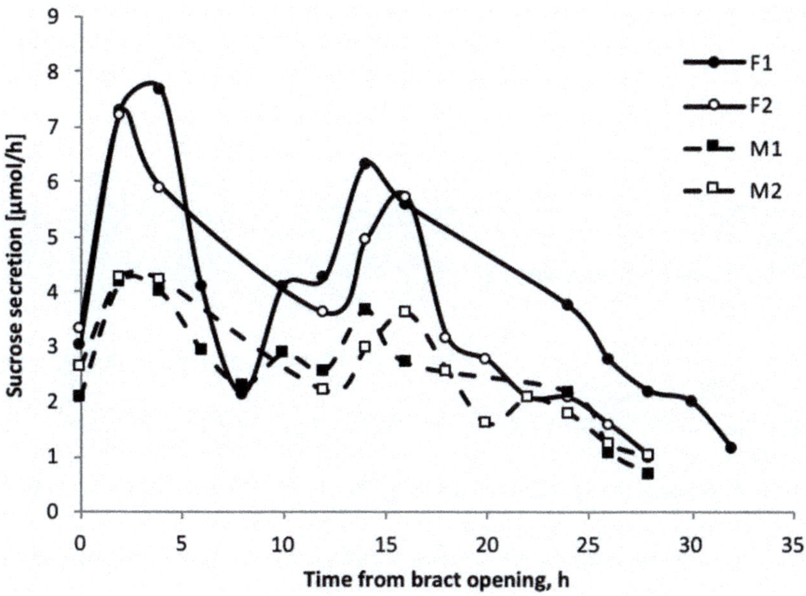

Figure 10 The secretion of sucrose in nectar from flowers of *Musa itinerans* at Kunming, China, in the hours from bract opening. F1 and M1 refer to female and male flowers that commence opening in the early morning. F2 and M2 refer to female and male flowers that commence opening in the late evening. Data from Liu et al. (2002).

(evening) or birds (morning) (Liu et al., 2002). A high correlation between the volume of nectar, particularly the amount of sugars secreted in nectar, and the receptivity of the stigma to pollination might be expected (Farkas and Orosz-Kovacs, 2003; Roy et al., 2017). Two peaks of sucrose secretion after the bracts opened were identified using the data of Liu et al. (2002) (Fig. 10). Sucrose was available to pollinators 2–5 h after bract opening, whether they were bats (night) or birds (day). A second but broader peak of sucrose secretion occurred about 12 h later.

If bract opening of *M. itinerans* is characteristic of bananas, then bracts of edible cultivars also are likely to open during darkness, particularly around sunset and before sunrise. Maximum receptivity of the stigma, if linked to sugar secretion in nectar, would follow 2–5 h later. This provides some rationale for focussing pollination efforts in banana breeding schemes to the early hours after sunrise, but overlooks those flowers opening in the evening, which are likely to be past their optimum receptivity the following morning.

3.3 Monitoring bract opening in the glasshouse and field

Diurnal behaviour of bracts was monitored in movies of bract opening on (i) individual bunches of two *Musa* (AAA) Cavendish cultivars, 'Grand Nain' and 'Williams' in a glasshouse and *Musa basjoo* in a garden in Belgium (51°N) over several weeks from June to September 2015 (i.e. summer), and (ii) four *Musa* (AA) 'Mchare' cultivars, 'Akondro Mainty', 'Mchare Mlelembo', 'Kisukari Mchare' and 'Huti White', in the field at Arusha, Tanzania (3°S) in December 2015 and January 2016 (i.e. dry season). The analysis of the movies focussed on the beginning of involute curling and first visible flower and the lifting of the bracts.

The movies are accessible at:

ftp://ftp.musabase.org/video/flowering/Movie_1_Grand_Nain_Belgium_3_Jun_2015.avi;
ftp://ftp.musabase.org/video/flowering/Movie_2_Williams_Belgium_7_Jul_2015.avi;
ftp://ftp.musabase.org/video/flowering/Movie_3_Williams_Belgium_28_Sep_2015.avi;
ftp://ftp.musabase.org/video/flowering/Movie_4_M_basjoo_Belgium_26_Jul_2015.avi);
ftp://ftp.musabase.org/video/flowering/Movie_5_Akondro_Mainty_Tanzania_21_Dec_2015.avi;
ftp://ftp.musabase.org/video/flowering/Movie_6_Mchare_Mlelembo_Tanzania_12_Jan_2016.avi;

ftp://ftp.musabase.org/video/flowering/Movie_7_Kisukari_Mchare_Tanzania_1_Dec_2015.avi;

ftp://ftp.musabase.org/video/flowering/Movie_8_Huti_White_Tanzania_25_Jan_2016.avi.

3.3.1 A link with sunrise and sunset?

The pattern of bract curl for *Musa* AA Mchare cultivars at Arusha, Tanzania, (Fig. 11) was consistent with the bimodal distribution observed by Liu et al. (2002) for *Musa itinerans* at Kunming, China. This was not the case for bracts of Cavendish cultivars in the glasshouse (KUL) or *M. basjoo* in the garden in Belgium (Fig. 12). While there was no significant difference in the response to sunrise in the population of bracts of *Musa* AAA Cavendish and *M. basjoo* (Table 1) these differed from the field in Arusha in that during long days all bracts curled during the day and there was no clear bimodal distribution of bract curling.

Further analysis suggested that bract curling was linked to a fixed time after sunrise, with bracts responding to the change of light at dawn and beginning to curl between 9 h and 12 h later irrespective of where they grew at latitudes 3°S, 21°N or 51°N (Fig. 13). Thus, whether or not the bracts curl in the day or the night depends on the day-length of the location in which bunches grew. The short day (11.5 h) in the tropics meant that bracts began involute curling mostly during the evening and night, but in Belgium, during daylight in the

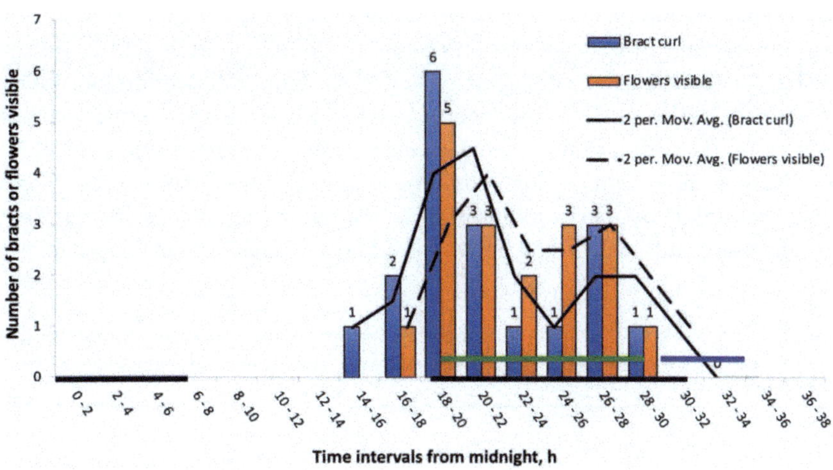

Figure 11 The horizontal green line indicates time over which moths visited open flowers (19–29h). Horizontal blue line indicates time of intense visitation of flowers by bees, wasps, and flies (30–34h). Time intervals are from midnight on day 1.

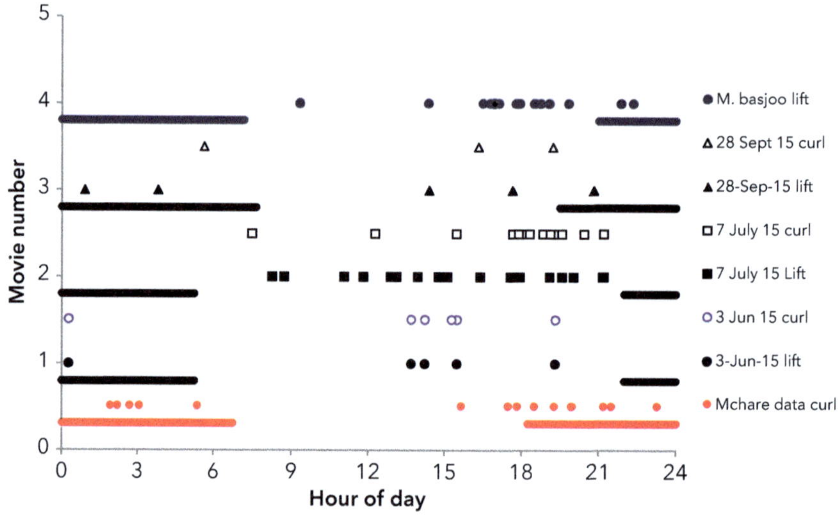

Figure 12 The time of day when bracts began to either lift from the floral bud or curl from the tip for three occasions (movie number) in a glasshouse in Belgium. Movie 1 - male hands of 'Grand Nain', movies 2 and 3 - 'Williams' bunches. Movie 2 and 3 included female and male hands. The beginning of bract lift is shown in closed black symbols for movies 1, 2 and 3 and in dark blue symbols for *M. basjoo* bunches in the garden in Belgium. The beginning of bract curl is shown in open symbols for movies 1, 2 and 3, and for the 'Mchare' bunches in the field in Arusha. Horizontal black, red or blue lines show night time for each location and time of year. Hours are from midnight.

long days of the summer (17 h) (Fig. 13). This phenomenon is consistent with a light signal received at dawn. However, a single light signal received at sunrise does not explain why in the field in Kunming, China (Liu et al., 2002), and in Arusha, Tanzania, bract opening or curling had a bimodal distribution during the hours of darkness. This may mean that two light signals are operating, one at sunrise and another at sunset. In each case curling of bracts begins about 9 h after the change in light regime, either from dark to light or the reverse. A light signal generated by a change from dark to light or the reverse seems likely, perhaps associated with a link between red light and ABA signalling (Lin and

Table 1 A comparison of the time from sunrise to bract lift in populations of bracts in the glasshouse and garden in Belgium. Mean values are not significantly different as determined by Student t-test (P = 0.129)

Time from sunrise to beginning of bract lift (h)	Mean	Standard error	Sample size (n)
Glasshouse, *Musa* AAA Cavendish cvs	12.29	0.88	22
Garden, *M. basjoo*	10.51	0.70	17

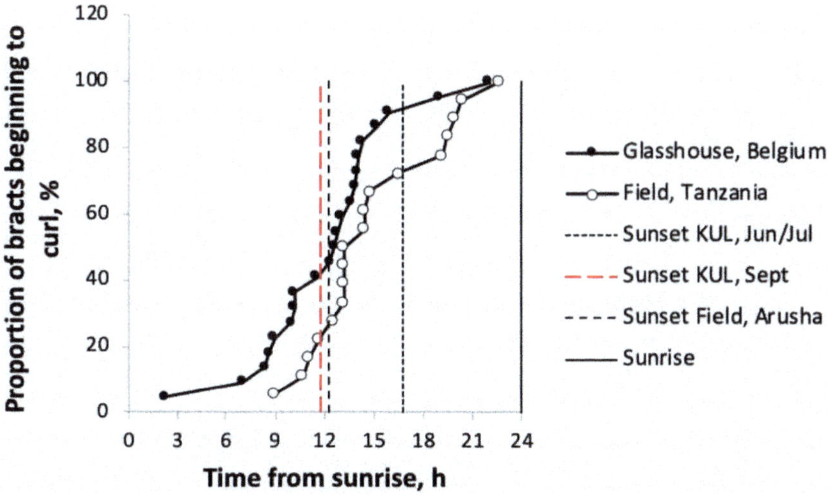

Figure 13 The cumulative start of bract involute curling for the glasshouse and field data in relation to sunrise (time zero). The solid symbols are for bract curling of the glasshouse data. The open symbols are for bract curling of the field data. Three vertical lines indicate time of sunset for the glasshouse movies (Belgium), in June or July (dotted black) and in September (hatched black), and for the field in Arusha (red). A solid black vertical line indicates the following sunrise.

Tang, 2014) as ABA is often linked to leaf (a bract is a modified leaf) abscission (Patterson, 2001).

3.3.2 Season-dependent seed set in plantain

If pollination procedures in banana breeding are guided by clock time throughout the year (e.g. 6.30 am to 7.30 am), then the relationship between the time of pollination and sunrise (or the previous sunset) will change throughout the year, according to photoperiod. They are likely to be closest twice each year when the day-length is 12 h. Swennen et al. (1991) observed two seasonal peaks of seed production, repeated over two years, in plantain cultivars at Onne, Nigeria (Lat 5°N), one in February and another in September to October when pollination began about 1.0–1.5 h after sunrise. In February, 60–70 seeds were set per bunch while in September to October 20–30 seeds were set per bunch. At other times of the year there were greater displacements between the start of pollination and the time of sunrise and seed set varied from 2 to 15 seeds per bunch (Swennen et al., 1991). There is a need to explore the relationship between receptivity, hours after sunrise and clock time to see whether this is associated with any change in the efficiency of pollination.

3.3.3 Bract lift, curl, water relations and energy supply

Observations on bract lift and curl in the glasshouse where three bracts lifted over three days, but curling of all three bracts occurred within an hour (Fig. 12, movie 2), suggested that while a signal at sunrise or sunset may result in bract curling 9–12 h later, bracts may not curl unless another condition is also met.

Flowering and pollination in plants requires energy to expose flowers to pollinators, excrete soluble carbohydrates, support pollen germination and movement. For lifting and curling of the bracts, energy is needed to move solutes from one part of the tissue to another so that there is a change of water distribution and cell turgor, providing a cantilever effect as bracts lift and curl. Exploring these relationships might shed some light on the variability in the sequence of bract lift and the readiness of the flower for pollination.

3.3.4 Bract opening and insect visitation

Rutikanga et al. (2016) observed nectar production in *Musa* AAA EAH bananas, noting that the number of visits by insects was not associated with the amount of nectar, but with the presence of nectar. Given the presence of insects at the opening bracts in the movies taken in the field, nectar production in 'Mchare' cultivars is likely to be the attractant. Insects at the inflorescence may be associated with maximum receptivity of the stigma to pollen. For bunches in the field at Arusha, this could mean stigma were receptive throughout the night and in the first hour or so after dawn. Understanding the relationship between which hands of flowers were visited by insects at which times and bract lifting and curling might help to identify times of maximum receptivity.

Across genotypes, bract behaviour was consistent with bract opening in banana being an energy-dependent process with at least two requirements. First, sufficient energy to move solutes between tissues creating changes in water distribution to promote gradients of turgor required for the mechanical aspects of the cantilever action of lifting and curling. Secondly, a signal initiated by a change in light regime, dark to light (sunrise) or light to dark (sunset), perhaps associated with a link between red light and ABA signalling (Lin and Tang, 2014), allowing flower opening to occur at the same time as pollinators visit. These possible mechanisms have implications for manipulating bract opening to coordinate pollination and stigma receptivity in banana breeding. Clearly, further experimentation is needed to test the hypotheses arising from these observations, specifically: (1) whether bract curling is initiated in part by a change of light regime – from dark to light (sunrise) or light to dark (sunset), (2) whether bract curling has two separate requirements, a light signal and a supply of energy to undertake the mechanics of curling and (3) whether maximum receptivity of the stigma is associated with nectar production and visiting insects may be indicative.

3.4 Pollination in Musa species before bract opening

Hermaphrodite flowers, which occur in some *Musa* species, have functional female and male gametes and can self-fertilise before the subtending bract opens (Simmonds, 1966). This feature influences the amount of out-crossing between *Musa* species and subspecies. For example, Argent (1976) discussed hybrids of *M. schizocarpa* and *M. acuminata*. ssp. *banksii* observed in the field in Papua New Guinea. *M. schizocarpa* consistently had hermaphrodite flowers in basal hands but in *M. acuminata* ssp. *banksii* this character showed variability. He concluded that this difference meant *M. acuminata* ssp. *banksii* was the most likely female parent of observed hybrids. Shepherd (1999) described experimental crosses between *M. schizocarpa* and *M. acuminata* ssp. *banksii* in Jamaica. While the inheritance of hermaphrodite flowers was not recorded, in practice there was no need for emasculation of flowers for successful pollination to produce backcross seeds. In this case Shepherd (1999) assumed that in the hybrids anthers were either absent or rare.

Häkkinen and Hong (2007) recorded hermaphrodite flowers on *M. yunnanensis* and *M. acuminata* ssp. *chinensis* in Yunnan, China, with no evidence of hybrids nearby. This supported the view that the plants of each species had self-fertilised. For breeding purposes, these observations suggest that successful pollination of hermaphrodite banana flowers may occur before bracts open and therefore without pollinators. Stigma may be receptive before bracts open even in male sterile edible cultivars used as female parents in banana breeding. Clearly further observations and experimental research is required to improve our understanding of floral biology and pollination among cultivated as well as *Musa* species.

4 Fertilization and seed set

In *Musa* there is little information about the journey of pollen tubes through the locules and their entry into ovules to prosecute fertilisation.

4.1 Pollination techniques

In banana breeding, crossing is an essential component in the process to combine desirable traits in new hybrids. Generally, hybridizations can be carried out through open or hand pollinations. Hand pollination is the main practice of carrying out targeted crosses for banana (Brown et al., 2017), although open pollination relying on natural pollinating agents has been suggested in the context of polycross breeding (Ortiz et al., 1995). Natural pollination in bananas is commonly carried out by bats, honey bees and birds (Ortiz and Crouch, 1997).

To make a controlled cross between two parents, unopened female and male flowers are covered with muslin bags before anthesis (Ortiz and Vuylsteke, 1995a; Ssebuliba et al., 2006). Bagging ensures that pollen contamination by bats and insects or other animal activity is prevented during pollination (Shepherd, 1954; Ray, 2002). In EAHB, flowers are ready for pollinatation when bracts are lifted away from the floral bud with an increasing angle from the tip of the bract and stigmas are exposed and appear fresh with a cream white color (Stage II and III) (Ssebuliba et al., 2006). After a series of pollinations at different times of the day on 'Gros Michel,' the highest seed set was obtained between 7 am and 10 am (Shepherd, 1954). This range of pollination time has thus been widely adopted without considering local and genetic variation across banana breeding programs with the understanding that pollen is normally shed in the morning and viability declines with time. Hand pollination is done by excising freshly exposed male flowers from the male bud and brushing sticky anthers onto stigmas of female flowers (Swennen and Vuylsteke, 1993). Pollinations are repeated daily on different hands on the same bunch for a period of 7–15 days depending on the number of hands present in the bunch (Ortiz and Vuylsteke, 1995a; Vuylsteke et al., 1997; Ssebuliba et al., 2006). The inflorescence usually bears 1–30 hands of fruit-forming female flowers followed by hands of non-fruit-forming flowers that might include 0–20 hands of transitional flowers and 100–300 male hands (White, 1928). These numbers depend on the genotype, environmental conditions and edaphic factors (Fortescue and Turner, 2011). On average, 1–3 female hands open over 24 h, often just after sunset or pre-dawn, and all open hands are pollinated (Shepherd, 1954). These values are approximate for a range of conditions and cultivars, and cannot be used to judge opening rates of female versus male hands.

The temporal and spatial separation of male and female flowers in bananas limit possibilities for self-fertilization between male and female flowers of the same inflorescence. But self-pollination is possible between plants of different cycles on the same mat (Fortescue and Turner, 2011). For breeding purposes, it is essential that by the time female flowers emerge from earmarked female parents, there is a ready source of pollen from male parents. To achieve this, male parents need to be planted earlier than female parents when establishing crossing blocks. Pollen preservation can also be explored to address the temporal and spatial separation of male and female flowers but protocols are yet to be established for pollen storage in bananas (Uma and Arun, 2016). Pollination can be carried out all year round depending on the availability of male and female flowers. Genotypes in crossing blocks may be planted at different times to allow staggered flowering and ensure availability of male and female flowers.

4.2 Understanding seed set

Some of the first experiments regarding fertilization and seed set in banana were conducted by Dodds (1945), Dodds and Simmonds (1948) and Simmonds (1952). They realised that some *M. acuminata* clones yielded less seed than expected as a result of genotypic influence but sometimes had greater than expected seed set. Simmonds (1952) also found that some diploids when selfed yielded less seeds than when other pollen sources were used. This was the first indication of the need for genetic research and optimisation of pollination to attain maximum seed set in banana. These two observations suggest optimization at the level of the female parent as well as the pollen source. There was also an observation of what appears to be a significant pollen source and female parent interaction in a number of experiments (Simmonds, 1952; Swennen and Vuylsteke, 1989; Ssebuliba et al., 2009). These interactions seem to be strongest when crosses between genomic groups are made. For example, some crosses between plantain and *M. balbisiana* yield less seed compared to *M. acuminata* as a male parent. This can be attributed to pollen-stigma interactions which lead to extracellular secretions from both pollen and the stigma as discussed in Section 2.5.1. Efforts to get around these challenges are still limited and banana breeders rather avoid futile crosses and only focus on successful crosses. Alternatively, the number of bunches to be pollinated can be increased to boost seed numbers. Consequently, banana improvement programs are working with limited numbers of progeny and a narrow genetic base despite the vast untapped natural diversity within the *Musa* genus. Simmonds (1952) also observed that crosses between diploids in the Eumusa section with those in the Rhodochlamys section were futile.

After successful pollen germination, pollen tube growth inhibition was identified as the next challenge. This was reported in 'Gros Michel' by Shepherd (1954) who observed arrested or slowed normal pollen tube growth and the consequent developmentof a swelling at the pollen tube tip. This is typically gametophytic self-incompatibility but there is limited knowledge of its presence and mechanism of operation in *Musa*. In other crops such as citrus, high temperatures are thought to inactivate proteins responsible for gametophytic self-incompatibility (Kawano et al., 2016). Normal *in vivo* pollen growth is therefore realised when temperatures are between 15°C and 25°C depending on the cross (Distefano et al., 2012). This could explain why higher seed set is observed when pollinations are made during high temperatures, though optimum temperature ranges have not been fully researched in *Musa* spp. Conversely, Fortescue and Turner (2011), focussing on some of Shepherd's observations, reported that pollen tube growth does not seem to be inhibited and more pollen tubes enter the ovary than the number of ovules. This suggests the uncertainty whether there are incompatibility systems at play. However, there is clear evidence that a necrosis forms at the tip of the ovary about a day after anthesis, especially in triploid edible bananas. This acts as a physical barrier preventing growth of pollen tubes into the ovary. It has also been observed that this necrosis develops much slower in diploid bananas (Soares et al., 2014). To optimise seed set, it is crucial that pollinations are made before this physical barrier develops.

It seems that seasonality plays a vital role in seed fertility in edible bananas but the link between season and seed fertility is complex and may involve a number of different pathways. Ortiz and Vuylsteke (1995a) reported a seasonal response in seed set, although this was variable across genotypes. Using 'Calcutta 4' as the male parent, low temperature, low irradiation and high relative humidity favoured seed set in *Musa* (AAA) 'Yangambi KM 5' and in *Musa* (ABB) 'Bluggoe', peaking in August to September. Seed set in *Musa* (AAB) plantain cultivars had two peaks, February and September. High temperature, high solar radiation and low relative humidity, characteristic of February, favoured production of tetraploid seed in plantain cultivars, possibly linked to the production of 2n gametes. These same conditions were associated with increased seed set in *Musa* (AAA) EAHB Matooke cultivars (Ssebuliba et al., 2009), although this seasonality was not apparent as determined by a long-term assessment of pollination success in a further study of Matooke cultivars (Batte et al., 2019). One has to keep in mind that the weather parameters of temperature, relative humidity and solar radiation are interrelated and the relative importance of each parameter is yet to be ascertained. For example, high temperature is positively correlated with high solar radiation and negatively correlated with relative humidity. Moreover, most r values reported by Ssebuliba et al. (2009) are less than 0.5, indicative of weak relationships suggesting the involvement of other factors.

4.3 Quantifying fertility in Musa

'Fertility' in *Musa* can have somewhat different meanings according to the scale adopted and whether ovules, pollen or seeds are considered. For broader studies in banana breeding, fertility is often expressed as the number of seeds per bunch. The more seeds the greater is the fertility. Seeds per bunch can also be used as a +/- comparison (either nil or at least one seed per bunch) to calculate the proportion of pollinated bunches containing at least one seed. This is considered a measure of 'pollination success' (Batte et al., 2019). In more detailed work with a focus on the biology of reproduction it is necessary to look at more definitive measures of fertility, such as seeds per fruit (Shepherd, 1954), or better, seeds/1000 ovules.

Seeds per bunch is a widely used measure of fertility in banana breeding. It has a long history (Cheesman, 1932; Shepherd, 1954, 1960; Swennen and Vuylsteke, 1993) and has been used to measure the 'fertility' of a clone (Ssebuliba et al., 2005) and the ability of a cross to produce seeds (Uma et al., 2011). Seeds per bunch, as a measure, has a number of components. Because a seed is the product of a successful fertilisation of an ovule, seeds per bunch (S_b) is made up of the number of seeds per thousand ovules (S_o), the number of ovules per ovary O_f, the number of ovaries per hand (F_h) and the number of female hands per bunch (H_b):

$$S_b = (S_o/1000) \cdot O_f \cdot F_h \cdot H_b. \quad (2)$$

The number of seeds per thousand ovules is a measure of the fertility of ovules. The number of ovules per ovary is a measure of the capacity of the ovary to produce seed. The number of ovaries or fruits per hand and the number of hands per bunch are measures of the architecture of the inflorescence. A similar value of seeds per bunch in two different genotypes, or treatments, can only be achieved if either all other components of Eqn 2 are the same, which is very unlikely, or if there are compensatory changes in the components.

These compensatory changes may reflect differences in the architecture of the inflorescence, or, more importantly in this case, the capacity of the ovules to set seed. Thus, at best, for the purposes of reproductive biology, seeds per bunch as a measure of fertility, has limited appeal.

The number of ovules per flower (O_f), also referred to as ovules per ovary, varies with genotype. Some examples are 130 ovules per flower in 'Calcutta 4' to >300 ovules in other diploids (Dodds, 1943). In triploid bananas, 260–350 ovules per ovary have been counted in genotypes of different clone sets of the *Musa* (AAA) EAHB (Ssebuliba et al., 2005). The major contributor to variation in the number of ovules per ovary is ovary length. From the data of Ssebuliba et al. (2005) for EAHB we can calculate the number of ovules per unit length of ovary, which ranges from 22 to 29 ovules per cm. Flowers per hand range from 5 to 20 and hands per bunch range from less than 5 to more than 17 in different plantain genotypes (Adheka et al., 2018).

The measure of fertility used in experimental studies needs to be considered in relation to the objectives of the study. Used alone, the number of seeds per bunch may be considered as a measure of success in breeding. However, it is unlikely to inform us of the reproductive strategies used to obtain seed set and so to increase breeding success for which seeds/1000 ovules is a more accurate measure.

5 Embryo dormancy: a cause for poor germination?

In addition to poor fertility and seed set a further hurdle faced by banana breeders is the low germination rates of seeds and even of cultured embryos. In general, germination of intact *Musa* seeds is variable and often not very successful. Both seed viability and seed dormancy can impact germination. In one study, less than half the seed harvested from 20 EAHB cultivars had an embryo (Ssebuliba et al., 2006), and so were viable, even before any impact of dormancy might be considered. Another example concerns the plantain breeding program in Onne, Nigeria, where only 1% of intact hybrid seeds germinated (Swennen et al., 1992). For plantain embryos extracted and cultured aseptically, the germination success rate was around 12% (Vuylsteke et al., 1990). So, for some seeds, embryo extraction and culture might have overcome a dormancy factor associated with the intact seed. However, the low success rate of embryo culture in this study raises the possibility of embryo dormancy.

Low germination of excised embryos *in vitro* in cultivars of *Musa* (AAA) EAHB (Ssebuliba et al., 2006) and other *Musa* hybrids (Bakry, 2008) is consistent with physiological dormancy of the embryo (embryo dormancy), although low germination of excised embryos is not always typical of *Musa* species. For example, more than 90% of embryos of *M. balbisiana* (BB), excised from fresh seeds (Bakry, 2008) and from seeds stored at room temperature for up to 18 months (Cox et al., 1960; Stotzky and Cox, 1962), developed into plantlets.

5.1 Is embryo dormancy heritable?

Breeding of bananas involves culturing the embryos of the seed of the hybrid progeny to grow new plants. This procedure overcomes limitations of using whole seed but the recovery of live plants from hybrid embryos undergoing culturing procedures is often low (Bakry, 2008), in the order of 20%–30% compared with 99% success for embryos of *M. balbisiana* (Bakry, 2008).

We hypothesised that embryos can express dormancy (D/d) and that this is inherited. We applied an elementary genetic analysis (Griffiths et al., 2000) to the data of Bakry (2008), because these data provided the parentage of the seeds and the success of embryos in culture.

We estimated the phenotypic ratio of the dormant:non-dormant character and then inferred possibilities of genetic composition (D/D, D/d, d/d) of parent(s) from the ratio. Simple Mendelian ratios were chosen in this exercise. We constructed a table indicating possible genotypes for dormancy for as many of the parents used by Bakry (2008) as possible (Table 2).

This provided some evidence that segregation of embryo dormancy in banana seed may follow Mendelian principles, although the patterns of

Table 2 Summary table showing hypothesised make up of dominant (D) or recessive (d) dormancy gene(s) of parent

Genome constitution, cultivar/selection of parents	Dominant (D) / recessive (d)
BB, 'Lal Velchi'; 'Pisang Klutuk Wulung'	d/d
AA, 'Pisang Madu'; 'Calcutta 4'; 'Tha 052'; 'Long Tavoy'; 'SF265'	D/d
AAAA, 'Pisang lilin' autotetraploid (Bakry et al. (2007)'.	D/d, D/d

inheritance are likely to be complex. Initial explorations suggest that the B genome appears to be recessive (d/d) while the A genome could be D/d or other combinations, perhaps depending on the origins of the A genome.

If it could be shown in a segregating population that embryo dormancy was inherited, then it could be selected against and increase the probability that cultured embryos will grow into plants. This would increase the efficiency of breeding by making more plants available for screening.

6 Parthenocarpy and sterility

Parthenocarpy is the growth of a fruit without fertilisation and seed development (Sinnott, 1960), or more broadly, producing fruit without seeds (Holmes, 1979). The broadly accepted hormonal mechanism of fruit formation and growth is that pollination itself produces hormones such as gibberellins and auxins, and the fertilised ovary subsequently produces hormones essential for fruit development. Fruit may be parthenocarpic without pollination or ovule fertilisation (vegetatively parthenocarpic), such as in banana, or it may be parthenocarpic because the embryo of the fertilised ovule aborts (stenospermocarpic), such as in sultana (*Vitis vinifera*), but the berry continues to grow (Atwell et al., 1999). While hormone mechanisms have dominated thinking about fruit growth, there are resource implications as well.

Pulp development in wild bananas is stimulated by seed development. Edible bananas are vegetatively parthenocarpic (Simmonds, 1962) and develop pulp autonomously from tissues on the ovary wall in the absence of pollination (Fig. 14). However, for edible bananas that set some seed, pulp development is more nuanced and will have a contribution not only from pulp that develops as a consequence of the autonomous process of vegetative parthenocarpy, but also from pulp that develops in association with seed set, if present (Simmonds, 1962). Partial parthenocarpy and partial sterility can be expressed in a single banana fruit at the same time. These observations contribute to the discussions in the literature about whether or not parthenocarpy is independent of sterility.

Sterility is the failure to form a seed and it has many causes. Female and/or male gametes may be defective because of faults in meiosis (Atwell et al., 1999) (see Section 2.2 on gametogenesis). Pollination and fertilisation are complex

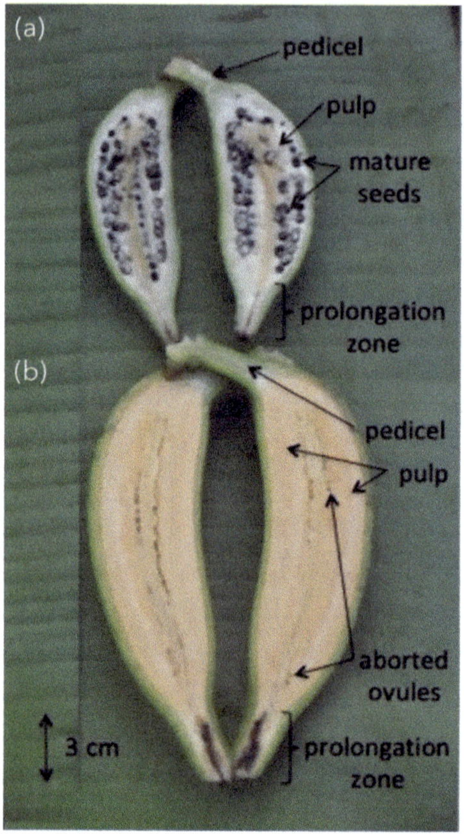

Figure 14 Longitudinal sections of mature fruit of a) seeded *Musa balbisiana* 'Tani' (BB) with mature seeds and a limited amount of pulp generated by the seeds, and b) parthenocarpic *Musa* (AAB) plantain 'Obino L'Ewai', with autonomous pulp originating from the ovary walls, and aborted ovules along the line of the placental tissue. In both fruits, flower parts have abscised, the prolongation zone shows necrotic tissues of the nectaries and the pedicel attaches the fruit to the female peduncle.

processes and failure in either may contribute to sterility in the absence of faulty meiosis. Edible banana clones generally do not form seeds, the latter reflecting in part their tendency to low pollen fertility (Simmonds, 1962). Provided with a fertile pollen donor, a low level of viable seed formation can occur (Shepherd, 1960). Some triploids, such as those from the commercially important *Musa* (AAA) Cavendish subgroup, are highly female sterile. Even in this highly female sterile sub-group, a few seeds have been obtained when very large numbers are crossed with pollen from Cavendish clones (Aguilar-Morán, 2013).

Simmonds (1962) warned of the confusion that has arisen in the literature from the use of the terms parthenocarpy and 'seedless' as if they are synonymous. He points out that a fruit in which pulp development has occurred can be seedless due to the abortion of fertilised ovules, that is, the fruit is seedless but not parthenocarpic. Indeed, there is a wide range of behaviours in pulp development, from seed-dependent to autonomous, and it can be incorrect and confusing to assume that a seedless fruit is parthenocarpic. Simmonds (1953) urges that to avoid confusion, the term 'parthenocarpy' is used in a defined and limited way, as for vegetative parthenocarpy, and concludes that the term 'seedless' is an ambiguous description for 'a condition that has several different causes' (Simmonds, 1962).

6.1 The relationship between parthenocarpy and sterility

In banana breeding, the relationship, if any, between parthenocarpy and female sterility is important because if they are independent, then to increase seed set, the focus needs to be on managing sterility. On the other hand, if parthenocarpy increases sterility then this needs to be borne in mind, especially for cultivars that appear almost completely female sterile. The concern is that any increase in parthenocarpy may eliminate seed set, contributing to the 'fertility crisis' in banana breeding.

In the past a case has been made that parthenocarpy and sterility are independent, involve different tissues and organs and result from two separate but interrelated genetic systems (Dodds and Simmonds, 1948; Simmonds,

	➔ Increasing parthenocarpy ➔		
↓ Increasing sterility ↓	A1. Many seeds Pulp volume proportional to seed number	B1. Many seeds Pulp volume more than expected from seed number	C1. No genotypes found
	A2. Some or few seeds Pulp volume proportional to seed number	B2. Some or few seeds Pulp volume more than expected from seed number	C2. Some or few seeds Pulp volume much more than expected from seed number
	A3. No seeds No pulp	B3. No seeds Some pulp	C3. No seeds Pulp volume generated without seeds

Figure 15 The phenotypic relationship between parthenocarpy (columns A -> C) and sterility (rows 1 -> 3) in different genotypes of bananas. A fertile genotype in A1 or A2 may show the behaviour of A3 if pollination fails or does not occur. Similarly, genotypes in B1 or B2 can appear as B3 and those in C2 can appear as C3. Based on Simmonds (1962).

1962). Parthenocarpy and sterility have a range of phenotypic expressions that give the appearance of a continuum (Fig. 15).

Simmonds (1962) concluded that the relationship between parthenocarpy and sterility in banana is not primarily causal, pointing out that failure of pollination and/or fertilization leads to sterility, but sterility does not lead to parthenocarpy (A3, Fig. 15). The logic behind his conclusion that parthenocarpy and sterility are independent systems is illustrated in Fig. 15. Simmonds (1962) acknowledged that this picture is not complete because there is no example of a highly parthenocarpic, highly fertile genotype (C1, Fig. 15). The absence of a genotype at C1 raises the possibility that a high level of parthenocarpy and a high level of fertility are not compatible. To take this possibility into account, Simmonds (1962) qualified his view on the independence of parthenocarpy and sterility with the statement that parthenocarpy might be responsible for failure to set viable seed, but if so, this effect is small. He commented that parthenocarpy and sterility are correlated in banana because 'they both have been subjected to human selection' (Simmonds, 1962).

Evidence from segregating populations indicated that parthenocarpy in banana is controlled by three independent but complementary dominant genes (P_i), a major switch gene (P_1) and two minor modifier genes (P_2 and P_3) (Dodds and Simmonds, 1948; Simmonds, 1953; Ortiz and Vuylsteke, 1995b). Seeded bananas, such as 'Calcutta 4' have double recessive genes for the P_1 switch gene involved in parthenocarpy, and so show no parthenocarpy (Ortiz and Vuylsteke, 1995b).

Sardos et al. (2016) referenced these earlier works in a genome-wide association study of *Musa* that focussed on 'seedlessness'. They noted that, unlike parthenocarpy, which is driven by a major dominant and several minor genes, sterility in banana is due to structural and genetic factors and reiterated that it is independent of parthenocarpy, although likely to have been co-selected with parthenocarpy by humans.

Parthenocarpy and sterility are of interest in fruits other than banana. In tomato, the relationship between parthenocarpy and sterility appears to parallel that of bananas, with observations suggesting that parthenocarpy and sterility in tomato are independently functioning systems, but with the qualification that parthenocarpy can lower seed set. For example, in parthenocarpic tomatoes, sterility lies with the pollen, not because it is sterile, but because the gene responsible controls the release of pollen from the anthers so the ovules are not pollinated and seeds are not formed (Gorguet, 2007). This is an example of structural and genetic factors acting together resulting in sterility. Independently, parthenocarpy gene(s) disturb the 'hormonal balance' of the wall tissues resulting in the development of pulp. In a separate study, paclobutrazol, which blocks biosynthesis of gibberellin, promoted seed production in a parthenocarpic tomato resulting in the production of a few seeds (Ohkawa et al., 2012), linking a change in the hormone balance associated with parthenocarpy with a change in sterility. Gorguet (2007) claimed that the *pat-1* gene, which is partly responsible for parthenocarpy in tomato, not only changed hormonal balance but disturbed meiosis in a proportion of female gametes. Takisawa et al. (2018) speculated that down-regulation of the *Pat-k* gene, which caused both parthenocarpy and low seed set in the parthenocarpic tomato cultivar MPK-1, could cause abnormal ovule formation, reducing the number of viable seeds formed.

6.1.1 How might parthenocarpy affect seed fertility?

An early study of parthenocarpy and sterility using edible diploids of banana concluded that, while many edible diploids produce little viable pollen, parthenocarpy did not lead to male sterility because at least one of the five parthenocarpic diploids studied, 'Pisang lilin', had abundant viable pollens that successfully pollinated seeded diploids (Dodds, 1943). 'Pisang lilin' was, however, highly female sterile demonstrating a separation of female and male fertility. This phenomenon was also observed in the parthenocarpic offspring of crosses between the seeded (female) and edible (male) diploids (Dodds, 1943). In all these cases both parthenocarpy and female sterility were inherited through the pollen. Dodds (1943) concluded a common factor acting in parthenocarpy and female sterility, suggesting that the auxin responsible for parthenocarpy might disrupt ovule development. This conclusion was later discounted by Simmonds (1962) on the basis that parthenocarpy can occur in seeded fruits (B1, Fig. 15) and that growth factors other than auxin are involved in pulp-filling. A later study using hybrids of 'Pisang lilin' and seeded diploids found evidence suggesting genetic modifiers of parthenocarpy diminished female fertility (Dodds and Simmonds, 1948).

So, the question remains: is there a fertility crisis in banana because it is parthenocarpic? To recap, Simmonds (1962) gave an example of high sterility with nil parthenocarpy (A3, Fig. 15) consistent with this not being the case. However, he could find no example of a highly fertile, highly parthenocarpic genotype (C1, Fig. 15) suggesting that while sterility can occur in the absence of parthenocarpy, parthenocarpy *per se* might have a negative impact on viable seed set, via its influence on female fertility.

Might the interpretation of a relationship between parthenocarpy and sterility be confounded by the possibility of a 'common factor' influencing both? Knowledge about parthenocarpy and sterility in bananas is derived mainly from studying pollen production and seed set in a wide range of genotypes in crossing experiments. When examining their relationship across genotypes, sometimes the two appear correlated, either negatively (A1 to B2 to C3 in Fig. 15) or positively (A3 to B2 in Fig. 15) and sometimes there is no association at all (A1 to A3 or B1 to B3 in Fig. 15). A hypothesis to account for these observations may indeed lie in a common factor that can affect the expression of parthenocarpy and sterility, but without any need for parthenocarpy to influence sterility or the reverse.

Whether parthenocarpy affects processes involved in viable seed sets will in part depend on:

1. Whether genes associated with parthenocarpy in banana influence or act on the expression of processes in the pathway to seed set. For example, genetic modifiers identified in diploid bananas by Dodds and Simmonds (1948), and the apparent influence of *pat*-genes linked with parthenocarpy on meiosis (Gorguet, 2007) and ovule formation (Takisawa et al., 2018) in tomato.
2. Whether proximate processes arising from the expression of parthenocarpy [autonomous pulp-filling] in the maternal parent negatively impact processes essential for viable seed set. For example, interaction and/or competition between pulp-filling and pollen-tube growth and guidance (Simmonds, 1962). For this to occur, pulp-filling processes would need to coincide with or precede seed set processes.

Genetic controls lie at the base of the phenological expressions of parthenocarpy and sterility. More investigations are needed to untangle their relationships in banana. We are left with the conclusion that in bananas parthenocarpy and sterility are independent, but the question of whether or not parthenocarpy increases sterility is still unresolved. We suggest that this is likely to be largely unimportant for breeding, except in the case of edible, highly female-sterile clones, where the high level of sterility might reflect the influence of parthenocarpy, and therefore some dependence between the processes. Although parthenocarpy and sterility are for the most part independent processes, an in-depth understanding of the genetics underlying both processes and their relationship is fundamental to improve seed set, particularly in edible, highly female sterile banana clones.

6.2 Parthenocarpy as an adaptive feature

Picarella and Mazzucato (2019) described parthenocarpy as an adaptive feature, suggesting it could exert an advantage in multi-seeded fruit species under sub-optimal pollination regimes where too few seeds are formed to support fruit growth. In this circumstance, parthenocarpy would promote the development of an attractive fruit with which to spread the few seeds formed, so providing a selective advantage.

If the hypothesis of Picarella and Mazzucato (2019) is supported, then a corollary to be expected would be high viability and capacity for germination in the few seeds per fruit. This seems not to be the situation in banana breeding programs based on current practices of seed germination. That parthenocarpy might depress viable seed set in banana seems at odds with the hypothesis of Picarella and Mazzucato (2019). Perhaps selection towards spread of viable seed contributes to adaption in some plant species, but in bananas, co-selection of parthenocarpy and sterility by humans has reinforced parthenocarpy, favouring pulp-filling over production of viable seed. We suggest that female sterility is the issue of importance to banana breeders, not parthenocarpy, because at this stage, breeding programs need more seeds.

6.3 A gene linked to female sterility?

The genome-wide association analysis conducted by Sardos et al. (2016) in bananas identified one strong candidate gene, a putative orthologous gene to histidine kinase CKI1, as being associated with female sterility, based on work with *Arabidopsis*. In *Arabidopsis*, the gene histidine kinase CKI1 is involved in gametophyte development, and, once silenced, resulted in female sterility (Deng et al., 2010). Subsequent analysis showed that the orthologous gene in banana was found on a region of the chromosome most significantly associated with 'seedlessness', strengthening its link to female sterility. Because of its association with sterility, it would be worthwhile looking at CKI1 expression in experiments manipulating the environment, such as soil water supply, modification of the atmosphere around the bunch, pollination or not, and using edible banana clones, including those with a reputation for no seeds and those with some seeds. This would allow for examination of genetic and environmental components of sterility in banana, although issues with sampling of fruit or fruit components would require some thought.

7 Conclusion and future trends

Banana breeding programs will need to play an increasing role in providing solutions to existing and emerging threats to production, such as the devastating threats of *Fusarium oxysporum* f. sp. *cubense* (*Foc*) Tropical race 4 (TR4) and Black Leaf Streak caused by *Pseudocercospora fijiensis*. This will require the integration of conventional breeding techniques as well as modern molecular breeding tools for more efficient hybrid development, selection and delivery. The success of breeding using these techniques will depend on conducting recombinations focussed on target traits and generating sufficient progeny for selection, which is largely affected by varying levels of sterility. An increase in the number of possible recombinations to create large hybrid progenies to select from requires overcoming the obvious fertility challenges. While considerable efforts have been made in this area which underpins the progress made so far in banana breeding, much is still left to be done.

As summarized in Fig. 1, successful fertilization and seed production are affected by several factors including pollen quantity, viability, potency and compatibility of pollen-pistil interactions as well as the ability of the female gametophyte to be fertilised and developed into a viable seed. These processes are in turn affected by genotype and the environment. Therefore, a sound understanding of these processes in the context of the wide variability in bananas is critical. Most of the limited work on floral biology has focussed so far on *Musa* species forming a solid basis, but for the future this should be extended to edible cultivars frequently used in crossing schemes as sources of desirable traits, particularly including the triploid category where infertility is most obvious. This can lead to a more targeted approach where critical processes can be optimized to improve seed set and meet crossing objectives.

Shepherd (1954) first established the best time to pollinate (between 7 am and 10 am), which has so far been adopted to maximize pollen viability and stigma receptivity. However, there may be variations in bract opening times and sequence with different genotype and environment interactions. Moreover, the short window in which to pollinate limits the number of crosses that can be made in a day. Further studies on understanding bract opening may inform further optimization of pollination time. On-going work on the application of pollen germination medium to enhance stigma receptivity has shown promising results, giving rise to increased seed set, at least with the use of one male parent, with potential for further exploration. Application of growth regulators and other stimulating substances on the stigma to stimulate pollen germination and pollen tube growth (Wu et al., 2008) may also be explored.

Establishing pollination blocks in environments that favor seed set as described by Ortiz and Vuylsteke (1995a) in plantain and Ssebuliba et al. (2009) in EAHB 'Matooke' cultivars may be pursued and may need to be separated from progeny testing. Progeny testing requires an ideal growing environment and/or environments in which certain biotic and abiotic factors are present to provide sufficient selection pressure. Similarly using chemicals like boron which is well known to enhance pollen fertility and stigma receptivity (Peñaloza and Toloza, 2018) also needs to be explored.

Beyond the mechanical manipulations, Sardos et al. (2016) identified a putative gene associated with 'seedlessness' in *Musa*. The use of RNA-silencing technologies with applications in other crop improvement programs (Guo et al., 2016) is dependent on such findings and can also be applied to banana. Ultimately the challenge is that fertility is desirable for breeding while sterility is desirable in final products to ensure edibility. Further characterization of regulatory genes or sequences implicated in these processes, in combination with breeding strategies, are required before modern tools such as gene editing using CRISPR/Cas9 (Tripathi et al., 2019) can be of use. A thorough understanding of the genes and processes involved in sterility and fertility is required to be able to effectively manipulate this using gene-editing techniques.

Hybridization and seed production are central to every crop breeding effort, and knowledge on these processes is unfortunately limited for bananas. In the era of modern molecular tools such as genomic selection and association genetics which hold considerable promise for banana (Nyine et al., 2017, 2019), multiple large progeny populations are required for their applications to increase breeding efficiency. Significant investments on advancing floral biology research in banana could revolutionize breeding efforts to deliver genetic gains in farmer's fields.

8 Where to look for further information

For a further detailed discussion on reproductive biology in *Musa,* see Fortescue and Turner (2011); for detailed anatomical studies of floral organogenesis in *M. velutina*, *M. ornata* and the triploid edible cultivar 'Go Sai Yung', see Kirchoff (1992) and Kirchoff (2017). For a functional approach to bunch formation including the occurrence of hermaphrodite flowers, see Turner and Gibbs (2018). For a description of the processes of parthenocarpy, Joldersma and Liu (2018) provide a recent review on the role of phytohormones in the induction of parthenocarpy, and the molecular and genetic basis of parthenocarpy. Mizuta and Higashiyama (2018) provide a very useful recent summary of the known mechanisms of pollen guidance. More information on banana flowers in general can be found on the ProMusa website, which is a platform for scientific discussions on banana http://www.promusa.org/Banana+flowers. The Breeding Better Bananas project (https://breedingbetterbananas.org/), which seeks to improve banana breeding in Tanzania and Uganda, also provides information on current research to address the fertility crisis in bananas.

9 Glossary

Androecium the stamens, the male reproductive unit.

Anther the part of a stamen that produces pollen grains.

Anthesis the opening of a flower, or in *Musa* the lifting of the bracts on an emerged bunch.

Archesporia cells that form the spore mother cells, which give rise to the egg cell in the embryo sacs.

Asynapsis lack of pairing of chromosomes in meiosis.

Autonomous independent.

Bract a modified leaf that on the female and male peduncle subtends a number of flowers. It usually dehisces after anthesis.

Bunch a horticultural term for the inflorescence or thyrse, or for the female peduncle carrying the edible fruit.

Carpel the female reproductive unit or gynoecium. In *Musa* the ovary consists of three fused carpels.

Cincinnus a type of inflorescence, a scorpioid cyme. In *Musa*, commonly known as a 'hand' of fruit.

Cultivar a term that blends the words 'cultivated + variety'. It is a term for cultivated plants and does not apply to those plants that occur in the wild. Cultivars include hybrids that are either deliberate or accidental, and cultivated plants selected from wild populations and recognised as a separate entity from the wild plants. *Musa* cultivars arise from the intended actions of people. They are parthenocarpic with various degrees of residual fertility.

Desynapsis separation of paired chromosomes in meiosis.

Dichogamous pistils and stamens mature at different times, in the case of banana on different flowers.

Dormancy the failure of a viable embryo to germinate despite being in a suitable environment.

Edible a commonly used term for a banana ovary with pulp of parthenocarpic origin and without seeds. Edible bananas can have a residual fertility and when pollinated, can set seed, albeit in very low quantities.

Embryo the product of a fertilised egg sac.

Embryo sac the megaspore of flowering plants within the ovule.

Energy-dependent a biological process requiring metabolic energy to proceed.

Fertilisation the combination of female and male gametes to produce an embryo. The process takes place within an ovule.

Fertility the capacity to produce offspring through sexual processes.

Flower a blossom made up of the ovary, petals, sepals, pistil and stamens.

Flowering the emergence of an inflorescence (bunch) from the pseudostem.

Gametogenesis the process of formation of female or male gametes for sexual reproduction.

Gynoecium (or Carpel) the pistils, comprising the stigma, style and ovary, the female reproductive unit.

Hand a horticultural term describing a cincinnus of flowers subtended by a single bract.

Hermaphrodite a flower with functional female and male parts capable of self-fertilisation.

Hybridity a plant derived from a cross between two different parents and which may express sterility.

Hybridization the process of cross pollination of two parents to produce progeny.

Infertility unable to produce offspring by sexual means.

Involute curling inwards, as involute curling of a bract.

Locule a cavity. The ovary of *Musa* normally has three locules.

Megaspore the larger spore in flowering plants that gives rise to the female gamete.
Meiosis cell division where the homologous chromosome number is reduced from diploid to haploid.
Microspore cells arising from the pollen mother cells that mature into pollen grains.
Monoecious female and male flowers produced on the same inflorescence.
Musa species taxa of the genus *Musa* occurring in the wild that can be classified into the species or sub-species taxon. Propagated by seeds and vegetatively.
Nectar sugary secretion arising from tissues at the apex of the ovary. It collects in the perianth.
Ovary the organ of female flowers that become the fruit in edible cultivars. Contains ovules that when fertilised, subsequently become seed in the species of *Musa*.
Parthenocarpy the growth of a fruit in the absence of seeds. In *Musa*, the most obvious expression is the autonomous development of pulp in edible cultivars. Parthenocarpy is a quantitative trait.
Parthenocarpy, vegetative the growth of a fruit without the stimulation of pollination or ovule fertilisation, characteristic of edible bananas.
Peduncle the external part of the true stem of a banana plant normally with three sections: sterile, female and male.
Pistil the stigma, style and ovary; the gynoecium.
Ploidy a set of chromosomes, hence haploid (one set), diploid and triploid.
Pollen mother cells the larger spores giving rise to pollen grains.
Pollen viability a measure of the portion of a population of pollen grains that are living.
Pollen tube an outgrowth of a germinating pollen grain. It contains the male gametic material.
Pollen tube guidance the anatomical and physiological mechanisms by which pollen tubes move from the surface of the stigma through the style to the locule and subsequently to the ovules. It includes the exclusion of pollen tubes from fertilised ovules.
Prolongation region the closure of the distal end of the locules. The stylar end of the ovary.
Receptivity the stage in stigma development when pollen on the surface can germinate and the pollen tube can grow into the stigmatic surface.
Restitution failure of a cell division in meiosis producing 2n gametes.
Restitution, double Failure of both cell divisions in meiosis, producing 4n gametes. Meiosis avoidance.
Scorpioid an inflorescence with a curved axis.
Seed a mature fertilised ovule where the integuments form the seed coat that contains the embryo and endosperm. Also **botanical seeds**, to distinguish

them from vegetative material such as suckers, also called 'seed' – a horticultural term.

Seedless, seedlessness fruit without seed, not to be confused with parthenocarpy since the fruit of *Musa* species can be seedless if either not fertilised or fertilisation has failed, and in the absence of parthenocarpy.

Staminodes imperfectly formed stamens with no or few pollen that are likely to be infertile.

Sterility the inability of gametes to form zygotes, seedless. Sterility is a quantitative trait in *Musa*.

Sterility, female in *Musa*, a plant with ovules unable to produce seed although the flower has been pollinated with fertile pollen.

Sterility, male in *Musa*, a plant the pollen of which is unable to fertilise an ovule and form seed.

Stigma the tissue that surmounts the style. When receptive, pollen grains germinate on the stigma.

Style the slender connecting organ between the locule and the stigma. The style contains stylar canals.

Sub-locular the basal region of the ovary beneath the locule.

Synapsis pairing of chromosomes during meiosis.

Transitional if present, transitional flowers appear between the basal female flowers and the distal male flowers. Transitional flowers are characterised by reduced the development of the gynoecium and androecium, but the non-functional ovaries are larger than in the male flowers. There are two types of transitional flowers: 'neuter' which are retained on the peduncle, and 'transitional' flowers that are functionally male and fall from the peduncle at flowering.'

Univalents unpaired chromosomes.

Viability a measure of the proportion of a population that is living.

10 References

Adeleke, M. T. V., Pillay, M. and Okoli, B. E. (2004). Relationships between meiotic irregularities and fertility in diploid and triploid *Musa* L., *Cytologia* 69(4): 387–393.

Adheka, J. G., Dhed'a, D. B., Karamura, D., Blomme, G., Swennen, R. and De Langhe, E. (2018). The morphological diversity of plantain in the Democratic Republic of Congo, *Scientia Horticulturae* 234: 126–133.

Aguilar-Morán, J. F. (2013). Improvement of Cavendish banana cultivars through conventional breeding, *Acta Horticulturae* 986: 205–208.

Argent, G. C. G. (1976). The wild bananas of Papua New Guinea, *Royal Botanical Garden Edinburgh Notes* 35: 77–114.

Atwell, B., Kriedemann, P. and Turnbull, C. (1999). *Plants in Action. Adaptation in Nature, Performance in Cultivation*, South Yarra: Macmillan Education Australia.

Bakry, F. (2008). Zygotic embryo rescue in bananas, *Fruits* 63(2): 111–115.

Bakry, F., Carreel, F., Jenny, C. and Horry, J. P. (2009). Genetic improvement of banana. In: Jain, S. M. (Ed.), *Breeding Plantation Tree Crops: Tropical Species*, New York: Springer, pp. 3–50.

Bakry, F., de la Reberdiere, N. P., Pichot, S. and Jenny, C. (2007). In liquid medium colchicine treatment induces non chimerical doubled-diploids in a wide range of mono-and interspecific diploid banana clones, *Fruits* 62(1): 3–12.

Batte, M., Swennen, R., Uwimana, B., Akech, V., Brown, A., Tumuhimbise, R., Hovmalm, H. P., Geleta, M. and Ortiz, R. (2019). Crossbreeding East African Highland bananas: lessons learnt relevant to the botany of the crop after 21 years of genetic enhancement, *Frontiers in Plant Science* 10: 81. doi: 10.3389/fpls.2019.00081.

Brown, A., Tumuhimbise, R., Amah, D., Uwimana, B., Nyine, M., Mduma, H ., Talengera, D. Karamura, D., Kuriba, J. and Swennen, R. (2017). The genetic improvement of bananas and plantains (*Musa* spp.). In: Campos, H. and Caligari, P. D. S. (Eds), *Genetic Improvement of Tropical Crops*, New York: Springer, pp. 219–240.

Cheesman, E. E. (1932). Genetic and cytological studies of *Musa*. 1. Certain hybrids of the Gros Michel banana, *Journal of Genetics* 26(3): 291–312.

Cheesman, E. E. and Dodds, K. S. (1942). Genetical and cytological studies in *Musa*, *Journal of Genetics* 43(3): 337–357.

Cox, E. A., Stotzky, G. and Goos, R. D. (1960). *In vitro* culture of *Musa balbisiana* Colla embryos, *Nature* 185(4710): 403–404.

Deng, Y., Dong, H., Mu, J., Ren, B., Zheng, B., Ji, Z., Yang, W. C., Liang, Y. and Zuo, J. (2010). Arabidopsis histine kinase CKI1 acts upstream of histidine phosphotransfer proteins to regulate female gametophyte development and vegetative growth, *Plant Cell* 22(4): 1232–1248.

Distefano, G., Hedhly, A., Las Casas, G., La Malfa, S., Herrero, M. and Gentile, A. (2012). Male-female interaction and temperature variation affect pollen performance in Citrus, *Scientia Horticulturae* 140(1): 1–7.

Dochez, C., Dusabe, J., Whyte, J., Tenkouano, A., Ortiz, R. and De Waele, D. (2006). New sources of resistance to *Radopholus similis* in *Musa* germplasm from Asia, *Australasian Plant Pathology* 35(5): 481–485.

Dodds, K. S. (1943). Genetical and cytological studies of *Musa*. 5. Certain edible diploids, *Journal of Genetics* 45(2): 113–138.

Dodds, K. S. (1945). Genetical and cytological studies of *Musa*. 6. The development of female cells of certain edible diploids, *Journal of Genetics* 46(2-3): 161–179.

Dodds, K. S. and Simmonds, N. W. (1946). Genetical and cytological studies of *Musa*. 8. The formation of polyploid spores, *Journal of Genetics* 47: 223–241.

Dodds, K. S. and Simmonds, N. W. (1948). Sterility and parthenocarpy in diploid hybrids of *Musa*, *Heredity* 2(1): 101–117.

Dumpe, B. B. and Ortiz, R. (1996). Apparent male fertility in *Musa* germplasm, *HortScience* 31(6): 1019–1022.

Fahn, A., Klarman-Kislev, N. and Ziv, D. (1961). The abnormal flower and fruit of May-flowering Dwarf Cavendish bananas, *Botanical Gazette* 123(2): 116–125.

Farkas, A. and Orosz-Kovacs, Z. (2003). Nectar secretion dynamics of Hungarian local pear cultivars, *Plant Systematics and Evolution* 238(1–4): 57–67.

Fortescue, J. A. and Turner, D. W. (2004). Pollen fertility in *Musa*: viability in cultivars grown in Southern Australia, *Australian Journal of Agricultural Research* 55(10): 1085–1091.

Fortescue, J. A. and Turner, D. W. (2005a). The association between low temperatures and anatomical changes in pre-anthetic ovules of *Musa* (Musaceae), *Scientia Horticulturae* 104(4): 433–444.

Fortescue, J. A. and Turner, D. W. (2005b). Growth and development of ovules of banana, plantain and Ensete (Musaceae), *Scientia Horticulturae* 104(4): 463–478.

Fortescue, J. A. and Turner, D. W. (2005c). The anatomy of ovule development of banana, plantain and ensat (Musaceae), *Scientia Horticulturae* 104(4): 479–492.

Fortescue, J. A. and Turner, D. W. (2011). Reproductive biology. In: Pillay, M. and Tenkauano, A. (Eds), *Banana Breeding: Progress and Challenges*, Boca Raton, FL: CRC Press, pp. 145–180.

Glinos, E. and Cocucci, A. A. (2011). Pollination biology of *Canna indica* (Cannaceae) with particular reference to the functional morphology of the style, *Plant Systematics and Evolution* 291(1–2): 49–58.

Goigoux, S., Salmon, F. and Bakry, F. (2013). Evaluation of pollen fertility of diploid and doubled-diploid clones of Mlali and their potential use for banana breeding, *Acta Horticulturae* 986: 195–204.

Gorguet, B. (2007). Parthenocarpy and functional sterility in tomato, PhD Thesis, Wageningen University, The Netherlands. ISBN 978-90-8504-762-9.

Griffiths, A. J. F., Miller, J. H., Suzuki, D. T., Lewontin, R. C. and Gelbart, W. M. (2000). *An Introduction to Genetic Analysis* (7th edn.), New York: WH Freeman.

Griffiths, A. J. F., Wessler, S. R., Carroll, S. B. and Doebley, J. (2015). *Introduction to Genetic Analysis* (11th edn.), New York: WH Freeman and Company.

Guo, Q., Liu, Q., Smith, N. A., Liang, G. and Wang, M. B. (2016). RNA silencing in plants: mechanisms, technologies and applications in horticultural crops, *Current Genomics* 17(6): 476–489.

Häkkinen, M. and Hong, W. (2007). New species and variety of *Musa* (Musaceae) from Yunnan, China, *Novon* 17(4): 440–446.

Harrison, C. J., Alvey, E. and Henderson, I. R. (2010). Meiosis in flowering plants and other green organisms, *Journal of Experimental Botany* 61(11): 2863–2875.

He, G., Hu, F., Ming, J., Liu, C. and Yuan, S. (2017). Pollen viability and stigma receptivity in Lilium during anthesis, *Euphytica* 213(10): 231.

Heslop-Harrison, J., Heslop-Harrison, Y. and Shivanna, K. R. (1984). The evaluation of pollen quality, and a further appraisal of the fluorochromatic (FCR) test procedure, *Theoretical and Applied Genetics* 67(4): 367–375.

Holmes, S. (1979). *Henderson's Dictionary of Biological Terms* (9th edn.), London: Longman.

IITA (1993). *Annual Report of the Plantain and Banana Improvement Program for 1992*, Ibadan, Nigeria: Crop Improvement Division of the International Institute of Tropical Agriculture International Institute of Tropical Agriculture.

Inta, W., Traiperm, P. and Swangpol, S. C. (2015). Floral micromorphology of the genus *Ensete* Bruce ex Horan. (Musaceae) in Thailand, *Taiwania* 60: 99–106.

Jeridi, M., Bakry, F., Escoute, J., Fondi, E., Carreel, F., Ferchichi, A., D'Hont, A. and Rodier-Goud, M. (2011). Homeologous chromosome pairing between the A and B genomes of *Musa* spp. revealed by genomic *in situ* hybridization, *Annals of Botany* 108(5): 975–981.

Joldersma, D. and Liu, Z. (2018). The making of virgin fruit: the molecular and genetic basis of parthenocarpy, *Journal of Experimental Botany* 69(5): 955–962.

Kawano, S., Li, Y., Yahata, M. and Kunitake, H. (2016). Effect of temperature on self-incompatibility in Citrus pistil and mature pollen culture systems, *Acta Horticulturae* 1135: 117–122.

Kirchoff, B. K. (1992). Ovary structure and anatomy in the Heliconiaceae and Musaceae (Zingiberales), *Canadian Journal of Botany* 70(12): 2490–2508.

Kirchoff, B. K. (2017). Inflorescence and flower development in *Musa velutina* H. Wendl. & Drude (Musaceae), with a consideration of developmental variability, restricted phyllotactic direction, and hand initiation, *International Journal of Plant Sciences* 178(4): 259–272.

Kronestedt, E. and Walles, B. (1986). Anatomy of the *Strelitzia reginae* flower (Strelitziaceae), *Nordic Journal of Botany* 6(3): 307–320.

Kronestedt, E., Walles, B. and Alkemar, I. (1986). Structural studies of pollen tube growth in the pistil of *Strelitzia reginae*, *Protoplasma* 131(3): 224–232.

Lin, R. and Tang, W. (2014). Cross talk between light and ABA Signaling. In: Zhang, D. P. (Ed.), *Abscisic Acid: Metabolism, Transport and Signaling*, Dordrecht: Springer Science+Business Media, pp. 255–269.

Liu, A. Z., Li, D. Z., Wang, H. and Kress, W. J. (2002). Ornithophilous and chiroterophilous pollination in *Musa itinerans* (Musaceae) a pioneer species in tropical rain forests of Yunnan, Southwestern China, *Biotropica* 34(2): 254–260.

Liu, B., De Storme, N. and Geelen, D. (2018). Cold-induced male meiotic restitution in Arabidopsis thaliana is not mediated by GA-DELLA signalling, *Frontiers in Plant Science* 9: 91.

Mizuta, Y. and Higashiyama, T. (2018). Chemical signalling for pollen tube guidance at a glance, *Journal of Cell Science* 131(2): jcs208447.

Nyine, M. and Pillay, M. (2007). Banana nectar as a medium for testing pollen viability and germination in *Musa*, *African Journal of Biotechnology* 6(10): 1175–1180.

Nyine, M., Uwimana, B., Akech, V., Brown, A., Ortiz, R., Doležel, J., Lorenzen, J. and Swennen, R. (2019). Association genetics of bunch weight and its component traits in east African highland banana (*Musa* spp. AAA group), *Theoretical and Applied Genetics* 132: 3295–3308.

Nyine, M., Uwimana, B., Swennen, R., Batte, M., Brown, A., Christelová, P., Hřibová, E., Lorenzen, J. and Doležel, J. (2017). Trait variation and genetic diversity in a banana genomic selection training population, *PLoS ONE* 12(6): e0178734.

Ohkawa, H., Sugahara, S. and Oda, M. (2012). Seed formation promoted by paclobutrazol, a gibberellin biosynthesis inhibitor, in *pat-2* parthenocarpic tomatoes, *Journal of the Japanese Society for Horticultural Science* 81(2): 177–183.

Ortiz, R. and Crouch, J. H. (1997). The efficiency of natural and artificial pollinators in plantain (*Musa* spp. AAB group) Hybridization and Seed Production, *Annals of Botany* 80(5): 693–695.

Ortiz, R., Ferris, R. S. B. and Vuylsteke, D. (1995). Banana and plantain breeding. In: Gowen, S. (Ed.), *Bananas and Plantains*, London: Chapman and Hall, pp. 110–146.

Ortiz, R., Ulburghs, F. and Okoro, J. (1998). Seasonal variation of apparent male fertility and 2n pollen production in plantain and banana, *HortScience* 33(1): 146–148.

Ortiz, R. and Vuylsteke, D. (1995a). Factors influencing seed set in triploid *Musa* spp. L. and production of euploid hybrids, *Annals of Botany* 75(2): 151–155.

Ortiz, R. and Vuylsteke, D. (1995b). Effect of parthenocarpy gene P_1 on fruit and bunch traits of plantain-banana hybrids, *Heredity* 75(5): 460–465.

Oselebe, H. O., Nnamani, C. V. and Ikeh, E. (2014). Pollen diversity, viability and floral structure of some *Musa* genotypes, *Nigerian Journal of Biotechnology* 27: 21–27.
Patterson, S. E. (2001). Cutting loose. Abscission and dehiscence in Arabidopsis, *Plant Physiology* 126(2): 494–500.
Peñaloza, P. and Toloza, P. (2018). Boron increases pollen quality, pollination, and fertility of different genetic lines of pepper, *Journal of Plant Nutrition* 41(8): 969–979.
Picarella, M. E. and Mazzucato, A. (2019). The occurrence of seedlessness in higher plants; insights on roles and mechanisms of parthenocarpy, *Frontiers in Plant Science* 9: 1997.
Ram, H. Y. M., Ram, M. and Steward, F. C. (1962). Growth and development of the banana plant. 3. A: The origin of the inflorescence and the development of the flowers. B: The structure and development of the fruit, *Annals of Botany* 26(4): 657–673.
Ray, P. K. (2002). *Breeding Tropical and Subtropical Fruits*, Pusa: Norosa Publishing House.
Rowe, P. R. and Richardson, D. L. (1975). *Breeding Bananas for Disease Resistance, Fruit Quality and Yield*. La Lima, Honduras, CA: Bulletin 2, Tropical Agriculture Research Services (SIATSA), p. 41.
Roy, R., Schmitt, A. J., Thomas, J. B. and Carter, C. J. (2017). Nectar biology: from molecules to ecosystems, *Plant Science* 262: 148–164.
Rutikanga, A., Tusiime, G., Night, G., Ocimati, W., and Blomme, G. (2016). Variation in nectar volume and sugar content in male flowers of *Musa* cultivars grown in Rwanda and their non-effect on the numbers of visiting key diurnal insect vectors of banana Xanthomonas wilt, *African Journal of Agricultural Research* 11(8): 607–623.
Sardos, J., Rouard, M., Hueber, Y., Cenci, A., Hyma, K. E., van den Houwe, I., Hribova, E., Courtois, B. and Roux, N. (2016). A genome- wide association study on the seedless phenotype in banana (*Musa* spp.) reveals the potential of a selected panel to detect candidate genes in a vegetatively propagated crop, *PLoS ONE* 11(5): e0154448.
Shepherd, K. (1954). Seed fertility of the Gros Michel banana in Jamaica, *Journal of Horticultural Science* 29(1): 1–11.
Shepherd, K. (1960). Seed fertility of edible bananas, *Journal of Horticultural Science* 35(1): 6–20.
Shepherd, K. (1974). Banana research at ICTA [Imperial College of Tropical Agriculture, West Indies], *Tropical Agriculture (Trinidad and Tobago)* 51: 482–490.
Shepherd, K. (1999). *Cytogenetics of the Genus Musa*, Montpellier: INIBAP.
Shepherd, K., Dantas, J. L. L. and Alves, E. J. (1987). Aspects of banana breeding at the Centro Nacional de Pesquisa de Mandioca e Fruticultura, Brasil. In: Galindo, J. J. and Jaramillo, R. (Eds), ACORBAT Memorias VII Reunion. Costa Rica 23–27 Set, 1985, Turrialba: CATIE, pp. 75–86.
Simmonds, N. W. (1952). Experiments on the pollination of seeded diploid bananas, *Journal of Genetics* 51(1): 32–40.
Simmonds, N. W. (1953). Segregations in some diploid bananas, *Journal of Genetics* 51(3): 458–469.
Simmonds, N. W. (1960). Megasporogenesis and female fertility in three edible triploid bananas, *Journal of Genetics* 57(2–3): 269–278.
Simmonds, N. W. (1962). *The Evolution of the Bananas*, London: Longman.
Simmonds, N. W. (1966). *Bananas* (2nd edn.), London: Longman.

Simmonds, N. W. and Dodds, K. S. (1949). Meiosis in seeded diploids of *Musa*, *Journal of Genetics* 49(3): 221–225.

Šimoníková, D., Nemecková, A., Karafiátová, M., Uwimana, B., Swennen, R., Doležel, J. and Hřibová, E. (2019). Chromosome painting facilitates anchoring reference genome sequence to chromosomes in situ and integrated karyotyping in banana (*Musa* spp.), *Frontiers in Plant Science* 10: 1503.

Sinnott, E. W. (1960). *Plant Morphogenesis*, New York: McGraw-Hill Book Company, Inc.

Soares, T. L., de Souza, E. H., de Carvalho-Costa, M. A. P., de Oliveira e Silva, S. and dos Santos-Serejo, J. A. (2014). *In vivo* fertilization of banana, *Ciencia Rural, Santa Maria* 44(1): 37–42.

Soares, T. L., de Souza, E. H., Sampaio, L. F. S., de Carvalho Costa, M. A. P., e Silva, S. D. O. and dos Santos-Sejero, J. A. (2015). Effect of collection time on the viability of banana pollen grains, *African Journal of Biotechnology* 14(14): 1207–1214.

Soares, T. L., Silva, S. O., Costa, M. A. P. C., Santos-Serejo, J. A., Souza, A. S., Lino, L. S. M., Souza, E. H. and Jesus, O. N. (2008). In vitro germination and viability of pollen grains of banana diploids, *Crop Breeding and Applied Biotechnology* 8(2): 111–118.

Specht, C. D., Yockteng, R., Almeida, A. M., Kirchoff, B. K. and Kress, W. J. (2012). Homoplasy, pollination, and emerging complexity during the evolution of floral development in the tropical gingers (Zingiberales), *Botanical Review* 78(4): 440–462.

Ssebuliba, R., Magambo, M., Talengera, D., Makumbi, D., Tenkouano, A., Rubaihayo, P. and Pillay, M. (2006). Biological factors affecting seed production in East African Highland bananas, *Journal of Crop Improvement* 16(1–2): 67–79.

Ssebuliba, R., Makumbi, D. and Pillay, M. (2009). Patterns of seed set in East African Highland banana (Musa spp.) hybrids, *Journal of New Seeds* 10(3): 160–170.

Ssebuliba, R. N., Rubaihayo, P., Tenkouano, A., Makumbi, D., Talengera, D. and Magambo, M. (2005). Genetic diversity among East African Highland bananas for female fertility, *African Crop Science Journal* 13: 13–26.

Ssebuliba, R. N., Tenkouano, A. and Pillay, M. (2008). Male fertility and occurrence of 2n gametes in East African Highland bananas (*Musa* spp.), *Euphytica* 164(1): 53–62.

Stotzky, G. and Cox, E. A. (1962). Seed germination studies in *Musa*. 2. Alternating temperature requirement for the germination of *Musa balbisiana*, *American Journal of Botany* 49(7): 763–770.

Stover, R. H. and Simmonds, N. W. (1987). *Bananas* (3rd edn.), London: Longman Group.

Swennen, R. and Vuylsteke, D. (1989). Aspects of plantain breeding at IITA. In: INIBAP Workshop on Sigatoka Leaf Spot Diseases (*Mycosphaerella* spp), San Jose, Costa Rica.

Swennen, R. and Vuylsteke, D. (1993). Breeding black sigatoka resistant plantains with a wild banana, *Tropical Agriculture (Trinidad)* 70: 74–77.

Swennen, R., Vuylsteke, D. and De Smet, K. (1991). Season dependent seed set in plantain. *Banana Newsletter* 14: 35–36.

Swennen, R., Vuylsteke, D. and Hahn, S. K. (1992). The use of simple biotechnological tools to facilitate plantain breeding. In: Thottapilly, G., Monti, L. M., Mohan, D. R. and Moore, A. W. (Eds), *Biotechnology: Enhancing Research on Tropical Crops in Africa*, Ibadan: International Institute of Tropical Agriculture, pp. 69–74.

Takisawa, R., Nakazaki, T., Nunome, T., Fukuoka, H., Kataoka, K., Saito, H., Habu, T. and Kitajima, A. (2018). The parthenocarpic gene *Pat-k* is generated by a natural mutation of *SlAGL6* affecting fruit development in tomato (*Solanum lycopersicon* L.), *BMC Plant Biology* 18(1): 72

Tenkouano, A., Pillay, M. and Ortiz, R. (2011). Breeding techniques. In: Pillay, M. and Tenkouano, A. (Eds), *Banana Breeding: Constraints and Progress*, Boca Raton, FL: CRC Press, pp. 181–202.

Tenkouano, A. and Swennen, R. (2004). Progress in breeding and delivering improved plantain and banana to African farmers, *Chronica Horticulturae* 44(1): 9–15.

Tripathi, J. N., Ntui, V. O., Ron, M., Muiruri, S. K., Britt, A. and Tripathi, L. (2019). CRISPR/Cas9 editing of endogenous banana streak virus in the B genome of *Musa* spp. overcomes a major challenge in banana breeding, *Communications Biology* 2: 46.

Turner, D. W. and Gibbs, D. J. (2018). A functional approach to bunch formation in banana. In: Kema, G. and Drenth, A. (Eds), *Achieving Sustainable Cultivation of Bananas: Cultivation Techniques*, Cambridge, UK: Burleigh Dodds Science Publishing, pp. 93–116.

Uma, S. and Arun, K. (2016). Understanding the diversity and reproductive biology of banana – for improvement through basic research, *Acta Horticulturae* 1114: 1–12.

Uma, S., Mustaffa, M. M., Saraswathi, M. S. and Durai, P. (2011). Exploitation of diploids in Indian banana breeding programmes, *Acta Horticulturae* 897: 215–224.

Vuylsteke, D., Ortiz, R., Ferris, R. S. B. and Crouch, J. H. (1997). Plantain improvement, *Plant Breeding Reviews* 14: 267–320.

Vuylsteke, D., Swennen, R. and De Langhe, E. (1990). Tissue culture technology for the improvement of African plantains. In: Fullerton, R. A. and Stover, R. H. (Eds), *Sigatoka Leaf Spot Diseases of Bananas*. Proceedings of the an International Workshop held at San José, Costa Rica. Montpellier, France: International Network for the Improvement of Banana and Plantain, pp. 316–337.

Vuylsteke, D., Swennen, R. and Ortiz, R. (1993a). Development and performance of Black Sigatoka-resistant tetraploid hybrids of plantain (*Musa* spp., AAB group),. *Euphytica* 65: 33–42.

Vuylsteke, D., Swennen, R. and Ortiz, R. (1993b). Registration of 14 improved tropical *Musa* plantain hybrids with Black Sigatoka resistance, *HortScience* 28(9): 957–959.

White, P. R. (1928). Studies on the banana: an investigation of the floral morphology and cytology of certain types of the genus *Musa* L., *Zeitschrift für Zellforschung und Mikroskopische Anatomie* 7(5): 673–733.

Wilson, G. B. (1946a). Cytological studies in the Musae. 1. Meiosis in some triploid clones, *Genetics* 31: 241–258.

Wilson, G. B. (1946b). Cytological studies in the Musae. 2. Meiosis in some diploid clones, *Genetics* 31: 475–482.

Wu, J., Qin, Y. and Zhao, J. (2008). Pollen tube growth is affected by exogenous hormones and correlated with hormone changes in styles in *Torenia fournieri* L., *Plant Growth Regulation* 55(2): 137–148.

Chapter 12

Targeted improvement of Cavendish clones

Eli Khayat, Rahan Meristem (1998) LTD., Israel

1 Introduction

Since the turn of the twentieth century the banana export industry has depended on monocultures of triploid and parthenocarpic cultivars. Most cultivars, unlike their wild precursors, are devoid of seeds and feature stagnant genomes. Clearly, sterility is a major hindrance to diversity of agro traits, especially tolerance to biotic pressures set off by microbial pathogens that rapidly adapt to changes in the environment. Moreover, the impacts of rapid climate changes provide a greater advantage to microorganisms that respond, adapt, and evolve quickly in contrast to the sterile banana genomes that remain stationary. Despite the fact that the gene pool of the genus *Musa* is highly diverse and contains a wealth of genes for resistance to the major diseases and pests (Khayat and Ortiz, 2011), the triploid and parthenocarpic nature of most cultivated varieties expose the plants to unsolved confrontations with biotic and abiotic challenges, especially in intense monoculture production systems. By no choice, most domesticated cultivars are vegetatively propagated, traditionally from corms, and in recent years from tissue culture. This review will provide an update of the methods that breeders have used to modify disease resistance, plant architecture and fruit ripening, circumventing constraints of sterility.

Cavendish cultivars were introduced to replace Gros Michel back in the 1960s mainly due to their resistance to Fusarium wilt caused by the so-called race 1 strains of *Fusarium oxysporum f.sp. cubense* (Foc), the causal agent of Panama disease, which have now been renamed as different *Fusarium* species (Maryani et al., 2019), following a severe outbreak in the late 1950s in Central America. But in addition to race 1 resistance, the Cavendish subgroup

http://dx.doi.org/10.19103/AS.2020.0070.15

exhibits excellent consumer traits and is by far the most popular imported banana in the marketplace, representing approximately 99% (Ploetz, 2006; Stover and Simmonds, 1987) of exported bananas. Several attempts were made to replace the Cavendish banana with various alternative cultivars, but until now consumers always preferred the Cavendish. In recent years, due to a new outbreak of Fusarium wilt, caused by *F. odoratissimum* tropical race 4 (TR4; Maryani et al., 2019) and viral diseases (mainly banana bunchy top virus (BBTV)), the option of replacing Cavendish with a man-created hybrid is being revisited by different research groups. Researchers have attempted to develop hybrids that will resemble Cavendish and/or receive consumer acceptance.

The main objective of this review is to outline the potential of breeding technologies that are presently available and has been utilized to improve the performance of Cavendish cultivars, and to overcome today's biological and environmental pressures. Genomic engineering and induced mutagenesis are the two options to achieve genetic modifications.

The Cavendish group of cultivars became the leading and nearly sole banana cultivar exported from banana-producing countries to Western markets. Paradoxically, despite sterility and lack of cross-hybridization, different Cavendish cultivars display an array of diverse morphologic and biochemical phenotypes. The cultivar Valery is approximately 5 m tall while Dwarf Cavendish is approximately 1.5 m from the ground to the inflorescence. Other Cavendish cultivars range in height between 2 m and 4.5 m (Khayat and Ortiz, 2011). For example, the Adi cultivar displays elevated levels of starch, sucrose and reducing sugars (Khayat, 2010).

2 Genetic background of bananas and plantains

Nearly all domesticated bananas and plantains are interspecific or intraspecific triploid hybrids that constitute all possible combinations of the A and B genomes representing *Musa acuminata* Colla. and *M. balbisiana* Colla., respectively. Triploids were preferred by farmers due to their seedless makeup. In this regard, cultivated bananas differ from their wild diploid ancestors (AA, BB genomes) that bear large and hard seeds. Ironically, the major banana constraints have been solved by genetic engineering for example augmentation of pro vitamin A (Paul et al., 2017).

Nearly in merely 100 years of banana and plantain breeding, very few qualitative or quantitative traits have been adopted in commercially important *Musa* cultivars. When compared to other crop species, in the most part, *Musa* breeding did not bear many beneficial products. But advances in genomics and related technologies that are fueled by advances in computation, spurred high expectations. Unfortunately, up to the present time, the use of transgenic plants is null in the global trade despite the successful development of plants resistant to TR4Panama disease (Dale et al., 2017), and extended shelf life (Elizure et al., 2016). Nevertheless, these insights and methodologies have driven a good understanding of the different banana genomes and to some extent the positional mapping of genes that control specific traits on virtual chromosomes. Still, the complexity of positional cloning in banana, combined with the lack of saturated linkage maps hinders gene-targeted hybridization breeding. The inability of researchers to use map-based cloning is a serious limitation for identifying strategic loci in segregating populations. However, the recent progress made in banana genomics will undoubtedly assist breeders to identify interesting loci and genes that govern important traits. To live up to such expectations will be a tall order due to remaining obstacles. The main barriers are a long cycling time span from seed to seed in the triploid lines, and the low fertility of most polyploid cultivars.

Several morphological and physiological traits were emphasized to distinguish domesticated crops from their wild ancestors. These characteristics are referred to as the domestication syndrome (Poncet et al., 1998). Normally, syndrome characteristics include a more compact growth habit, increased earliness, reduction/loss of seed dispersal and dormancy, gigantism and increased morphological diversity in the consumed portion of the plant (Frary and Doganlar, 2003). Studies on the domestication syndrome and process have revealed that numerous traits that distinguish crop plants from their wild relatives are often governed by a relatively small number of loci having effects of unequal magnitude. The differences between domesticated bananas and their wild relatives involve parameters that govern not only disease and pest resistance but also plant architecture, yield and fruit characteristics. In the most part, growers tended to select only hybrids exhibiting parthenocarpic fruit. The pressures of agro-technical prerequisites add to the complexity of developing a clone fit for commercial use.

3 In vitro mutagenesis: somaclonal variation

The components of yield are well-defined by banana and plantain producers and breeders. These include bunch weight, number of marketable hands in a bunch, number of fingers in a hand, finger size, cycling time and number of stems per hectare. In the export industry (mainly Cavendish) the most commonly used yield parameter is box per stem ratio. This appraisal combines weight and fruit quality. These traits are widely used for selection of hybrids (Osuji et al., 1997; Krikorian et al., 1992; Crouch et al., 2000) in plantains, as well as somaclones of Cavendish (Khayat et al., 2004; Khayat, 2010). The merit of selecting yield components is limited due to masking of the traits by diseases and pests. It was therefore argued that subtle quantitative traits are best When the selection is performed, usually devoid of diseases and pests (Khayat et al., 2004). To this end, somaclonal selections of Cavendish were highly productive in one of the most marginal climates for banana growth, Northern Israel. The yield performance of these selections was proven superior not only in subtropical areas such as in Israel and South Africa (Eckstein et al., 1998), but also for tropical climates such as the Davao region of the Philippines (Khayat et al., 2004; Khayat, 2010), Africa and Latin America (Khayat and Ortiz, 2014). The shift from diploid to polyploid karyotypes greatly increased the bunch weight and fruit size of both banana and plantain hybrids (Ortiz and Vuylsteke, 1995). Banana and plantain breeders should quantify additive and non-additive components of genetic variance to choose appropriate selection methods to improve quantitative traits. Furthermore, narrow-sense heritability, which is the ratio between the additive genetic variance and the total phenotypic variance, is a scale-independent quantity that plays an important role in the theory of selection methods because the response to selection can be predicted if the breeder knows the target-trait heritability. Triploid parent-offspring regression analysis suggests that bunch weight, number of fruits and fruit length had intermediate to low heritability in tetraploid plantain-banana hybrids from 3x-2x crosses (Ortiz, 2000). This indicates that they could not be predicted using the phenotype of the triploid parent and that progeny testing will be required to assess and select the best parents for population improvement of such traits. Progeny testing in secondary triploid hybrids (from 4x-2x crosses) shows, however, that general combining ability was much greater than specific combining ability for yield and associated traits (Tenkouano et al., 1998). This result suggests therefore that the 4x-2x breeding scheme should aim to accumulate favorable alleles in potential tetraploid and diploid parents through recurrent selection.

The frequency of off-types in tissue culture-propagated banana cultivars is several orders of magnitude above the expected level of random mutations. Interestingly, most mutations fall within distinct recurrent groups, namely dwarf, leaf variegation ('Massada'), tall and twisted hand in the top of the fruit bunch and tall and slender plants (Fig. 1). The frequency of these mutations is exacerbated when plants regenerate from extensive tissue culture (Khayat et al., 2004), especially when high levels of plant growth regulators (cytokinins or auxins) are utilized in the process of in vitro propagation. The term 'somaclonal variation' has been used to define somatic mutations triggered by the in vitro conditions, but the mechanism of these mutations is diverse in various crop species and in many cases is poorly understood.

Regardless of the source of mutagenesis, the question of genetic stability is always a concern when selecting somaclonal mutants. To this end, we have used a long and narrow leaf mutant to prove a true genomic polymorphic

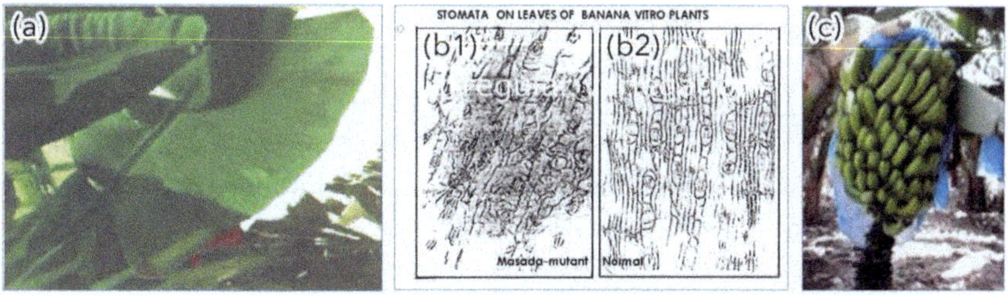

Figure 1 Phenotypes expressed in Massada (variegated) mutants. (a) Variegated leaf. (b) Irregularity in the arrangement of the stomata (b1) - Mutant; (b2) - Wt. (c) Fruit bunch of mutated plants.

change in the variant. Figure 2 exhibits an AFLP comparison of DNA banding pattern between a parental-line and its in vitro propagules originating from the same plant. The polymorphic pattern in one of the segregants was stable for several generations in the field. This experiment opened the path for the discovery of a new method for diversifying Cavendish clones.

Numerous genetic pathways have been associated with somaclonal variations. Higher ploidy number and aneuploidy were found in some cases (Mujib et al., 2007; Leva et al., 2012) including the Massada off-type in bananas where the number of chromosomes is 34 instead of 33 (Khayat et al., 2004). Other types of mutations include point mutations (Ngezahay et al., 2007), somatic crossing over and sister chromatid exchange (Bairu et al., 2011),

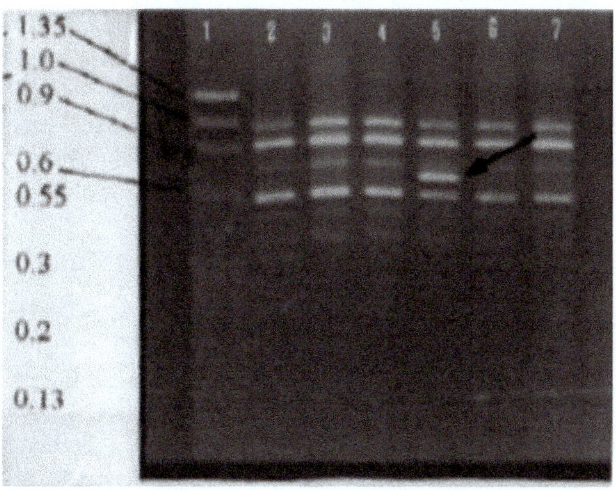

Figure 2 AFLP marker-analysis of genomic DNA, comparing the amplified products of the mother plant placed in culture (lane 2) and derived plants after eight cycles of subcultures of the meristem (lanes 3-7). A polymorphic band appeared in a single line (lane 5).

chromosome breakage and rearrangement (Alvarez et al., 2010), somatic gene rearrangement, DNA amplification (Karp, 1995; Tiwari et al., 2013), changes in organelle DNA (Cassels and Curry, 2001), DNA methylation (Stelpflug et al., 2014), epigenetic variation (Kaeppler et al., 2000; Guo et al., 2007; Smulders and de Klerk, 2011), histone modifications and RNA interference (Miguel and Marum, 2011), segregation of pre-existing chimeral tissue (Brar and Jain, 1998; Vázquez, 2001; Ravindra et al., 2012; Nwauzoma and Jaja, 2013) and insertion or excision of transposable elements (TEs; Gupta, 1998; Sato et al., 2011). In particular, retrotransposable elements are among the major causes of genetic rearrangements during in vitro culture (Hirochika et al., 1996; Sato et al., 2011).

A major source of genetic variation in many plant species entails transcriptional activation of REs. Although REs were initially regarded as 'junk DNA', with little or no effect on plant traits, it has been demonstrated that they play key roles in chromosome structures, gene expression, and regulation, as well as adaptation to biotic and abiotic stresses and evolution (Orozco-Arias et al., 2019). They are particularly important in the evolution of sterile plants that propagate asexually. These elements are mobile and transpose through transcription and reintegration in the genome. REs are ubiquitous to all plant species, but differ among species in their abundance and distribution along chromosomes. The abundance of long terminal repeat (LTR) retrotransposons differs among species, but can cover a large percentage of the total genome, for example, the rice genome approximates 389 Mb and comprises ~22% LTR retrotransposon sequences (Ma and Bennetzen, 2004), and no less than 74.6% of the 2045 Mb maize genome (Baucom et al., 2009). Many of these sequences are typically concentrated in centromeres. Their structure is made of LTRs flanking the element and two open-reading frames encoding a Gag protein and a single ORF coding for a reverse transcriptase, RNAs H, a protease and an integrase (Fig. 3) (Kumar and Bennetzen, 1999). Many plant genera have provided a view of the dynamic nature of retrotransposon distribution in plant genomes and the sporadic post-speciation bursts of retrotransposition that

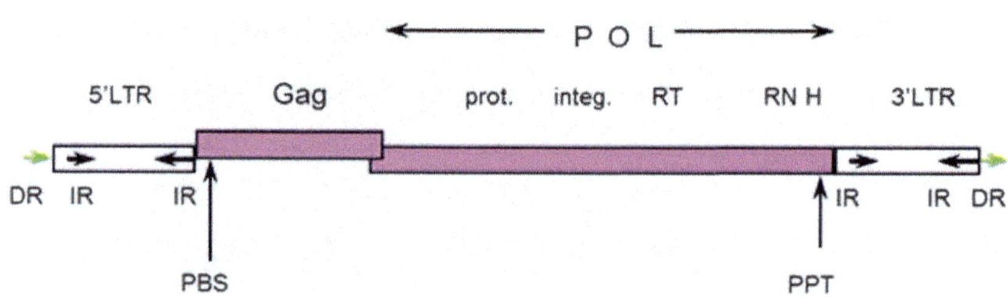

Figure 3 Structure of the BAN-1 retrotransposable element coding to a GAG protein and a single open-reading frame containing (Prot) a protease, (integ) integrase, (RT) reverse transcriptase and (RN H) RNAs H. The element is flanked by two long terminal repeats (LTR).

have shaped plant chromosomes, including those of crop plants (Bennetzen, 1998; Bennetzen and Kellogg, 1997; Bennetzen et al., 1993).

In a previous publication (Khayat et al., 2004), we have shown a clear linkage between somaclonal mutations and activation of retrotransposable elements. These elements are especially abandoned in the banana genome. The large majority belong to the T-1 *copia* type although the T-3 gypsy type is also abundant. Both types belong to the LTR-containing retrotransposons. The LTR sequences encode both a promotor and a terminator sequence. In consensus to other monocots, the promotor sequence is CpG-rich and heavily methylated. Recent data show DNA methylation through small interfering (Si)RNAs (Paszkowski, 2015) in cell replication. Cytosine methylation has been associated with genetic inactivity and a heterochromatic state in many higher eukaryotes, including plants where de novo methylation was first detected during the inactivation of DNA transposable elements of the mutator (Bennetzen et al., 1993). This cytosine methylation has been associated with both transcriptional inactivity and a higher rate of C to T transitions, giving rise to enhanced transcriptional and mutational silencing.

Despite many years of investigation, in many species, it is not clear whether the methylation itself establishes the inactive state of a retrotransposon, or if they incite a cascade effect leading to new traits. On a chromatin level the outcome of inactivation in is caused by an alteration of chromatin structure. The first evidence of a role of DNA methylation in the regulation of retrotransposition came from a study on *Tto1* by Hirochika et al. (2000). Here it was observed that after the initial active retrotransposition in the *Arabidopsis* genome, *Tto1* was silenced. But this was only observed in transgenic *Arabidopsis* lines where the *Tto1* copy number was increased by active transposition during the culture-mediated transformation process. Thus the silencing appears to mimic repeat-induced gene silencing (Matzko and Matzko, 1998).

Using AFLP markers for polymorphic analysis of the common banana mutant, long and narrow leaves (LNL) and its precursor line, we have shown a banding pattern consisting of a unique fragment (Fig. 4). The nucleotide sequence of the polymorphic band (BR1) encodes a 340 bp ortholog of a reverse transcriptase of the rice TOS 17 RTE (Hirochika et al., 1996).

Using the BR-1 (banana retrotransposon) as a probe in a DNA blot-hybridization analysis revealed strong amplification of specific bands in the mutant compared to the DNA pattern of the original mother plant (Fig. 4).

To confirm activation of the retro-element, we used (RT)-PCR reverse transcriptase. A single amplicon of approximately 350 bp was generated only from RNA samples containing the RT enzyme (Fig. 5).

These data demonstrate that the BR-1 element in Cavendish banana is transcriptionally regulated. It was also evident that extensive tissue culture activates transcription of BR-1 leading to modulation of plant architecture.

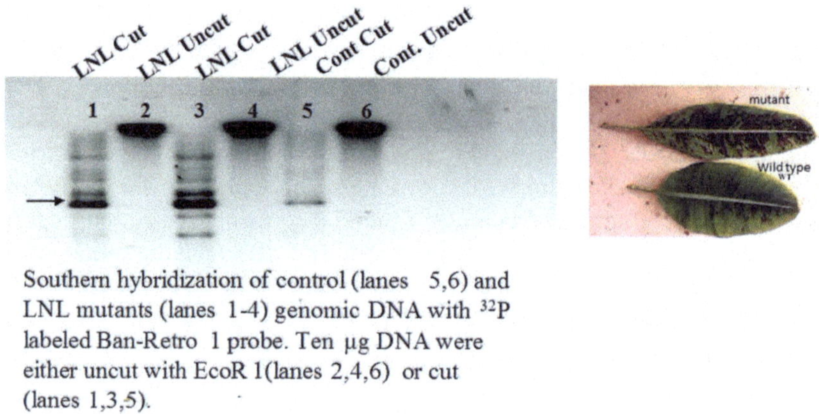

Figure 4 Southern hybridization of control (lanes 5-6) and LNL mutants (lanes 1-4) genomic DNA with ^{32}P-labeled Ban-retro 1 probe. Ten μg DNA were either uncut with EcoR 1 (lanes 2, 4, 6) or cut (lanes 1, 3, 5).

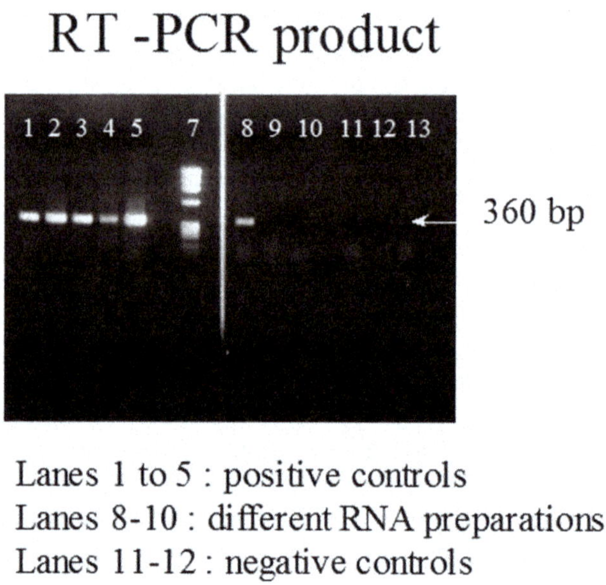

Figure 5 RT-PCR product. Lanes 1-5: positive controls. Lanes 8-10: different RNA preparations. Lanes 11-12: negative controls.

Among Cavendish cultivars there are major differences in height and girth, ranging from the tall and slender Valery to short and thick Dwarf Cavendish. These polymorphic growth characteristics have an influence on yield and fruit quality. Using a BAC library and genome walking we have found a cluster of BR-1 sequences residing in the vicinity of a *Gai* gene, a receptor of the plant-growth regulator, gibberellic acid (GA3). Activation of BR-1 by demethylation resulted in plants with a wide diversity of heights including very short mutants (Adi mutants), of 2.3 m compared to the 4.5 m wild-type mother line (Fig. 6) (Khayat, 2010).

These mutants have been selected in vitro in the presence of GA3 in the medium. Plants insensitive to GA3 did not elongate. The chromatin distance between the edges of the landing pad of the BR-1 cluster and the GAI receptor was approximately 200 Kbp upstream of the promotor (Fig. 7).

Depending on the cycling number, hormonal regiments and frequency of using the demethylation agent, it is possible to distinguish between lines of plants that exhibit a different profile of banding pattern. These represent mutations in different QTLs. The DNA hybridization blots represent an array of 'landing pads' for the retro-element cascades. Through the different profiles

Figure 6 Comparison of stature between Adi (photo in the left panel), Grand Neine (photo in the center panel) and Jaffa (photo in the right panel).

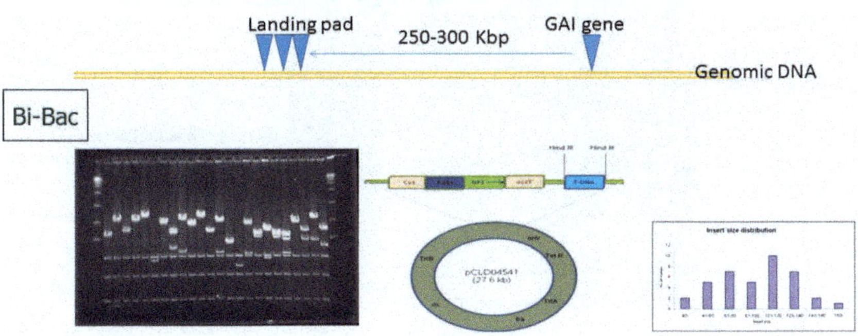

Figure 7 Physical mapping of a major landing pad of retrotransposable element residing 200 Kbp upstream from a GA receptor. Mutated plants in this QTL express plants with different heights. The distance was calculated from an assembly of two clones in a Bi-BAC library (shown below) with an average insert size of 101–120 Kbp.

we were able to segregate mutant lines according to the locus of mutation. By genome walking it is possible to identify genes in the vicinity of the mutated QTLs.

We have found that using the demethylation compounds 5-Aza-2′-deoxycytidine it is possible to control DNA methylation of plant meristems. This induced the activation of RTE and the generation of an array of mutants (Fig. 8). Given the new insertions of the retro-elements (BR-1 activation) the mutations remain stable. Employing this technique, we have recently Mutated GAL (a Cavendish cultivar) and tested the plants for resistance to TR4. From a population of 9640 in vitro-mutated plants that were inoculated with TR4, we selected 514 lines that exhibited an asymptomatic phenotypes, after challenging with the pathogen. These lines are currently being evaluated in a TR4-infested area in the Philippines (Fig. 9).

Major efforts were put in by the export industry to overcome the challenge of controlling fruit ripening. Being a climacteric fruit, growers are forced to harvest at early stages and suffer a significant loss of yield due to harvesting prior to full weight gain. The commerce of Cavendish banana entails long periods of shipping to remote markets. Extended shelf life is a highly desired trait that can be achieved by genetic modifications. Climacteric ripening is an inherent natural feature in all banana and plantains and given the distant markets it deserves special attention when selecting a new cultivar. Ironically, excellent genetic tools do exist to prolong the green life of the fruit (Elitzur et al., 2016). Simple gene silencing significantly augmented the shelf life. But, due to lack of acceptance of genetically modified products the industry is forced to

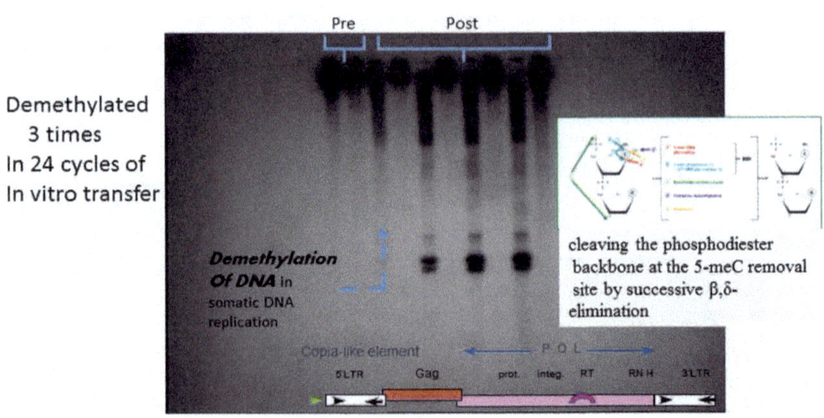

Figure 8 In-vitro activation of Retrotransposable elements by in vitro demethylation of DNA in GAL plants. Ex vitro plants of the mutated lines were screened for TR4 resistance in a greenhouse (after inoculation) and in the field. Pre - is DNA extracted from plants prior to de-methylation with 'AZA'; Post - is DNA extracted from plants post the de-methylation treatment. Polymorphic genotypes were detected in the post-treated lines.

Figure 9 Field trial of mutated plants. Resistant GAL clones in Dabong-Dabong, Valencia City Bukidnon, the Philippines. Free trial: Number of TC plants in the trial: 3000; number of treatments (clones): 130; number of replications (blocks): 25; number of Foc-sensitive cases in selected clones: 10; first Foc incidence: 18 WAP; clone, 1585; plant stage: between first to second harvest.

deploy ripening-prevention protocols requiring exorbitant packing, shipping and handling practices. Meticulous protocols have been deployed by the industry to assure the fruit is offered at a perfect stage of ripening. Transitioning to a new cultivar from a hybrid will require immense efforts on the part of the industry. On the other hand nearly all the biochemical components involved in banana conversion of starch to sucrose, chlorophyll degradation and ethylene biosynthesis have been described (for review see Khayat and Ortiz, 2014). This information is valuable to genetic engineers, and may be used through transgenic, intergenic/cisgenic breeding or genome editing.

4 Conclusion and future trends

The Cavendish banana is a highly valuable product globally. The various Cavendish cultivars manifest a range of phenotypes that have been adapted by farmers in various parts of the world. But these differences are not obvious to the consumer. The TR4 epidemic, initially confined to Southeast Asia develops into a pandemic (Ordóñez Román et al., 2015), particularly since the occurrence of TR4 in the La Guajira peninsula, Colombia, (García-Bastidas et al., 2019), which again alarmed the industry and consumers worldwide. Luckily, it was demonstrated that excellent resistance exists in the *Musa* genus (Dale et al., 2017). Furthermore, the gene that confers resistance in the wild *Musa acuminata* ssp. *malaccensis* is a single dominant gene that exists in Cavendish as well but its expression level is low. Genetic modification or genome-editing tools currently available are very effective and could dramatically change this trait by either transgenic plants

(GMO) or genome editing. At the same time, the Cavendish subgroup has a remarkable competency to express novel phenotypes through the process of somaclonal variation, but not for all traits. For example Black Sigatoka seems to be controlled by several genes in resistant varieties (Ortiz and Vuylsteke, 1992). It seems at least in part this genetic mechanism is retro-element activation. In vitro demethylation of DNA is very effective in creating populations of mutants that can be examined according to their banding pattern using hybridization assays. This method is useful not only for expressing or suppressing genes, but also for mapping mutations to a specific genomic position.

Clearly as long as 'Cavendish' are available, they will be the choice of both the industry as well as consumers. The topic of retro-transposable elements involvement in gene expression and chromatin structure merits a deeper look. The wealth of retro elements in the *Musa* genome contributes to a certain plasticity in traits. A better understanding of activation and stability of the changes could open possibilities in mitigating new biological and environmental burdens. Finally other sources of variability for example activation of epigenetic changes should be explored by banana breeders. New breeding technologies (NBT), are currently being attempted by several groups. This line of research should expose new opportunities in the challenges facing 'Cavendish' under mono culturing and changes in the global climatic environment.

5 References

Alvarez, M. E., Nota, F. and Cambiango, D. A. (2010). Epigenetic control of plant immunity. *Mol. Plant Pathol.* 11:563–576.

Bairu, M. W., Aremu, A. O. and Staden, J. V. (2011). Somaclonal variation in plants, causes and detection methods. *Plant Growth Regul.* 63:147–173.

Baucom, R. S., Estill, J. C., Chaparro, C., Upshaw, N., Jogi, A., Deragon, J. M., Westerman, R. P., Sanmiguel, P. J. and Bennetzen, J. L. (2009). Exceptional diversity, non-random distribution, and rapid evolution of retroelements in the B73 maize genome. *PLOS Genet.* 5:e1000732.

Bennetzen, J. L. (1998). The structure and evolution of angiosperm nuclear genomes. *Curr. Opin. Plant Biol.* 1:103–108.

Bennetzen, J. L. and Kellogg, E. A. (1997). Do plants have a one-way ticket to genomic obesity? *Plant Cell* 9:1509–1514.

Bennetzen, J. L., Springer, J. L. and Crease, A. D. (1993). Specificity regulation of the Mutator transposable elements system in maize. *Crit. Rev. Plant Sci.* 12:57–95.

Brar, D. S. and Jain, S. M. (1998). Somaclonal variation: mechanisms and applications in crop improvement. In: Jain, S. M., Brar, D. S. and Ahloowalia, B. S. (Eds), *Current Plant Science and Biotechnology in Agriculture*, pp. 15–37, Springer, USA.

Cassels, A. C. and Curry, R. F. (2001). Oxydative stress and physiological epigenetic and genetic variability in plant tissue culture. Implications for micropropagation and genetic engineering. *Plant Cell Tissue Organ Cult.* 64:145–157.

Crouch, H. K., Crouch, J. H., Madsen, S., Vuylsteke, D. R. and Ortiz, R. (2000). Comparative analysis of phenotypic and genotypic diversity among plantain landraces (*Musa* spp., AAB group). *Theor. Appl. Genet.* 101:1056–1065.

Dale, J., James, A., Paul, J. Y., Khanna, H., Smith, M., Peraza-Echeverria, S., Garcia-Bastidas, F., Kema, G., Waterhouse, P., Mengersen, K. and Harding, R. (2017). Transgenic Cavendish bananas with resistance to Fusarium wilt tropical race 4. *Nat. Commun.* 8:1496.

Eckstein, K., Fraser, C., Husselmann, J. and Temple Murray, N. (1998). The evaluation of promising new banana cultivars. *Acta Hortic.* 490:57–70.

Frary, A. and Doganlar, S. (2003). Comparative genetics of crop plant domestication and evolution. *Turk. J. Agric. For.* 27:59–69.

García-Bastidas, F. A., Quintero-Vargas, J. C., Ayala-Vasquez, M., Schermer, T., Seidl, M. F., Santos-Paiva, M., Noguera, A. M., Aguilera-Galvez, C., Wittenberg, A., Hofstede, R., Sørensen, A. and Kema, G. H. J. (2019). First report of Fusarium wilt in Cavendish bananas caused by Fusarium odoratissimum Tropical Race 4 in Colombia. *Plant Dis.* 104. doi:10.1094/PDIS-09-19-1922-PDN.

Guo, W., Gong, L., Diang, Z., Li, Y., Li, F., Zhao, S. and Lin, B. (2007). Tissue culture induced locus specific alteration in DNA methylation and correlation with genetic variation in *Codonopsis lanceolata* Benth et Hook f. *Plant Cell Rep.* 26:1297–1307.

Gupta, P. K. (1998). Chromosomal basis of somaclonal variation in plants. *Curr. Plant Sci. Biotechnol. Agric.* 32:149–168.

Hirochika, H., Okamoto, H. and Kakutani, T. (2000). Silencing of retrotransposons in *Arabidopsis* and reactivation by the ddm1 mutation. *Plant Cell* 12(3):357–369.

Hirochika, H., Sugimoto, K., Otsuki, Y., Tsugawa, H. and Kanda, M. (1996). Retrotransposons of rice involved in mutations induced by tissue culture. *Proc. Natl. Acad. Sci. U.S.A.* 93:7783–7788.

Kaeppler, S. M., Kaeppler, H. F. and Rhee, Y. (2000). Epigenetic aspects of somaclonal variation in plants. *Plant Mol. Biol.* 43(2–3):179–188.

Karp, A. (1995). Somaclonal variation as a tool for crop improvement. *Euphytica* 85(1–3):295–302.

Khayat, E. (2010). Banana plant named Adi. USPTO (United States Patent Office) PP20,645 P3.

Khayat, E., Duvdevani, A., Lahav, E. and Balisteros, B. A. (2004). Somaclonal variation in banana (*Musa acuminata* Grande Naine). Genetic mechanism, frequency, and application as a tool for clonal selection. In: *Musa Biotechnology*. Swennen, R. and Jain, M. (Eds). Klawer Academic Publishing, pp. 97–109.

Khayat, E. and Ortiz, R. (2011). Genetics of important traits in *Musa*. In: *Banana Breeding, Progress and Challenges*. Pillay, M. and Tenkouano, A. CRC Press, Boca Raton, FL, pp. 71–84.

Krikorian, A. D., Irizarry, H., Caronauer-Mitra, S. S. and Rivera, E. (1992). Clonal fidelity and variation in plantain (*Musa* AAB) regenerated from vegetative stem and floral axis tips in vitro. *Ann. Bot.* 71:519–535.

Kumar, A. and Bennetzen, J. L. (1999). Plant retrotransposons. *Annu. Rev. Genet.* 33:479–532.

Leva, A. R., Petruccelli, R. and Rinaldi, L. M. R. (2012). Somaclonal variation in tissue culture: a case study with olive. *Recent Adv. Plant In Vitro Cult.7:*10–17.

Ma, J. and Bennetzen, J. L. (2004). Rapid recent growth and divergence of rice nuclear genomes. *Proc. Natl. Acad. Sci. U.S.A.* 101:12404–12410.

Maryani, N., Lombard, L., Poerba, Y. S., Subandiyah, S., Crous, P. W. and Kema, G. H. J. (2019). Phylogeny and genetic diversity of the banana Fusarium wilt pathogen *Fusarium oxysporum* f. sp. cubense in the Indonesian centre of origin. *Stud. Mycol.* 92:155–194.

Matzko, A. J. M. and Matzko, M. A. (1998). Position effects and epigenetic silencing of plant transgenes. *Curr. Opin. Plant Biol.* 1:142–148.

Miguel, C. and Marum, L. (2011). An epigenetic view of plant cells cultured in vitro: somaclonal variation and beyond. *J. Exp. Bot.* 62:3713–3725.

Mujib, A., Banerjee, S. and Dev Ghosh, P. (2007). Callus induction, somatic embryogenesis and chromosomal instability in tissue culture raised hippeastrum (Hippeastrum hybridum cv. United Nations). *Propag. Ornam. Plants* 7:169–174.

Ngezahay, F., Dong, Y. and Liu, B. (2007). Somaclonal variation at the nucleotide sequence level in rice (*Oryza sativa* L.) as revealed by RAPD and ISSR markers, and by pairwise sequence analysis. *J. Appl. Genet.* 48(4):329–336.

Nwauzoma, A. B. and Jaja, E. T. (2013). A review of somaclonal variation in plantain (Musa spp): mechanisms and applications. *J. Appl. Biol. Sci.* 67:5252–5260.

Ordóñez Román, N. I., Seidl, M. F., Waalwijk, C., Drenth, A., Kilian, A., Thomma, B. P. H. J., Ploetz, R. C. and Kema, G. H. J. (2015). Worse comes to worst: bananas and Panama disease - when plant and pathogen clones meet. *PLoS Pathog.* 11:e1005197.

Orozco-Arias, S., Gustavo, I. and Romain, G. (2019). Retrotransposons in plant genomes: structure, identification, and classification through bioinformatics and machine learning, nt. *J. Mol. Sci.* 20:3837.

Ortiz, R. (2000). Understanding the Musa genome: an update. *Acta Hortic.* 54:157–168.

Ortiz, R. and Vuylsteke, D. (1992). Inheritance of black Sigatoka resistance and fruit parthenocarpy in the triploid AAB plantain. In: *Agronomy Abstracts*. American Society of Agronomy – Crop Science Society of America – Soil Science Society of America, Madison, WI.

Ortiz, R. and Vuylsteke, D. (1995). Effect of the parthenocarpy gene P1 and Ploidy on fruit and bunch traits of plantain-banana hybrids. *Heredity* 75:460–465.

Osuji, J. O., Bosa, E. O., Vuylsteke, D. and Ortiz, R. (1997). Multivariate pattern of quantitative trait variation in triploid banana and plantain cultivars. *Sci. Hortic.* 71:197–202.

Paul, J. Y., Khanna, H., Kleidon, J., Hoang, P., Geijskes, J., Daniells, J., Zaplin, E., Rosenberg, Y., James, A., Mlalazi, B., Deo, P., Arinaitwe, G., Namanya, P., Becker, D., Tindamanyire, J., Tushemereirwe, W., Harding, R. and Dale, J. (2017). Golden bananas in the field: elevated fruit pro-vitamin A from the expression of a single banana transgene. *Plant Biotechnol. J.* 15:520–532.

Paszkowski, J. (2015). Controlled activation of retrotransposition for plant breeding. *Curr. Opin. Biotechnol.* 32:200–206.

Ploetz, R. C. (2006). Fusarium wilt of banana is caused by several pathogens referred to as Fusarium oxysporum f. sp. *cubense*. *Phytopathology* 96:653–656.

Poncet, V., Lamy, F., Enjalbert, J., Joly, H., Sarr, A. and Robert, T. (1998). Genetic analysis of the domestication syndrome in pearl millet (*Pennisetum glaucum* L., Poaceae): inheritance of the major characters. *Heredity* 81:648–658.

Sato, M., Hosokawa, M. and Doi, M. (2011). Somaclonal variation is induced *de novo* via the tissue culture process: a study quantifying mutated cells in *Saintpaulia*. *PLoS ONE* 6:e23541.

Smulders, M. J. M. and de Klerk, G. J. (2011) Epigenetics in plant tissue culture. *Plant Tissue Cult.* 63:137–146.

Stelpflug, S. C., Eichten, S. R., Hermanson, P. J., Springer, N. M. and Kaeppler, S. M. (2014). Consistent and heritable alterations of DNA methylation are induced by tissue culture in maize. *Genetics* 198:209–218.

Stover, R. H. and Simmonds, N. W. (1987). *Bananas*. Longman Scientific and Technical, John Wiley and Sons, New York.

Tenkouano, A., Ortiz, R. and Vuylsteke, D. (1998). Combining ability for yield and plant phenology in plantain-derived populations. *Euphytica* 104:151–158.

Tiwari, J. K., Chandel, P., Guptas, S., Goptal, J., Singh, B. P. and Bhardwaj, V. (2013) Analysis of genetic stability of in vitro propagated potato microtubers using DNA markers. *Physiol. Mol. Biol. Plants* 19:587–595.

Vázquez, A. M. (2001). Insight into somaclonal variation. *Plant Biosyst.* 135:57–62.

Chapter 13

Developing hybrid banana varieties with improved properties

Edson Perito Amorim, Vanusia Batista de Oliveira Amorim, Manassés dos Santos Silva, Fernando Haddad, Claudia Fortes Ferreira and Janay Almeida dos Santos Serejo, Embrapa, Brazil

1 Introduction

Banana production in Brazil includes different types of banana subgroups, with the different *Prata* clones accounting for 70% of the crop area, followed by the Cavendish varieties, *Maçã*, and plantain varieties. Other cultivars are grown at a smaller scale, including Bluggoe and Pisang Mas. Within the *Prata* banana clones, we highlight the *Prata Catarina*, *Prata Gorutuba*, and *Prata Rio* cultivars, which were obtained from a positive selection in populations of the *Prata-Anã* cultivar, which was recommended by Embrapa in 1985 to Brazilian farmers (Rebouças et al., 2018). The primary banana-producing states in Brazil are Bahia, Minas Gerais, São Paulo, and Santa Catarina, which account for approximately 50% of the fruit available for the consumer market (IBGE, 2019).

In 2019, Brazil produced approximately 7 million tons of bananas in an area of 465 000 ha, which makes it one of the largest banana-producing countries in the world, together with India, China, and Indonesia (FAOSTAT, 2020). The economic resources generated by banana production account for 4% of the net value of the Brazilian production, that is, $3 billion annually. Banana crops

http://dx.doi.org/10.19103/AS.2020.0070.16

stand fifth among the most important crops, behind such commodities as soybean, corn, sugarcane, and coffee (IBGE, 2019).

The main challenges of banana production in Brazil include fungal diseases, namely the black Sigatoka and Fusarium wilt race 1. Moreover, the weevil borer and several species of nematodes also are problematic for banana production in certain regions. It is worth noting that the *Prata* cultivars are susceptible to these diseases.

Considering that the best strategy to mitigate the effects of these plant diseases is the use of resistant cultivars, the Brazilian Agricultural Research Corporation (Embrapa) created a program for the genetic improvement of banana production in 1976, which is still in operation today. The first step was the establishment of a comprehensive collection of germplasm, starting with collection expeditions conducted in the banana's center of origin in Southeast Asia, as well as in Africa and Central and South America. These expeditions were supervised and undertaken by researchers Kenneth Shepherd and Francisco Ricardo Ferreira, who specialized in the Musaceae family of plants. This collection now has 400 accessions, some of which are not held in other collections, including diploids.

The banana germplasm bank of Embrapa is field-maintained, with five specimens of each accession. Conservation in the field has some disadvantages, due to the requirement of large experimental areas, high maintenance cost, and the exposure of accessions to unfavorable weather conditions and pathogen and pest attack, which may result in a high risk of accession loss. The advantage of field-maintained collections is that they allow for ongoing characterization and assessment of the germplasm with a view of obtaining information that is useful to the breeding program. To avoid the risks associated with field maintenance, Embrapa keeps 12–24 specimens of each accession as duplicates in vitro, under minimum growth conditions, to reduce the risk of loss.

2 Released hybrids

Embrapa has released 11 hybrid banana cultivars, which are all products of crosses between improved diploids, resistant to the main pests and diseases, and commercial cultivars, especially *Prata* and *Maçã* (Table 1 and Fig. 1). The cultivar have a three-letter prefix BRS (Embrapa code) followed by the cultivar name. The vast majority of these hybrids are used by smallholder farmers in the Amazon region where agricultural technologies, such as the use of fungicides, are incipient. Three of the eleven cultivars have shown high levels of acceptance among farmers, the *BRS Platina*, *BRS Tropical*, and *BRS Princesa* cultivars. The first is a *Prata* cultivar hybrid, and the other two are *Maçã* cultivar hybrids. They are all resistant to yellow and black Sigatoka and Fusarium wilt race 1. *BRS Princesa* was tested in Africa, in collaboration with the IITA, and

Table 1 Characteristics of eleven commercial banana cultivars developed by Embrapa

	Embrapa BRS cultivars										
Characteristics	Tropical	Princesa	Pacovan Ken	Platina	Japira	Preciosa	Pioneira	Caprichosa	Garantida	Vitória	Pacoua
Type	Silk	Silk	Prata	Prata	Prata	Prata	Prata	Prata	Prata	Prata	Prata
Height	AH	AH	H	AS	H	H	AS	H	H	H	H
Plant cycle (days)	400	387	385	347	390	400	390	335	330	390	300
Bunch weight (kg)	19	19	23	20	15	19	16	28	22	15	25
Number of fruits	125	121	105	114	122	106	100	117	120	122	90
Number of hands	7	8	7	8	8	7	7	8	10	8	10
Fruit length (cm)	15	8	19	16	18	15	14	12	13	18	13
Fruit weight (g)	136	125	215	160	180	121	123	198	200	180	101
Yield (t/ha/cycle) (NI/I)	15/30	15/30	20/50	20/40	15/-	15/30	20/-	30/40	21/32	15/-	13/25
Yellow Sigatoka	R	R	R	R	R	R	R	R	R	R	R
Black Sigatoka	MS	MS	R	MR	R	MS	S	R	R	R	R
Fusarium wilt (race 1)	MR	MR	R	R	R	MR	MS	R	R	R	R
Moko	S	S	S	S	S	S	S	S	S	S	S
Nematodes	-	-	MS	R	MS	-	MR	MR	MS	MS	T
Weevil borer	-	-	MS	R	MR	-	MR	MR	MS	MR	T

AH: average to high; AS: average to short, A: average, H: high, S: short. NI/I: non-irrigated/irrigated. S: susceptible; MS: moderately susceptible; MR: moderately resistant; R: resistant, T: tolerant.

Figure 1 Commercial banana hybrids developed by Embrapa.

had high sensory acceptability, with better results than the local *Maçã* cultivar. Comprehensive testing of *BRS Princesa* is being discussed with international partners.

3 Seeds as the basis of the genetic improvement program of Embrapa

Banana plant improvement in Embrapa is based on seed production from crosses between commercial cultivars and different improved diploids. Embrapa started developing improved diploids more than 30 years ago. Targeted crosses are made daily, and embryos from the seeds obtained are cultured in vitro to form a new plant for subsequent field and greenhouse agronomic assessment. Up to 2018, more than 200 000 seeds have been produced, and their embryos excised and grown under tissue-culture conditions (Tables 2 and 3).

4 Improved diploids and their importance in the improvement of Musa spp.

Considering that the different agronomic traits of interest for banana plant improvement are scattered among several wild diploids, it is essential to concentrate the highest number of traits in a single or a few diploid genotypes. Moreover, the hybrid resulting from the 3x/2x cross has the complete genome of the female parent and traits of the pollen donor, because 25% of the alleles are associated with the male parent, and 75% of the alleles are inherited from the female parent. The selection of the male parent (diploid) is of extreme importance for this strategy to be successful.

Table 2 Main female and male parents used by the Embrapa banana genetic-breeding program including number of crosses and numerical ratio of seeds per number of crosses from January 1982 to 1995

Parents ♂\♀	Fruits crosses (F)	Seeds (S)	S/100F
		1982-1992	
		Pacovan	
Wild type			
Calcutta 4	29.942	3.348	11.18
Buitenzorg	6.946	259	3.73
Madang	5.544	380	6.85
Pahang	762	204	26.77
Cultivar			
Lidi	30.419	772	2.54
Madu	7.723	68	0.88
S/N°2	2.284	5	0.22
Hybrid			
003004	22.591	1.170	5.18
M53	97.812	1.874	1.92
Total	204.023	8.080	$\overline{X}$ = 3.96
		Prata-Anã	
Wild type			
Calcutta 4	7.314	68	0.93
Buitenzorg	2.524	13	0.52
Madang	3.818	12	0.31
Pahang	-	-	-
Cultivar			
Lidi	17.851	37	0.21
Madu	2.934	0	-
S/N°2	-	-	-
Hybrid			
003004	14.003	23	0.16
M53	136.666	138	0.10
Total	185.110	291	$\overline{X}$ = 0.93
		1993-1995	
		Diploids	
1993	741	2.914	25.43
1994	4.785	17.929	26.69
1995	6.549	85.154	7.69
Total	12.075	105.997	$\overline{X}$ = 11.39
		Prata*/Silk**	
1993	5.184	51	0.98
1994	8.145	175	2.15
1995	19.382	268	1.38
Total	32.711	494	$\overline{X}$ = 1.50

*Prata Comum, Prata Ponta Aparada, Pacovan e Prata Sta. Maria. **Yangambi N°2.

Embrapa operates a parallel breeding program that focuses on the development of improved diploids with the following traits: low height, >100 fruits per bunch, loose bunch architecture, male fertility, good offspring production, resistance to the main diseases (yellow and black Sigatoka and Fusarium wilt race 1 and more recently tropical race 4), and tolerance to pests, especially nematodes and weevil borer. Commercial varieties are used in crosses with wild diploid ancestors as a strategy for the development of new diploids, in addition to combinations of genotypes resistant to yellow and black Sigatoka and Fusarium wilt race 1 and tropical race 4. Table 4 shows some of the new, improved diploids developed in recent years by Embrapa, and Fig. 2 shows photographs of bunches. Recent results obtained from genotyping using SNPs by Embrapa, in collaboration with the University of Queensland (UQ), indicated that six new-improved diploids may have the genes that confer resistance to Fusarium tropical race 4. This finding will be confirmed in Colombia through tests conducted in an area where the disease is present, in a partnership with AGROSAVIA and AUGURA, starting in 2020.

5 Phenotyping for resistance to Fusarium wilt race 1

The clones in the *Prata* subgroup are susceptible to Fusarium wilt race 1 (Foc 1), a disease that has been limiting production, especially in Brazilian agriculture areas under irrigation. The combination of severe winter conditions, common in Southern Brazil, and Fusarium wilt has a significant influence on banana production as stress due to cold increases the number of plants affected by the disease. In this context, and because there is no known chemical control for Fusarium wilt, unlike black Sigatoka, the control through genetic resistance to this disease is a priority for Embrapa. To identify resistant genotypes, Embrapa has developed specific selection protocols for Fusarium wilt, both for field and greenhouse conditions, using artificial inoculation using the isolate CNPMF 218 (Gonçalves et al., 2019). Embrapa has a field site artificially infested with Foc race 1, in which the inoculum load was further increased through successive crops of the *Maçã* clone which is highly susceptible to this disease. This field site has been used for over 20 years for genotype selection and has been shown to enable extremely efficient discrimination between resistant and susceptible genotypes. For genotypic selection, the genotypes are distributed in this field site in a completely random fashion, with ten replicates of each genotype. Plants of the *Maçã* banana are used as controls and are planted around each new hybrid. At the time of flowering, a cross-sectional cut is made in the pseudostem for the assessment of internal symptoms of Foc 1. Assessment is performed using a scoring scale developed by Embrapa (Fig. 3).

Table 3 Number of crosses, number of seeds, ratio of seeds per number of crosses, and number and percentage of crosses made with di-, tri- and tetraploids by the Embrapa banana and plantain genetic-breeding improvement program from 1996 to 2018

Year	Number of seeds (S)	Number of crosses (C)	S/C	Number of crosses (%)		
				di- x di-	tri- x di-	tetra- x di-
1996	5.209	298	17.48	51 (17.11)	247 (82.89)	-
1997	903	293	3.08	98 (33.44)	195 (66.56)	-
1998	280	202	1.39	25 (12.37)	177 (87.63)	-
1999	689	6	114.83	4 (66.67)	2 (33.33)	0 (0.00)
2000	6.756	82	82.39	62 (75.61)	1 (1.22)	19 (23.17)
2001	28.845	205	140.71	130 (63.41)	20 (9.76)	55 (26.83)
2002	32.021	231	138.62	181 (78.35)	34 (14.72)	16 (6.93)
2003	13.340	125	106.72	88 (70.40)	12 (9.60)	25 (20.00)
2004	23.782	168	141.56	107 (63.69)	7 (4.17)	54 (32.14)
2005	10.362	96	107.94	47 (48.96)	41 (42.71)	8 (8.33)
2006	11.490	156	73.65	107 (68.59)	31 (19.87)	18 (11.54)
2007	20.009	163	122.75	85 (52.15)	44 (26.99)	34 (20.86)
2008	1.127	39	28.90	19 (48.72)	0 (0.00)	20 (51.28)
2009	4.427	32	138.34	20 (62.50)	1 (3.13)	11 (34.38)
2010	3.230	36	89.72	21 (58.33)	1 (2.78)	14 (38.89)
2011	3.385	115	29.43	9 (7.83)	17 (14.78)	89 (77.39)
2012	1.360	80	17.00	14 (17.50)	3 (3.75)	63 (78.75)
2013	6.541	244	26.81	19 (7.79)	19 (7.79)	206 (84.43)
2014	11.082	372	29.79	27 (7.26)	24 (6.45)	321 (86.29)
2015	4.729	228	20.74	23 (10.09)	19 (8.33)	186 (81.58)
2016	298	35	8.51	1 (2.86)	2 (5.71)	32 (91.43)
2017	3.306	174	19.00	2 (1.16)	17 (9.83)	154 (89.01)
2018	1.482	72	20.59	0 (0.00)	5 (6.95)	67 (93.05)
Total	196.481	3.561	55.18	1.144 (32.12)	927 (26.03)	1.490 (41.85)

Considering the wide variability of reaction of banana genotypes to Foc 1, Embrapa has been continuously collected isolates in the main banana-producing regions of Brazil to monitor the changes in the fungal population and ensure selection of genotypes with effective resistance against the pathogen population present in Brazil. One of the collected isolates, known as CNPMF 218, was shown to be highly aggressive and virulent, both in the *Prata* and in Cavendish clones. Therefore, this isolate is used as a standard for the selection of resistant genotypes under greenhouse conditions. For the screening of different pathogen strains with regards to aggressiveness and

Table 4 New generation of improved diploids developed by Embrapa

Codes	Genealogies Improved diploids
CNPMF0038	[(M53 × Madu)] × [(Malaccensis × Tjau Lagada)]
CNPMF0496	[(M61 × Pisang Lilin)] × [(Terrinha × Calcutta 4)]
CNPMF0513	[(M61 × Pisang Lilin)] × [(M53 × Kumburgh)]
CNPMF0519	Self-fertilization (wild diploid Tambi)
CNPMF0534	[(Calcutta 4 × Madang)] × [(Borneo × Guyod)]
CNPMF0536	[(Calcutta 4 × Madang)] × [(Borneo × Guyod)]
CNPMF0542	[(SH3263)] × [(Malaccensis × Sinwobogi)]
CNPMF0557	[(M61 × Pisang Lilin)] × [(Malaccensis × Tjau Lagada)]
CNPMF0565	[(Calcutta 4 × Pahang) × (Borneo × Madang)] × Khae
CNPMF0572	[(Khai × (Calcutta 4 × Madang)] × [(Calcutta 4 × Madang)]
CNPMF0612	[(M53 × Madu) × Madu)] × SH3263
CNPMF0731	[(Malaccensis × Madang)] × [(Tuugia × Calcutta 4)]
CNPMF0767	[(Malaccensis × Madang)] × [(Khai × (Calcutta 4 × Madang)]
CNPMF0811	[(Khai × (Calcutta 4 × Madang)] × [(Calcutta 4 × Pahang) × (Borneo × Madang)]
CNPMF0978	[(Calcutta 4 × Madang)] × [(Terrinha × Calcutta 4)]
CNPMF0993	[(Borneo × Guyod) × (Tuugia × Calcutta 4)] × [(Khai × (Calcutta 4 × Madang)]
CNPMF0998	[(Borneo × Guyod)] × [(Borneo × Guyod) × SH3263]
CNPMF1102	[(Jari Buaya × (Calcutta 4 × Madang)] × [(Borneo × Guyod) × (Tuugia × Calcutta 4)]
CNPMF1105	[(Borneo × Guyod) × (Calcutta 4 × Heva)] × [(Calcutta 4 × Madang)]
CNPMF1171	[(Malaccensis × Madang)] × [(M53 × (Tuugia × Calcutta 4)]
CNPMF1272	[(Borneo × Guyod) × (Calcutta 4 × Heva)] × [(Tuugia × Calcutta 4)]
CNPMF1286	[(Calcutta 4 × Madang)] × [(Terrinha × Calcutta 4)]
CNPMF1323	[(Malaccensis × Sinwobogi)] × [(Calcutta 4 × Heva)]

virulence, inoculation protocols and a scoring scale have been developed by Embrapa researchers (Fig. 4).

6 The Embrapa breeding cycle

The banana genetic improvement program of Embrapa has a flowchart to generate and validate the new hybrids developed for Brazilian conditions.

The main traits targeted for improvement are the following: quality of the fruit (e.g. size, texture, and flavor), resistance to pests and diseases (nematodes, rhizome weevil, Sigatoka, Fusarium wilt race 1 and tropical race 4), low height,

Figure 2 New improved diploids developed by Embrapa.

early cycle, high productivity, and bunch profile (e.g. loose architecture, cylindrical shape, and high fruit yield), among other traits with less direct impact on production.

The main phenotypes of plants selected for the breeding program are the ones having traits similar to *Prata*, Pacovan, *Maçã*, plantain, and more recently

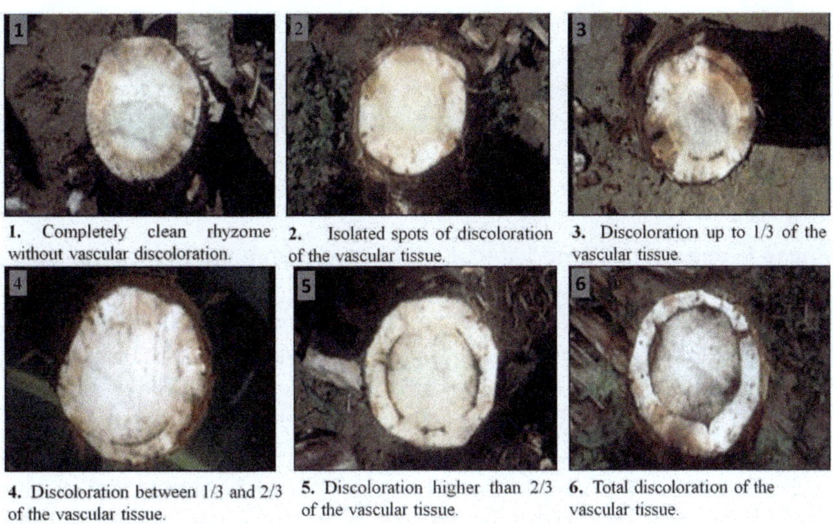

Figure 3 Longitudinal cut of the rhizome to analyze the vascular discoloration caused by the presence of Foc.

Figure 4 External and internal symptoms of Foc. External symptoms: (0) no symptoms; (1) initial yellowing of old leaves; (2) yellowing of old leaves and initial discoloration of young leaves; (3) intense yellowing of all leaves; and (4) plant dead. Internal symptoms: (0) absence of rhizome discoloration; (1) initial discoloration at the junction of the root and rhizome up to 5%; (2) rhizome discoloration between 6% and 25%; (3) rhizome discoloration between 26% and 50%; and (4) rhizome discoloration more than 50%.

to Cavendish clones. It is worth highlighting that a greater focus is placed on the genotypes similar to *Prata* clones because there is higher consumer demand for this variety in Brazil. Breeding involving Cavendish clones has been conducted in collaboration with companies and international institutions, under confidentiality agreements.

The Embrapa breeding process starts with crosses between commercial triploid cultivars that have one or more undesirable agronomic traits and improved diploids selected according to their genetic background. Currently, on a larger scale crosses between tetraploids and improved diploids are also made. Embryos are recovered from the seeds through embryo rescue and germinated in vitro culture immediately after the crosses in the Tissue Culture Laboratory of Embrapa. The resulting plants are acclimatized before being planted in the field.

Phase I - seeds: In this step, the plants obtained from the seeds are planted in the field and evaluated individually during two cycles regarding the traits described at the beginning of this section. In addition, preliminary sensory tests are performed on the genotypes whose traits are aligned with the market's requirements, particularly flavor and aroma.

Phase II - clonal: The hybrids selected in Phase I are submitted to micropropagation, and five clones of each plant are again agronomically characterized. This step includes replicates, and it is possible to observe the stability of the measured traits. Phase II includes two production cycles. Moreover, the fruits are submitted to sensory analysis. At the end of this step, the selected genotypes are maintained in the field for another six months to be able to increase the number of offspring for in vitro establishment.

Phase III - evaluation of resistance to Fusarium wilt race 1 (Foc 1): The genotypes selected in Phase II, already established in vitro and multiplied, are planted in a field site artificially infested with Foc 1 and assessed as indicated in Section 5 of this chapter. Some Fusarium wilt tests are currently also performed on small plants under greenhouse conditions as a form of preselection to reduce the size of the populations that need to be planted in the field. These genotypes are also assessed for their resistance to tropical race 4 of Foc using SNP markers. In addition, genotypes are assessed in the field with the presence of the pathogen outside of Brazil.

Phase IV - trial network: The plants selected in Phase III are sent to collaborating producers for final validation prior to commercial release. The network thus includes small, medium, and large producers located in the major centers of banana production in Brazil and the Embrapa unit, where more detailed studies are conducted. This step is performed during two or three production cycles, and in addition to the evaluation of the traits of interest to producers/consumers, agronomic studies are conducted (e.g., planting density, point of harvest) for the development of efficient production systems specific for each new cultivar.

Phase V - the release of cultivars: The genotypes selected in Phase IV are provided license to private tissue culture facilities for large-scale in vitro multiplication. After this process, field days are held in areas belonging to partner producers for the official release of the new cultivars. Several promotion materials and activities are also organized, including brochures, press releases, presentations, and interviews.

Phase VI - post-release: This step consists in the follow-up with several reference producers who adopted the new cultivars to identify potential problems during production. This monitoring is conducted for over three years.

7 Alternatives to breeding based on crosses

The banana-breeding program of Embrapa uses alternatives to genetic improvement based on crosses, both as a support per se to the crosses and as a parallel route for the development of new cultivars.

7.1 Duplication of chromosomes

The aim of chromosome duplication is to obtain tetraploids that can be used in crosses with improved diploids, with the final product being secondary triploids for commercial production, developed in less time than that required for the conventional improvement process. Several wild diploids have been duplicated, and autotetraploid populations have been characterized in the field for a series of agronomic traits. Crosses with improved diploids have been made, and several secondary triploid genotypes are currently under evaluation, with some already in advanced stages of selection. The priority was the duplication of wild diploids viewed as potential ancestors with a genetic background associated with the commercial cultivars, including *Akondro Mainty, Malbut, Niyarma Yik, Jari Buaya, Pisang Mas, Tuu Gia, Mambee Thu, Thong Dok Mak, Banksii, Mlali*, and *Pisang Lilin* (Costa et al., 2011; Perrier et al., 2009; Hippolyte et al., 2012; Amaral et al., 2015; Pio et al., 2018; Christelová et al., 2017; Carvalho Santos et al., 2019). The focus was also on the duplication of wild diploids resistant to Fusarium wilt tropical race 4, which were identified by Zuo et al. (2018) and are also rich in carotenoids, thus being a biofortification tool (Amorim et al., 2009).

7.2 Induction of somaclonal variation

Embrapa has been working on the induction of in vitro somaclonal variants of the *Prata* and Cavendish clones. All protocols in use were adapted to take into account the Brazilian conditions of the originals used by the Taiwan Banana Research institute (TBRI), in Taiwan. Several modifications have been made to these protocols resulting in a significantly higher frequency of useful somaclonal variants. The main objective of using induction is the development of genotypes of existing clones resistant to black Sigatoka and Fusarium wilt (race 1 and tropical race 4, Foc 1 and 4).

Fusaric acid is used during the process as an agent for the in vitro preselection of somaclones potentially resistant to Foc 1. The selected somaclones are subsequently subjected to the Fusarium pathogen under controlled conditions in a greenhouse. At this point, the isolate CNPMF 218 of Foc 1 is used for inoculation, being highly aggressive and virulent both to the *Prata* and Cavendish cultivars (Fig. 5). The plants considered resistant to Foc race 1 are also tested for resistance to black Sigatoka through inoculation in a greenhouse.

Some somaclonal variants resistant to the isolate CNPMF 218 of Foc race 1 were identified among plants of the cultivars *Grande Naine* and *Prata-Anã* after treatment with plant regulators. All plants used as controls exhibited symptoms of the disease and were deemed susceptible. The somaclonal variants are under the process of agronomic evaluation to determine whether other traits have been modified during in vitro variation induction. The promising somaclonal variants will be tested for their resistance to tropical race 4 of Foc after the evaluation of the agronomic and sensory characteristics through the international collaborations established by Embrapa.

Figure 5 Selection of banana somaclonal variants to be screened in water tanks with isolate race 1 CNPMF 218 of *Fusarium oxysporum* f. spp. *cubense*.

8 Future trends and conclusion

The identification of genes associated with resistance to Fusarium wilt race 1 is the current focus of advanced biology studies conducted in Embrapa, as well as the quantification of the genetic diversity of populations of Foc race 1 and black Sigatoka in the major banana-producing regions of Brazil (Peixouto et al., 2015; Costa et al., 2014). Moreover, SSR molecular markers for assisted selection have been developed (Passos et al., 2012; Amorim et al., 2013; Silva et al., 2015).

Embrapa has developed cell suspensions for use in genetic transformation or potential application of gene editing with the CRISPR-Cas9 technique

(Morais-Lino et al., 2016; Silva et al., 2016). More recently, Embrapa has been working on the identification of genes associated with tolerance to drought in *Musa* spp. (Mattos-Moreira et al., 2018; Santos et al., 2018). All generated knowledge is applied in the genetic improvement program to accelerate the development of new cultivars and improve the durability of genetic resistance to pests and diseases in the new hybrids.

The Brazilian banana and plantain genetic-breeding program at Embrapa is in strict relationship with producers since research demands are aligned with problems seen at farm level especially associated with pests and diseases focusing in Fusarium and Sigatokas. The different hybrids developed by Embrapa are used by growers mainly from the north of the country where black Sigatoka limits the production of susceptible cultivars. The BRS Platina, BRS Princesa, BRS Pacoua, and BRS Tropical hybrids are used in many production hubs in Brazil. It is estimated that 30% of the cultivated area with bananas in Brazil use some of the hybrids developed by Embrapa in their production systems together with other cultivars. Embrapa's hybrids make up almost the totality of plantations in some regions or farms.

9 References

Amaral, C. M., Santos-Serejo, J., Silva, S., Ledo, C. A. and Amorim, E. P. (2015). Agronomic characterization of autotetraploid banana plants derived from Pisang Lilin (AA) obtained through chromosome doubling. *Euphytica* 202:435–443.

Amorim, E. P., Lessa, L. S., Ledo, C. A. S., Amorim, V. B. O., Reis, R. Vd, Santos-Serejo, J. Ad and Silva, SdOe (2009). Caracterização agronômica e molecular de genótipos diploides melhorados de bananeira. *Rev. Bras. Frutic.* 31:154–161.

Amorim, E. P., Santos-Serejo, J. A., Amorim, V. B. O., Ferreira, C. F. and Silva, S. O. (2013). Banana breeding at Embrapa Cassava and Fruits. *Acta Hortic.* 986:171–176.

Carvalho Santos, T. T., Amorim, V. B., Santos-Serejo, J. A., Ledo, C. A., Haddad, F., Ferreira, C. F. and Amorim, E. P. (2019). Genetic variability among autotetraploid populations of banana plants derived from wild diploids through chromosome doubling using SSR and molecular markers based on retrotransposons. *Mol. Breed.* 39:95.

Christelová, P., De Langhe, E., Hřibová, E., Čížková, J., Sardos, J., Hušáková, M., Van den houwe, I., Sutanto, A., Kepler, A. K., Swennen, R., Roux, N. and Doležel, J. (2017). Molecular and cytological characterization of the global Musa germplasm collection provides insights into the treasure of banana diversity. *Biodivers. Conserv.* 26(4):801–824.

Costa, F. H. S., Pasqual, M., Silva, S. O., da Silva Neto, H. P., Amorim, E. P. and dos Santos-Serejo, J. A. (2011). Poliploidização em ápices caulinares de bananeira e seus efeitos morfofisiológicos *in vitro*. *Pesq. Agropec. Bras.* 46:805–813.

Costa, S. N., Bragança, C. A. D., Ribeiro, L. R., Amorim, E. P., Oliveira, S. A. S., Dita, M. A., Laranjeira, F. F. and Haddad, F. (2014). Genetic structure of Fusarium oxysporum f. sp. cubense in different regions from Brazil. *Plant Pathol.* 64:137–146.

FAOSTAT (2020). Food and agricultural organization of the United Nations. Available at: http://faostat3.fao.org/home/E. Accessed on: January 7, 2020.

Gonçalves, Z. S., Haddad, F., de Oliveira Amorim, V. B., Ferreira, C. F., de Oliveira, S. A. S. and Amorim, E. P. (2019). Agronomic characterization and identification of banana genotypes resistant to Fusarium wilt race 1. *Eur. J. Plant Pathol.* 155:1093-1103.

Hippolyte, I., Jenny, C., Gardes, L., Bakry, F., Rivallan, R., Pomies, V., Cubry, P., Tomekpe, K., Risterucci, A. M., Roux, N., Rouard, M., Arnaud, E., Kolesnikova-Allen, M. and Perrier, X. (2012). Foundation characteristics of edible Musa triploids revealed from allelic distribution of SSR markers. *Ann. Bot.* 109(5):937-951.

IBGE (Instituto Brasileiro de Geografia e Estatística) (2019). LSPA, Levantamento sistemático da Produção Agrícola. Available at: http://www.ibge.gov.br/home/estatistica/indicadores/agropecuaria/lspa/lspa_201212.pdf. Accessed on: January 7, 2020.

Mattos-Moreira, L. A., Ferreira, C. F., Amorim, E. P., Pirovani, C. P., de Andrade, E. M., Filho, M. A. C. and da Silva Ledo, C. A. (2018). Differentially expressed proteins associated with drought tolerance in bananas (Musa spp.). *Acta Physiol. Plant.* 40(3):60.

Morais-Lino, L. S., Santos-Serejo, J. A., Amorim, E. P., Santana, J. R. F., Pasqual, M. and Silva, S. O. (2016). Somatic embryogenesis, cell suspension, and genetic stability of banana cultivars. *In Vitro Cell. Dev.-Pl.* 52:99-106.

Passos, M. A., Oliveira Cruz, V., Emediato, F. L., de Camargo Teixeira, C., Souza, M. T., Matsumoto, T., Rennó Azevedo, V. C., Ferreira, C. F., Amorim, E. P., de Alencar Figueiredo, L. F., Martins, N. F., de Jesus Barbosa Cavalcante, M., Baurens, F. C., da Silva, O. B., Pappas, G. J., Pignolet, L., Abadie, C., Ciampi, A. Y., Piffanelli, P. and Miller, R. N. (2012). Development of expressed sequence tag and expressed sequence tag-simple sequence repeat marker resources for Musa acuminata. *AoB Plants* 2012:pls030.

Peixouto, Y. S., Dórea Bragança, C. A., Andrade, W. B., Ferreira, C. F., Haddad, F., Oliveira, S. A., Darosci Brito, F. S., Miller, R. N. and Amorim, E. P. (2015). Estimation of genetic structure of a Mycosphaerella musicola population using inter-simple sequence repeat markers. *Genet. Mol. Res.* 14:8046-8057.

Perrier, X., Bakry, F., Carreel, F., Jenny, C., Horry, J. P., Lebot, V. and Hippolyte, I. (2009). Combining biological approaches to shed light on evolution of edible bananas. *Ethnobot* 7:199-216.

Pio, L. A. S., Pasqual, M., Silva, S. O., Rocha, H. S. and Magalhães, H. M. (2018). Inducing and identifying articially-induced polyploidy in bananas. *Afr. J. Biotechnol.* 13:3748-3758.

Rebouças, T. A., Haddad, F., Ferreira, C. F., Oliveira, S. A. S. and Ledo, C. A. S. (2018). Identification of banana genotypes resistant to Fusarium wilt race 1 under field and greenhouse conditions. *Sci. Hortic.* 239:308-313.

Santos, A. S., Amorim, E. P., Ferreira, C. F. and Pirovani, C. P. (2018). Water stress in Musa spp.: a systematic review. *PLoS ONE* 13(12):e0208052.

Silva, P. R. O., de Jesus, O. N., Bragança, C. A. D., Haddad, F., Amorim, E. P. and Ferreira, C. F. (2016). Development of a thematic collection of Musa spp accessions using SCAR markers for preventive breeding against Fusarium oxysporum f. sp cubense tropical race 4. *Genet. Mol. Res.* 15(1):15017765.

Silva, P. R. O., Jesus, O. N. J., Creste, S., Figueira, A., Amorim, E. P. and Ferreira, C. F. (2015). Evaluation of microsatellite loci from libraries derived from the wild diploid 'Calcutta 4' and 'Ouro' banana cultivars. *Genet. Mol. Res.* 14:11410-11428.

Zuo, C., Deng, G., Li, B., Huo, H., Li, C., Hu, C., Kuang, R., Yang, Q., Dong, T., Sheng, O. and Yi, G. (2018). Germplasm screening of *Musa* spp. for resistance to *Fusarium oxysporum* f. sp. cubense tropical race 4 (FocTR4). *Eur. J. Plant Pathol.* 151:723-734.

Chapter 14

Genetic modification of bananas: the long road to farmers' fields

James Dale, Queensland University of Technology, Australia; Wilberforce Tushemereirwe, National Agricultural Research Organisation, Uganda; and Robert Harding, Queensland University of Technology, Australia

1 Introduction

1.1 The problem with bananas

There is now a large and growing concern that the bananas we know are going to be decimated by diseases. Fusarium wilt or Panama disease tropical race 4 (TR4) is highest on this disease list followed by Black Sigatoka and banana bunchy top disease. Because of these disease problems, there is now also a renewed concern about the lack of genetic diversity in bananas that are grown on a large scale, such as Cavendish originating from a single clone which is now being used almost exclusively for the export market.

The reality is, or at least could be, very different. There is huge diversity in bananas both within and between the cultivated bananas and importantly their wild progenitors, the diploids. Further, there is resistance or tolerance in this broad germplasm to virtually all the banana diseases and pests. The real problem with bananas is that world production and consumption relies on, and export production completely depends on, just one subgroup of bananas, the Cavendish, a high-yielding banana that transports well and has widely acceptable flavour and texture. Its resistance to Fusarium wilt race 1 led to Cavendish by replacing the susceptible Gros Michel as the world's export banana as a result of the Fusarium wilt race 1 epidemic in Central and South America through to the 1950s. Cavendish, however, is susceptible to TR4 as well as Black Sigatoka and banana bunchy top disease. There does not seem to be

http://dx.doi.org/10.19103/AS.2020.0070.19

a naturally occurring TR4-resistant cultivar available at present that can replace Cavendish as the primary dessert banana suitable for the export market.

Outside the export market, there are more than 1000 different banana cultivars and landraces (Heslop-Harrison and Schwarzacher, 2007). As a result, there is a much greater diversity of bananas grown for local markets. In Uganda, for instance, the favourite dessert bananas are not Cavendish but Gros Michel and Sukali Ndizi; in Australia, Lady finger commands about 5% of the market; in the Philippines, a major exporter of Cavendish, the more popular banana for the domestic market is Lakatan. Indonesia contains hundreds of varieties with many local preferential selections. In Micronesia, the orange-fleshed Fe'i bananas are an important part of the local diets as they are rich in pro-vitamin A (PVA). Where bananas are a staple food, particularly in Africa, cooking bananas such as the East African Highland banana (EAHB) clones dominate in East Africa whereas in West Africa a range of different plantain cultivars dominate. It is highly likely that all these cultivars and landraces will be vulnerable to at least one major disease or have negative fruit or yield characteristics that diminish their utilisation for broader cultivation and acceptability and importantly 'transportability' to provide alternatives for landraces or replace Cavendish in the export markets.

Technological advances have changed the landscape with regards to genetic improvement of our crops. We now have mechanisms to combine germplasm with proven agronomic characteristics with disease resistances and other desirable traits very effectively. Conventional breeding of bananas is becoming much more sophisticated and as a more efficient process to generate new cultivars. In parallel, genetic modification has, while genome editing has the potential, to provide the ability to introduce disease resistances, as well as other valuable traits, into the current accepted banana cultivars, thus building upon the amazing biodiversity already available.

1.2 The Europe problem

The other problems of bananas (and most other crops)that specifically relate to genetically modified and genome-edited crops including bananas is the Europe problem. Europe has incredible restrictive regulations regarding the growth or importation of modified crops and food and this now includes crops and foods that have been or will be genome-edited. The practical effect of this is that it is virtually impossible to gain access to the European market with a genetically modified product other than, for instance, corn and soybeans for animal feed and cotton as a fibre. Europe is the largest importer of bananas, the vast majority of which are Cavendish. However, only around 15% of the total banana production is exported. Therefore, of the approximately 140 million tonnes of bananas produced per annum, Europe imports about 6 million tonnes, less than 5% of world production. The reality is that the global influence of Europe's position on modified crops in relation to bananas is far greater than their proportion of the market should justify.

The rapidly developing technologies for genome sequencing and analysis are laying the foundations to enable the quick identification of genes encoding valuable traits. This information can be used as markers for conventional breeding and provide the genes to be moved through genetic modification or edited in situ.

2 Banana improvement by genetic modification and gene editing

2.1 Genetic modification

As noted previously, the two major approaches in developing new or improved banana cultivars are conventional breeding and genetic modification, either through introgression of genes or editing existing genes in the banana genome. The key technical differences between these two approaches are that with conventional breeding the output will be a different cultivar as it is virtually impossible to introgress new traits such as resistance to a particular disease into a current banana cultivar without linkage drag resulting in many other changes to the original cultivar. There are two reasons for this: (i) repeated backcrossing to the original cultivar is not possible because of the high level of sterility in almost all domesticated female parents and (ii) the majority of cultivated bananas, including Cavendish, are highly heterozygous polyploids. Thus, conventional crosses with current cultivars will likely result in wide segregation and heterozygosity giving rise to new gene combinations and cultivars quite different from the progenitors.

In contrast, any banana cultivar can be genetically modified as long as the gene responsible for a particular phenotype is available and it is possible to generate suitable cells that can be transformed and regenerated. The addition of one or a few new genes to a current cultivar should not significantly alter the background characteristics of that cultivar. Efforts to genetically modify bananas have been highly successful. The early transformation efforts started with Bluggoe (ABB), a cooking banana, using microprojectile bombardment (Sagi et al., 1995). Following Bluggoe came the transformation of Cavendish by *Agrobacterium*-mediated transformation (May et al., 1995; Khanna et al., 2004) and microprojectile bombardment (Becker et al., 2000). The list of transformed banana cultivars now exceeds 18 (Table 1) including Cavendish and Bluggoe as well as Rasthali (AAB) (Ganapathi et al., 2001), Lady finger (AAB) (Khanna et al., 2004), Taijiao (AAA) (Pei et al., 2005), a range of plantains (AAB) including Agbagba (Tripathi et al., 2005), East African Highland banana (AAA) (Tripathi et al., 2008), Sukali Ndizi (AAB) (Tripathi et al., 2010), and Gros Michel (AAA)

Table 1 Summary of the published reports on the generation of transgenic bananas

Banana cv	Gene	Trait	Reference
Grand Nain (AAA)	uidA		May et al. (1995)
Bluggoe (ABB)	uidA		Sagi et al. (1995)
Grand Nain	uidA		Becker et al. (2000)
Rasthali (AAB)	uidA		Ganapathi et al. (2001)
Rasthali	Frog magainin	Resistance to Foc race 1	Chakrabarti et al. (2003)
Grand Nain	Rice cystatin	Nematode resistance	Atkinson et al. (2004)
Grand Nain; Lady finger (AAB)	uidA, GFP		Khanna et al. (2004)
Taijiao (AAA) China	Human lysozyme	Resistance to Foc race 4	Pei et al. (2005)
Rasthali	HBsAg	Edible vaccine	Kumar et al. (2005)
Plantain cv Agbagba (AAB)	uidA		Tripathi et al. (2005)
Rasthali	Soybean β-1,3-glucanase	Resistance to Foc race 1	Maziah et al. (2007)
EAHB Mpologoma (AAA), Nakitembe (AAA)	uidA		Tripathi et al. (2008)
Plantain (Three Hand Planty; AAB)		Promoter identification	Santos et al. (2009)
Robusta (AAA)	uidA		Ghosh et al. (2009)
Plantain (Navolean AAB)	B-1,3-endoglucanase (PR-2) and AP24 osmotin (PR5)	Resistance to Black Sigatoka	Kosky et al. (2010)
EAHB Mpologoma Sukali Ndizi (AAB)	Sweet pepper hrap	Resistance to BXW	Tripathi et al. (2010)
Pei Chiao (AAA); Gros Michel (AAA)	Sweet pepper pflp	Resistance to Foc race 4	Yip et al. (2011)
Williams (AAA)	BBTV-S	BBTV resistance	Ismail et al. (2011)
Rasthali	Soybean ferritin	Iron biofortification	Kumar et al. (2011)
Dwarf Brazilian (AAB)	BBTV Rep	BBTV resistance	Borth et al. (2011)
Rasthali	Musa DHN-1	Abiotic stress tolerance	Shekhawat et al. (2011)
Lady finger	Nematode CED-9, human Bcl-xL, human Bcl-2 3′ UTR	Resistance to Foc race 1	Paul et al. (2011)

(Continued)

Table 1 Summary of the published reports on the generation of transgenic bananas (*Continued*)

Banana cv	Gene	Trait	Reference
Grand Nain	Fungal endochitinase (ThEN-42), grape stilbene synthase, tomato superoxide dismutase gene (Cu,Zn-SOD)	Black Sigatoka resistance	Vishnevetsky et al. (2011)
Rasthali	Petunia defensin	Resistance to Foc race 1	Ghag et al. (2012)
Rasthali	Musa SAP	Abiotic stress tolerance	Sreedharan et al. (2012)
Dwarf Cavendish (AAA)	uidA		Chong-Perez et al. (2012)
Sukali Ndizi EAHB Nakinyika (AAA)	Sweet pepper pflp	Resistance to BXW	Namukwaya et al. (2012)
Plantain (Gonja Manjaya AAB)	uidA		Tripathi et al. (2012)
Plantain (Gonja Manjaya)	Maize cystatin, synthetic peptide	Nematode resistance	Roderick et al. (2012)
Rasthali		BBTV resistance	Shekhawat et al. (2012)
Sukali Ndizi	Synthetic mCED9	Resistance to Foc race 1	Magambo (2012)
Rasthali	Onion Ace-AMP	Resistance to Foc Race 1	Mohandas et al. (2013)
Gros Michel	Rice chitinase	Black sigatoka resistance	Kovács et al. (2013)
Rasthali	Musa WRKY71	Abiotic stress tolerance	Shekhawat and Ganapathi (2013)
Rasthali	Musa PIP1;2	Abiotic stress tolerance	Sreedharan et al. (2013)
Furenzhi (AA)	Fungal endochitinase gene chit42	Resistance to Foc TR4	Hu et al. (2013)
Rasthali	VEL, FTF1	Resistance to Foc race 1	Ghag et al. (2014a)
Rasthali	Sm-AMP-D1	Resistance to Foc race 1	Ghag et al. (2014b)
Rasthali	Musa DAD1, BAG1 and BI1	Resistance to Foc race 1	Ghag et al. (2014c)
Plantain (Gonja Manjaya)	Rice Xa21	Resistance to BXW	Tripathi et al. (2014a)
Sukali Ndizi EAHB Nakinyika	Sweet pepper pflp and hrap	Resistance to BXW	Tripathi et al. (2014b)

(*Continued*)

Table 1 Summary of the published reports on the generation of transgenic bananas (*Continued*)

Banana cv	Gene	Trait	Reference
Rasthali	Musa bZIP53	Abiotic stress tolerance	Shekhawat and Ganapathi (2014)
Rasthali	Musa PIP2:6	Abiotic stress tolerance	Sreedharan et al. (2015)
Matti (AA)	Peanut salinity-induced pathogenesis-related class 10 protein gene	Salt and drought tolerance	Rustagi et al. (2015)
Williams, Gros Michel, Sukali Ndizi	uidA		Tripathi et al. (2015a)
Plantain (Gonja Manjaya)	Maize cystatin Synthetic peptide	Nematode resistance	Tripathi et al. (2015b)
Rasthali	Musa VND1	Stress response and developmental regulation	Negi et al. (2015a)
Rasthali	Musa VND2,3	Stress response and developmental regulation	Negi et al. (2015b)
Brazil	Musa MpMYBS3	Abiotic stress tolerance	Dou et al. (2016)
Rasthali	Musa NAC042	Abiotic stress tolerance	Tak et al. (2017a,b)
Dwarf Cavendish	Musa Psy	pVA biofortification	Paul et al. (2017)
Rasthali	Musa NAC68	Plant development	Negi et al. (2015a,b)
Rasthali	Musa ferritin	Biofortification, stress tolerance	Yadav et al. (2017)
Rasthali	Musa MYB31	Lignin biosynthesis study	Tak et al. (2017a)
Rasthali	Musa NAC042	Abiotic stress tolerance	Tak et al. (2017b)
Grand Nain	Nematode CED-9, Musa RGA2	Resistance to Foc TR4	Dale et al. (2017)
Ambon Lumut (AAA)	GFP		Kusumastuti et al. (2017)
Baxi (AAA cv Brazilian)	uidA		Liu et al. (2017)
Gongjiao (AA cv Mas)			
Red banana (AAA, Musa corniculata)			
Rose banana (AA)			
Xinglongnaijiao (AAB)			

(*Continued*)

Table 1 Summary of the published reports on the generation of transgenic bananas (*Continued*)

Banana cv	Gene	Trait	Reference
Williams	CRISPR/Cas9	Inactivate PDS	Naim et al. (2018)
Rasthali	CRISPR/Cas9	Inactivate PDS	Kaur et al. (2018)
Rasthali	Musa SNAC1	Drought tolerance	Negi et al. (2018)
Rasthali	Musa NAC67	Chlorophyll catabolism	Tak et al. (2018)
Plantain (Gonja Manjana)	CRISPR/Cas9	Inactivate eBSOLV	Tripathi et al. (2019)
Rasthali	Musa NAC68	Xylem wall thickness	Negi et al. (2019)

(Yip et al., 2011). Diploids have also been transformed, including Rose banana (AA) (Liu et al., 2017) and Matti (AA) (Rustagi et al., 2015).

Results from multiple experiments indicate that genetic transformation of bananas is not highly genotype-specific in contrast to many other crop species (Rai and Shekhawat, 2014). To date, cultivars with AA, AAA, AAB and ABB genomes have been transformed. It is likely, therefore, that almost any cultivar can be transformed given sufficient effort. This makes the technology ideal for improving the vast range of genetically diverse banana germplasm. This is very important as it is highly likely that there will be no further generations of transgenic banana lines after the T_0 generation. The exception to this is the transformation of germplasm to be used in conventional breeding programmes. Importantly, this lack of genotype dependency provides the opportunity to genetically improve the vast array of banana germplasm for traits such as disease resistance or 'transportability', the two characteristics that probably limit the expansion of the diversity of bananas for the export market. There are currently no reports of transformation of the fertile diploids to be used in conventional breeding programmes, but this is likely to come. Genetically modified fertile diploids could be very useful as parents in conventional breeding programmes particularly as a mechanism to introduce traits that are not available in the accessible banana gene pool such as, for example, banana bunchy top virus resistance.

Genetic modification virtually assures that cultivars such as Cavendish and EAHB will not disappear, but their future will most likely include modified genomes. It is also likely that modified cultivars such as Gros Michel with resistance to Fusarium wilt race 1 and TR4 could be brought back to the market to broaden the genetic base of internationally and locally traded dessert bananas. There could be many other similar examples, including Lady finger in Australia, Sukali Ndizi in East Africa and Prata ana in Brazil with resistance to Fusarium wilt race 1 and probably TR4; Lakatan in the Philippines with TR4 resistance; Cavendish in Malawi with BBTV and ideally TR4 resistance.

2.2 Genome editing

The situation regarding genome editing in bananas is becoming more complex. The first challenge is of a regulatory nature. There was hope that international regulations would be harmonised across the world where specific applications of genome editing would not be classified as genetic modifications and therefore not require extensive regulation and possibly labelling. This has not eventuated and there is a general lack of agreement about the need to regulate the products of genome editing (Eckerstorfer et al., 2019). The extreme position is that of Europe and New Zealand where all genome editing is now defined as a form of genetic modification. Japan and Australia will consider only plants edited using the SDN-1 (site-directed nuclease-1) approach as not genetically modified. SDN-1 does not involve any additional insertion of DNA and therefore is limited to knock-outs or base substitutions. This excludes moving genes or parts of genes from, for instance, one banana to another by homologous recombination.

In the United States, the agency responsible for regulating GM plants and crops is the United States Department of Agriculture (USDA). Until recently, the regulatory status of crops generated through genome-editing technologies was decided on a case-by-case basis. Under this system, the USDA allowed the non-regulated cultivation of CRISPR-edited non-browning mushrooms and waxy corn (Waltz, 2016). More recently, the USDA indicated that crops modified by CRISPR or other genome-editing techniques would not be regulated (USDA, 2018), thus greatly reducing the time for commercialisation and compliance costs. As an example, Yield10 Bioscience (formerly Metabolix) stated that at least six years and $30–50 million had been saved by the USDA decision to approve its *Camelina sativa* genome-edited for enhanced omega-3 oil (Waltz, 2018).

Japan appears likely to adopt a similar regulatory stance on genome editing to that of the United States, with a recent statement from a Japanese government panel supporting the non-regulation of genome-edited organisms (https://www.the-scientist.com/news-opinion/japanese-authorities-recommend-not-regulating-gene-editing-64675).

Argentina was the first country to adopt specific regulations for genome-editing technologies (Resolution 173/2015). Under this policy, the Argentinian regulatory authority (CONABIA) decides the regulatory status of genome-edited plants on a case-by-case basis, with deregulation approved for several crops (Whelan and Lema, 2015; OECD, 2018). A similar resolution was recently issued by Brazil in 2018 (Resolution 16/2018) and it is expected that similar guidelines will be introduced in Chile. The scientists, policymakers and regulatory authorities in Argentina, Brazil, Chile, Paraguay and Uruguay are currently working towards the harmonisation of their genome-editing regulations (Metje-Sprink et al., 2019).

Unlike countries such as the United States, the regulatory scheme in Canada is focussed on the product (Canadian Novel Products Act) rather than the process. As such, crops developed by genome editing could fall under the criteria 'plants with novel traits' which would require regulation. This has been the case with genome-edited, herbicide-resistant crops being subjected to evaluation prior to approval (http://www.inspection.gc.ca/plants/plants-with-novel-traits/approved-under-review/decision-documents/eng/1303704378026/1303704484236).

In South Africa, GMO legislation is currently applied to all plants developed using genome editing although there are ongoing discussions regarding policy amendments (Eckerstorfer et al., 2019).

Genome-edited bananas are already a reality. There are three examples of gene editing in bananas all using CRISPR/Cas9. Naim et al. (2018) knocked out the phytoene desaturase (PDS) gene in Cavendish and tri-allelic knock-outs had an albino phenotype as expected. Tripathi et al. (2019) used genome editing to delete sequences of Banana streak virus (BSV) in the plantain Gonja Manjaya. Finally, Kaur et al. (2018) also knocked out PDS but in Rasthali. These results suggest that genome editing, like genetic modification, will not be strongly genotype-specific in bananas. However, all these edited plants were transgenic as they all contained an integrated cassette containing a selectable marker gene, the Cas9 gene and the guide RNA gene. In seed-propagated crops, it is possible to 'breed away' from the integrated CRISPR/Cas9 cassette. This is not possible with bananas unless the edited banana is to be used in a conventional breeding programme. Therefore, to develop an edited 'non-GM' banana, it will probably be necessary to only transiently express, for instance, the guide RNA and the Cas9 protein, perhaps as a ribonucleoprotein (RNP) and the edited cell will need to be regenerated in the absence of a selectable marker. A possible approach would be to develop a protoplast generation and regeneration protocol to ensure that resulting edited plants are regenerated from a single-edited cell. This will be challenging.

2.3 Traits, transgenes and genomes

Since the first report in 1995, there has been a steady increase in the number of published papers on the generation of transgenic bananas. A comprehensive list of these reports is presented in Table 1. There have been more than 50 publications to early 2019 of which only 14 were pre-2010. Many of these, particularly the early reports, involved the development of transformation systems and protocols but, once reproducible and efficient transformation protocols became available, the emphasis quickly turned to applications.

Interestingly, one of the early practical targets for genetically modified bananas was to develop bananas as a vehicle for 'edible vaccines'. The term 'edible vaccines' was first adopted by Charles Arntzen in 1992 (Mason et al., 1992). Arntzen identified bananas as a very attractive vehicle for producing edible vaccines and he and his team pioneered the transformation of bananas (May et al., 1995) as part of the progress towards this goal. Unfortunately, expression levels in banana fruit at that time were found to be too low to be practically exploited (Arntzen, 2015). With our current knowledge, however, this limitation may be overcome. Kumar et al. (2005) continued the attempts to develop bananas as an edible vaccine vehicle and expressed the hepatitis B surface antigen in Rasthali, but with maximum expression of only ~20 ng/g in the leaves of glasshouse-grown plants. Since that time, there have been no further published reports on the development of edible vaccines in bananas. Instead, the emphasis has changed to the use of confined high level expression systems such as *Nicotiana benthamiana* and subsequent purification of candidate vaccines.

Once genetic modification of bananas became available and relatively efficient, developing disease resistance in bananas has been the primary target of most programmes. This was an obvious target considering the impact that diseases have on banana production worldwide in both commercial plantations and smallholder farmers. The target diseases have been the 'usual suspects', Fusarium wilt race 1 and TR4, Black Sigatoka, banana Xanthomonas wilt (BXW), banana bunchy top and nematodes. The strategies have been either expression/overexpression of broad-spectrum resistance genes or RNA interference (RNAi) (Table 1) with one exception of the successful introduction of a resistance gene to TR4 from the wild progenitor *M. acuminata* spp. *malaccensis* into Cavendish bananas (Dale et al., 2017).

For all of the following transgenes, resistance either in the glasshouse or in the field has been reported. The majority of the broad spectrum resistance transgenes were derived from non-banana sources or non-plant sources. Included in the plant-derived genes were those encoding cystatins (Atkinson et al., 2004; Roderick et al., 2012; Tripathi et al., 2015a,b), an endoglucanase (Maziah et al., 2007), defensins (Ghag et al., 2012,b), antimicrobial peptides (Mohandas et al., 2013), a chitinase (Kovács et al., 2013), a thaumatin-like protein (Mahdavi et al., 2012) and the hypersensitive response-assisting protein and the plant ferredoxin-like protein (Tripathi et al., 2010, 2014a,b; Namukwaya et al., 2012; Muwonge et al., 2016; Yip et al., 2011). The non-plant-derived genes included mammalian-derived anti-apoptosis genes (Paul et al., 2011), nematode-derived anti-apoptosis genes (Paul et al., 2011; Magambo, 2012; Dale et al., 2017), mammalian-derived lysozyme (Pei et al., 2005), frog-derived magainin (Chakrabarti et al., 2003) and fungal-derived endochitinases (Hu et al., 2013; Vishnevetsky et al., 2011).

There are only two examples of the modification of bananas for disease resistance involving banana-derived genes: Dale et al. (2017) introduced

RGA2 from *M. acuminata* spp. *malacenssis* into Cavendish for Fusarium wilt TR4 resistance and Ghag et al. (2014c) introduced the programmed cell death-related genes, *Musa DAD1, Musa BAG1* and *Musa BI* for Fusarium wilt race 1.

There have also been a number of reports of the use of pathogen-derived RNAi-inducing transgenes for disease resistance. Four of the reports targeted the banana bunchy top virus (Borth et al., 2011; Ismail et al., 2011; Shekhawat et al., 2012; Mware, 2016) and one report which targeted Fusarium wilt race 1 (Ghag et al., 2014b).

The development of abiotic stress tolerance in bananas has been less successful. The primary targets have been drought, cold and salt tolerance with the group from the Bhaba Atomic Research Centre, India, most active (Shekhawat et al., 2011; Sreedharan et al., 2012, 2013, 2015; Shekhawat and Ganapathi, 2013, 2014; Tak et al., 2017b) with other contributions by Rustagi et al. (2015) and Dou et al. (2016). Interestingly, all but one of the transgenes were derived from bananas. While in each instance there were positive indications at the glasshouse level, none have progressed to field trial.

Finally, biofortification has been a significant target for banana genetic modification. Both PVA and iron-biofortified bananas have progressed through to the field (Paul et al., 2017; Matovu, 2016) subsequent to glasshouse evaluations (Kumar et al., 2011).

2.4 Progression to field trials

Only a few projects have progressed through to field trials and, as yet, no genetically modified bananas have been released for general production and consumption. Anecdotal reports suggest probable early field trials of genetically modified bananas in Honduras possibly investigating disease resistance and delayed fruit ripening. However, the details and the outcomes of these trials have not been published and the projects seem not to have progressed.

Since those early unpublished trials, there has been a significant number of field trials involving genetically modified bananas and the indications are that at least some of these projects will generate transgenic lines that will be progressed through to release to farmers.

The earliest report of a modified banana field trial was by Pei et al. (2005) in China. They transformed Taijiao, apparently a Cavendish type, with a constitutively expressed human lysozyme gene in an attempt to generate resistance to 'Foc R4', probably Fusarium wilt TR4 which is now widely distributed across the Chinese banana belt. A total of 51 transgenic plants were generated and challenged with the fungus in the glasshouse. Of those challenged, 24 remained healthy and were transferred to the field. In the field, 22 plants developed Panama disease symptoms while two plants remained healthy. It is possible that these two plants were resistant but, since the lines were not replicated, it is more likely that they had escaped infection. There have been no further reports of these plants in expanded field trials.

In 2009, the first biofortification field trial was planted in Australia in 2009 (Paul et al., 2017). This was the first of a series of trials along the path to develop biofortified cooking bananas with elevated levels of PVA. This example is covered later in detail as the Case Study. That same year, a field trial was planted in Uganda with bananas genetically modified at the Katholieke Universiteit Leuven (Belgium) as a collaboration with the National Agricultural Research Organisation of Uganda (Pillay and Tenkouano, 2011). The modification using a chitinase gene was aimed at producing bananas resistant to *Pseudocercospora fijiensis,* the fungus that causes Black Sigatoka. The results of this trial have not been published.

The following year, 2010, the first field trial of modified bananas with potential resistance to banana Xanthomonas wilt (BXW) was planted in Uganda. BXW is a major disease problem in East Africa. It is caused by a soil-borne bacterium, *Xanthomonas campestris* pv *musacearum*. Originally first recorded in Ethiopia in Enset, it is now widespread in East Africa including Uganda, Kenya, Democratic Republic of Congo, Tanzania, Rwanda and Burundi. Bananas grown in infested soil easily become infected and the bacteria can be transmitted in infected planting material, via movement of soil and contaminated tools and by insects. The major bananas grown in the region, including EAHB, plantains, Gros Michel, Cavendish and Sukali Ndizi, are all highly susceptible. Despite agronomic protocols that can control the disease, these are often poorly implemented in smallholder settings. In 2014, Tripathi et al. (2014b) reported the first field trial of bananas resistant to BXW in Uganda. This was a collaboration between the International Institute of Tropical Agriculture and the National Agricultural Research Organisation of Uganda. For this trial, the local dessert banana, Sukali Ndizi (AAB), was transformed with either constitutively expressed Hrap (hypersensitive response-assisting protein) or Pflp (plant ferredoxin-like protein). These two transgenes had been previously isolated from *Capsicum annuum* by others and demonstrated in other crops to provide resistance to bacterial diseases (Ger et al., 2002; Pandey et al., 2005; Huang et al., 2004, 2007; Liau et al., 2003; Yip et al., 2007; Tang et al., 2001). The Sukali Ndizi lines were pre-screened in the glasshouse for resistance to BXW and 40 lines with Hrap and 25 lines with Pflp were progressed through to the field. The plants were artificially inoculated and the trial was conducted over two crop cycles. Seven of the 40 Hrap and three of the Pflp lines showed complete resistance to BXW. This project has now progressed significantly with the two transgenes stacked and genetically modified lines of the EAHB clone Nakitembe and the dessert banana Cavendish generated. Resistant lines have been identified through glasshouse screening and are now proceeded to field trial in either Uganda (Nakitembe) or Kenya (Cavendish).

There has only been one field trial targeting the extremely damaging disease, Black Sigatoka, caused by *Pseudocercospora fijiensis* (previously *Mycosphaerella fijiensis*). This leaf disease causes serious losses in both commercial plantations and smallholder gardens with the major bananas Cavendish, EAHBs and West African plantains severely affected. Many commercial plantations are sprayed with fungicides more than 50 times a year to control the disease, a strategy not usually available to smallholders. Clearly, cultivar resistance would be of considerable benefit to commercial producers to particularly reduce costly fungicide applications and to smallholders to reduce yield losses. To this end, Vishnevetsky et al. (2011) generated transgenic Cavendish (Grand Nain) bananas expressing three different transgenes: the endochitinase gene *ThEN-42* from *Trichoderma harzianum* under the control of the constitutive CaMV 35S promoter, the grape stilbene synthase gene *StSy* under the control of the inducible PR-10 promoter and the tomato superoxide dismutase gene *Cu,Zn-SOD* under the control of the constitutive maize polyubiquitin promoter. The field trial was conducted in Florida, USA, over four years with three consecutive plantings and a total of 71 lines with 12 plants per line. The level of disease was scored using an 'index of non-spotted leaves (INSL)'. Unfortunately, none of the lines was immune and only two lines had a significantly lower INSL than the controls. There have been no further reports of these lines or the approach in the field.

Early in 2012, we planted the first field trial of modified Cavendish banana with potential resistance to Fusarium wilt TR4 (Dale et al., 2017). Fusarium wilt is probably the disease that banana producers most fear. The soil-borne fungus, previously called *Fusarium oxysporum* f.sp. *cubense*, has recently been renamed in a series of *Fusarium* spp. causing Panama disease (Maryani et al., 2019). They can survive in the soil for decades and are easily spread in infected planting material, infested soil attached to boots, clothing, machinery and vehicles and also in water including irrigation and flood water. They infect the roots and invade the plants' vascular tissues and invariably the plants die. Two races are extremely important: race 1 caused one of the most devastating epidemic plant diseases ever recorded (Ploetz, 2005). Early last century, Gros Michel (AAA) was the primary exported banana and was grown widely in South and Central America. Almost certainly, Fusarium wilt race 1 was introduced into the Americas likely from the centres of origin of bananas. Gros Michel is highly susceptible to Fusarium wilt race 1; the fungus and the disease spread rapidly through the plantations and production was greatly curtailed. Cavendish, however, is resistant to Fusarium wilt race 1 and, because of this and other characteristics such as yield, robustness and consumer acceptability, it supplanted Gros Michel as the global export banana.

Within the last 30 years, a separate species, Fusarium wilt tropical race 4 (TR4) recently renamed as *F. odoratissimum* (Maryani et al., 2019), has risen from initial concerning reports from Taiwan to currently the major threat to global banana production. TR4 has very similar characteristics to Fusarium wilt race 1 with the major addition of many banana cultivars, including Cavendish as a host. It has probably been present in Southeast Asia for a significant period of time and has been spread from there to most countries of Southeast and South Asia (Zheng et al., 2018). It is already having a major impact on the export-oriented production of Cavendish in the Philippines as well as significantly impacting the domestic production of Cavendish in China. The fungus is now present in several States in India (Thangavelu et al., 2019), the world's largest producer of bananas, but it is not clear how widespread the impact of the disease is there. Of great concern is that the fungus has been introduced into two commercial plantations in Mozambique. These plantations are the potential source populations of the fungus to move out of Mozambique to the neighbouring banana-producing countries of Malawi, South Africa and Zambia and beyond as there is no obvious impediment to stop the disease infiltrating further into sub-Saharan Africa. It must also be remembered that TR4 is not only a disease of Cavendish, but also infects a wide range of cultivars as well as some wild diploids. It is probable that most of the banana cultivars are susceptible to either TR4, race 1 or both. Of particular concern in Africa is that the susceptibility of two major food-security crops in sub-Saharan Africa, the EAHBs and the West African plantains has not been determined. The recent confirmation of TR4 in Colombia, South America (García-Bastidas et al., 2019), now poses a significant threat to banana production in this, and other major exporting countries, in Latin America.

We had previously isolated RGA2 from a TR4-resistant wild diploid banana, *Musa acuminata* ssp. *malaccensis*, which was highly expressed in resistant progeny but not in susceptible progeny (Peraza-Echeverria et al., 2008). RGA2 is a typical NB-LRR resistance gene and is located on Chromosome 3 of the *Musa acuminata* ssp. *malaccensis* genome (D'Hont et al., 2012). We transferred this gene, under the control of the Agrobacterium nopaline synthase (nos) promoter, to the Cavendish cultivar Grand Nain. The resultant transgenic lines were not challenged in the glasshouse due to local biosecurity TR4 restrictions. Consequently, six RGA2 transgenic lines were trialled at Lambells Lagoon in the Northern Territory in Australia over three years in a field previously growing Cavendish which had been decimated by TR4 infections. At the end of the trial, the non-GM controls were heavily infected with rates of 67% in Cavendish Williams, 69% in GCTCV218 and 88% in Cavendish Grand Nain. In contrast, none of the RGA2-3 line was infected and a further three lines, RGA2-2, RGA2-4 and RGA2-5 had only 20%, 20% and 14% infection, respectively (Dale et al., 2017). Two of the six lines did not demonstrate significant TR4 resistance. Two other important outcomes from this field trial were that the level of resistance was highly correlated with the level of expression of RGA2, and RGA2 is also present in the Cavendish genome but was either not expressed or was expressed at a very low level.

A larger Phase 2 field trial commenced in 2018 which includes new RGA2 lines. Data is being collected on agronomic performance, disease resistance and various biosafety parameters that could be included in a dossier for deregulation.

Nematodes are a major production constraint in virtually all banana-growing areas. The key nematodes of bananas are *Radopholus similis*, *Pratylenchus goodeyi* and *P. coffeae*, *Helicotylenchus multicinctus* and *Meloidogyne* spp. Nematode resistance in transgenic bananas has previously been demonstrated through glasshouse assessment (Roderick et al., 2012), but not in the field. In 2015, Tripathi et al. (2015b) reported on a one-cycle-field trial of transgenic plantains in Uganda. This was a collaboration between the International Institute of Tropical Agriculture, the University of Leeds, UK, and the National Agricultural Research Organisation of Uganda. The plantain involved was the popular Gonja manjaya. This plantain was transformed with either a cysteine protease-inhibitor gene (for an anti-feeding protein) or a synthetic peptide gene (for a secreted anti-root invasion peptide) or a combination of the two (Roderick et al., 2012). From 250 lines that were screened in the glasshouse, 12 lines progressed to the field and were statistically more resistant to particularly *R. similis* at harvest and one line, which contained the synthetic peptide gene alone, had a 99% resistance level. Importantly, for this line, the resistance level translated into about a 25% increase in yield. These results are extremely encouraging. The field trial is continued to a first and second ratoon. If the resistance is maintained through multiple cycles of banana production, this will be a major step to potentially control nematodes in both bananas and plantains.

Bunchy top is considered one of the most destructive diseases of bananas. Caused by the single-stranded DNA virus, banana bunchy top virus (BBTV), the disease is well-established in Asia but, very importantly, is expanding its distribution to new countries in Africa and is causing very significant damage to the point where in Malawi, for example, it is extremely difficult to grow bananas such as Cavendish. Four groups have attempted to use RNAi as a transgenic strategy to generate resistance to BBTV in banana. Borth et al. (2011) and Shekhawat et al. (2012) reported glasshouse resistance with an inverted repeat derived from the BBTV replication-associated gene, the Rep gene (DNA-R). Ismail et al. (2011) targeted the BBTV coat protein gene (DNA-S) while Mware (2016) targeted the Rep gene, the movement protein gene (DNA-M) or the nuclear shuttle protein gene (DNA-N), as well as combinations of these. Mware (2016) recorded the highest level of resistance in the glasshouse where DNA-M was the target. In a collaboration between the Queensland University of Technology, the Malawi Department of Agricultural Research Services and the International Institute of Tropical Agriculture, the transgenic Cavendish lines of Mware (2016) were planted in a field trial in Malawi in 2016. After two years trialling there are a number of lines which show significant lower levels of infection and these lines are progressing into a larger field trial in Malawi together with plantain lines currently being assessed in the glasshouse.

3 Case study: biofortified East African Highland bananas

Probably the most advanced GM bananas are the PVA-enhanced EAHBs. These bananas are about to enter multi-location, regulatory field trials in Uganda.

EAHBs are the staple food of Uganda. These are cooking bananas and are harvested green and usually steamed for a number of hours. They are the primary starch source across Uganda with daily consumption around 0.5 kg per person. This food, known as Matooke, is very much part of the Ugandan culture. EAHB are also grown widely in other countries of East Africa including Tanzania, Kenya, Rwanda, Burundi and Democratic Republic of Congo. Uganda, with a population of approximately 40 million, produces annually in excess of 9 million tonnes of bananas. As with most developing countries with heavy dependence on a single starch source, Uganda has high levels of micronutrient deficiencies particularly vitamin A deficiency (VAD) and iron deficiency anaemia (IDA). For instance, the 2016 Uganda Department of Health survey indicated that about 9% of children had clinical VAD with similar levels in women of child-bearing age. The levels of IDA are higher. These unacceptable levels persist despite interventions such as food fortification and supplements.

The concept of biofortification, developing staple foods that contain elevated levels of micronutrients, has been expanding as an additional intervention strategy with low recurrent costs. It can be achieved through conventional breeding, for instance, high iron beans, where the necessary traits are available in the accessible gene pool or by genetic modification, the best example of which is Golden Rice (Ye et al., 2000; Paine et al., 2005) that has been biofortified to produce PVA in the seed.

The 'golden banana' project commenced in 2005 funded as part of the Bill & Melinda Gates Foundation's Grand Challenges in Global Health with subsequent additional funding from the UK Department for International Development. The project goal is to alleviate vitamin A deficiency in Uganda by elevating the levels of PVA in the staple food of Uganda, namely bananas. PVA, a combination of α-carotene and β-carotene, naturally occurs in banana fruit at varying levels depending on the cultivar and is converted to vitamin A by the liver. PVA is non-toxic even at high dosages. The target is to provide 50% of the estimated average requirement (EAR) of vitamin A which translates to a level of 20 μg/g dry weight of β-carotene equivalents (βCE) in banana fruit flesh. The 20 µg/g dwt βCE was determined from the EAR, the average consumption of cooked bananas in Uganda by both children under 60 months and women of child-bearing age, a ratio for the conversion of β-carotene equivalents in bananas to vitamin A in mammals and the losses of βCE during cooking and storage. It was determined during the project that the bioavailability of PVA in bananas was low in raw fruit but high in cooked fruit (Bresnahan et al., 2012).

The project has been a collaboration between Queensland University of Technology (QUT) in Australia and the National Agricultural Research Organisation (NARO) of Uganda. The strategy from the commencement of the project was for two parallel sub-projects where the technology would be developed in Australia using Cavendish and the technology, but not the plants, would be continuously transferred to Uganda such that the final product would be produced in Uganda. QUT assembled a range of promoters and transgenes into more than 48 different promoter/transgene combinations including single transgene and multi-transgene cassettes which were then transformed into Cavendish. The resultant transgenic lines were transferred to the field without prior characterisation in the glasshouse. In the Phase 1 field trials, planted from 2009, more than 2000 individual transgenic lines were planted in North Queensland with up to 30 transgenic lines per cassette but only one plant per line. From these early trials, four promoter/transgene combinations were determined to result in lines with the highest levels of fruit PVA. These were either the maize phytoene synthase 1 gene (*ZmPsy1*) or the banana (Asupina) phytoene synthase 2a gene (*MtPsy2a*) each under the control of either the constitutive maize polyubiquitin promoter (Ubi) or the fruit-preferred banana ACC oxidase promoter (ACO) (Paul et al., 2017). The *ZmPsy1* gene had been previously used to generate Golden Rice 2 (Paine et al., 2005). The *MtPsy2a* gene was isolated from a Fe'i type banana, Asupina, which naturally has high fruit PVA (Mlalazi et al., 2012; Buah et al., 2016).

Multiple plants of each of the most promising lines were planted in the Phase 2 trials in Australia. From this trial, the highest performing line was a Ubi>*MtPsy2a* line with an average of 55 µg/g dwt βCE. There were also multiple lines with ACO>*MtPsy2a*, Ubi>*ZmPsy1* and ACO>*ZmPsy1* with higher than the target 20 µg/g dwt βCE. However, lines containing *ZmPsy1* invariably had phenotypic abnormalities (Paul et al., 2017). As a consequence, only the Ubi>*MtPsy2a* and ACO>*MtPsy2a* cassettes were selected for transfer to Uganda for the generation of the biofortified cooking bananas. The best lines were monitored over successive years in Australia. There was no evidence of the transgenes becoming silenced; however, the PVA levels varied both up and down between harvests. This variation correlated strongly with the time from flowering (bunch emergence) until harvest; the longer this period the higher the PVA levels (Paul et al., 2017).

Figure 1 Field trial of PVA-biofortified genetically modified banana plants in Uganda.

For Phase 3, the Ubi>*MtPsy2a* and ACO>*MtPsy2a* cassettes were each transformed into two genotypes in Uganda, Nakitembe, a popular EAHB clone, and Hybrid M9, a conventionally bred cooking banana with resistance to Black Sigatoka, Fusarium wilt race 1 and weevils. A total of 518 transgenic lines were planted in the elite line-selection trial at a site near Kampala in Uganda (Fig. 1). The two primary target criteria for lines in this trial were PVA levels greater than 20 μg/g dwt βCE and bunch weights within 10% of the non-transgenic controls. Multiple lines of both genotypes and either cassette have met these criteria, some with PVA levels four times the target (Fig. 2). The project has now commenced Phase 4, which includes the multi-location regulatory field trials at four different agroecological zones across Uganda. Five lines each of Nakitembe and Hybrid M9 have been selected to progress to these trials based on multiple additional criteria including transgene copy number and the integrity of the insertion locus as well as fruit composition and bunch characteristics. It is planned that these trials will be completed in the second half of 2021 for Hybrid M9 and the first half of 2022 for Nakitembe with the regulatory dossiers submitted soon after the completion of the trials.

It is likely that these will be the first GM bananas released to farmers: a staple crop with a trait derived from a single banana gene to alleviate vitamin A deficiency in a developing country and grown by smallholder farmers.

Figure 2 Longitudinal section of fruit from a PVA-enhanced (left) and wild-type (right) cooking banana (M9).

4 Future trends and conclusion

Genetically modified bananas are at a dynamic point in their progress towards release to farmers. There are at least two projects that are likely to result in deregulated bananas within the next three to four years, the biofortified cooking bananas in Uganda and the BXW-resistant cooking bananas also in Uganda and in Kenya. These could be closely followed by Fusarium wilt TR4-resistant Cavendish in Australia and BBTV-resistant Cavendish in Malawi as well as nematode-resistant plantains in Africa.

The path of modified bananas from the laboratory to the field and, in the future, to the farmers and ultimately to consumers could not have been more different to the path taken for the first wave of modified crops particularly maize, soybean, cotton and canola. These latter crops were developed and financed by large multinational companies such as Monsanto, Bayer (now including Monsanto), DuPont Pioneer, Dow AgroSciences and BASF. The farmers were broadacre producers initially in North America. The traits were, to begin with, herbicide-tolerant and insect-resistant, both using bacterial-derived transgenes, contrast this history with that of modified bananas. While there was some interest from one or two of the large multinationals early, this soon evaporated. Rather, the development from laboratory to the field has been undertaken by national departments of agriculture, universities and international and national research centres. These projects have been funded primarily from philanthropic sources such as the Bill & Melinda Gates Foundation, aid organisations such as USAID and UK DFID, national governments primarily in Africa and Australia and a small group of Australian banana growers and marketers. The large multinational banana-marketing companies have been conspicuously absent from supporting the development of genetically modified bananas. The Europe problem has almost certainly been a major inhibiting factor to invest in modified bananas. However, one prominent 'banana' company, Rahan Meristem, has invested in the development of modified bananas and has conducted field trials.

In contrast to the broadacre-modified crops, the roll-out of modified bananas will not be in North America, but most likely in Africa and will be to smallholders and not broadacre farmers. The second continent is likely to be Australia. Importantly, these geographies either do not export bananas to Europe or the volumes are extremely small. Thus, the Europe problem is not relevant. Finally, the regulatory environment for modified crops is continuing to develop in Africa. Australia has had a biosafety Bill for nearly two decades. In sub-Saharan Africa, South Africa, Nigeria, Malawi and Kenya, all have functioning biosafety Bills and Uganda is close to developing and implementing that. All these countries produce bananas. The first modified bananas on the market will certainly be genetically modified in the 'conventional' sense. They will all contain additional DNA as transgenes and, in most cases, these transgenes will be derived from plants, including bananas. The first trait will likely be a consumer trait, biofortified cooking bananas with elevated PVA, again in stark contrast to the first wave of genetically modified field crops. The next group of released bananas will have disease resistance.

What will follow? With the plunging costs of DNA and RNA sequencing, the amount of genomic and transcriptomic information becoming available for an array of different banana genotypes is increasing rapidly. As a consequence, genes for a range of different traits will be identified and will become available initially for genetic modification. However, very quickly, protocols for developing non-GM genome-edited bananas will be developed.

Within a decade, it is likely that genetic modification and genome editing of bananas will lead to current cultivars with multiple disease resistances, enhanced fruit quality, abiotic stress tolerance and increased transportability. Most importantly, these traits will be incorporated into a wide range of different banana genotypes from dessert bananas such as Cavendish, Gros Michel, Lakatan, Sukali Ndizi, Rasthali and Prata ana to the cooking bananas of East Africa and the plantains of West Africa and the Pacific to name just a few. These advances will have a significant positive impact on food security through reduced disease losses, enhanced nutrition and adaption to climate change and for the environment with the reduction in use of pesticides. Importantly, the biodiversity of cultivated bananas will be greatly increased and wonderful cultivars and landraces not lost. Finally, genome-edited bananas are likely to enter the export trade.

5 References

Arntzen, C. (2015). Plant-made pharmaceuticals: from 'Edible Vaccines' to Ebola therapeutics, *Plant Biotechnol. J.* 13(8), 1013–1016.

Atkinson, H. J., Grimwood, S., Johnston, K. and Green, J. (2004). Prototype demonstration of transgenic resistance to the nematode Radopholus similis conferred on banana by a cystatin, *Transgenic Res.* 13(2), 135–142.

Becker, D. K., Dugdale, B., Smith, M. K., Harding, R. M. and Dale, J. L. (2000). Genetic transformation of Cavendish banana (Musa spp. AAA group) cv 'Grand Nain' via microprojectile bombardment, *Plant Cell Rep.* 19(3), 229–234.

Borth, W., Perez, E., Cheah, K., Chen, Y., Xie, W. S., Gaskill, D., Khalil, S., Sether, D., Melzer, M., Wang, M., Manshardt, R., Gonsalves, D. and Hu, J. S. (2011). Transgenic banana plants resistant to banana bunchy top virus infection, *Acta Hortic.* 897, 449–457.

Bresnahan, K. A., Arscott, S. A., Khanna, H., Arinaitwe, G., Dale, J. L., Tushemereirwe, W., Mondloch, S., Tanumihardjo, J. P., De Moura, F. F. and Tanumihardjo, S. A. (2012). Cooking enhances but the degree of ripeness does not affect provitamin A carotenoid bioavailability from bananas in Mongolian gerbils, *J. Nutr.* 142(12), 2097–2104.

Buah, S., Mlalazi, B., Khanna, H., Dale, J. L. and Mortimer, C. L. (2016). The quest for golden bananas: investigating carotenoid regulation in a Fe'i group Musa cultivar, *J. Agric. Food Chem.* 64(16), 3176–3185.

Chakrabarti, A., Ganapathi, T. R., Mukherjee, P. K. and Bapat, V. A. (2003). MSI-99, a magainin analogue, imparts enhanced disease resistance in transgenic tobacco and banana, *Planta* 216(4), 587–596.

Chong-Perez, B., Reyes, M., Rojas, L., Ocaña Díaz, B., Pe'rez, B., Gómez-Kosky, R. and Angenon, G. (2012). Establishment of embryogenic cell suspension cultures and Agrobacterium-mediated transformation in banana cv. "Dwarf Cavendish" (*Musa* AAA): effect of spermidine on transformation efficiency, *Plant Cell Tiss. Org. Cult.* 111, 79–90.

Dale, J. L., James, A., Paul, J. Y., Khanna, H., Smith, M., Peraza-Echeverria, S., Garcia-Bastidas, F., Kema, G., Waterhouse, P., Mengersen, K. and Harding, R. (2017). Transgenic Cavendish bananas with resistance to Fusarium wilt tropical race 4, *Nat. Commun.* 8(1), 1496.

D'Hont, A., Denoeud, F., Aury, J. M., Baurens, F. C., Carreel, F., Garsmeur, O., Noel, B., Bocs, S., Droc, G., Rouard, M., Da Silva, C., Jabbari, K., Cardi, C., Poulain, J., Souquet, M., Labadie, K., Jourda, C., Lengellé, J., Rodier-Goud, M., Alberti, A., Bernard, M., Correa, M., Ayyampalayam, S., Mckain, M. R., Leebens-Mack, J., Burgess, D., Freeling, M., Mbéguié-A-Mbéguié, D., Chabannes, M., Wicker, T., Panaud, O., Barbosa, J., Hribova, E., Heslop-Harrison, P., Habas, R., Rivallan, R., Francois, P., Poiron, C., Kilian, A., Burthia, D., Jenny, C., Bakry, F., Brown, S., Guignon, V., Kema, G., Dita, M., Waalwijk, C., Joseph, S., Dievart, A., Jaillon, O., Leclercq, J., Argout, X., Lyons, E., Almeida, A., Jeridi, M., Dolezel, J., Roux, N., Risterucci, A. M., Weissenbach, J., Ruiz, M., Glaszmann, J. C., Quétier, F., Yahiaoui, N. and Wincker, P. (2012). The banana (Musa acuminata) genome and the evolution of monocotyledonous plants, *Nature* 488(7410), 213–217.

Dou, T. X., Hu, C. H., Sun, X. X., Shao, X. H., Wu, J. H., Ding, L., Gao, J., He, W., Biswas, M., Yang, Q. and Yi, G. (2016). MpMYBS3 as a crucial transcription factor of cold signaling confers the cold tolerance of banana, *Plant Cell Tissue Organ Cult.* 125(1), 93–106.

Eckerstorfer, M. F., Engelhard, M., Heissenberger, A., Simon, S. and Teichmann, H. (2019). Plants developed by new genetic modification techniques – comparison of existing regulatory frameworks in the EU and non-EU countries, *Front. Bioeng. Biotechnol.* 7, 26.

Ganapathi, T. R., Higgs, N. S., Balint-Kurti, P. J., Arntzen, C. J., May, G. D. and van Eck, J. M. (2001). Agrobacterium-mediated transformation of the embryogenic cell suspensions of the banana cultivar "Rasthali" (AAB), *Plant Cell Rep.* 20(2), 157–162.

García-Bastidas, F. A., Quintero-Vargas, J. C., Ayala-Vasquez, M., Schermer, T., Seidl, M. F., Santos-Paiva, M., Noguera, A. M., Aguilera-Galvez, C., Wittenberg, A., Hofstede, R., Sørensen, A. and Kema, G. H. J. (2019). First report of Fusarium wilt Tropical Race 4 in Cavendish bananas caused by Fusarium odoratissimum in Colombia, *Plant Dis.* 104(3), 994.

Ger, M. J., Chen, C. H., Hwang, S. Y., Huang, H. E., Podile, A. R., Dayakar, B. V. and Feng, T. Y. (2002). Constitutive expression of hrap gene in transgenic tobacco plant enhances resistance against virulent bacterial pathogens by induction of a hypersensitive response, *Mol. Plant Microbe Interact.* 15(8), 764–773.

Ghag, S. B., Shekhawat, U. K. S. and Ganapathi, T. R. (2012). Petunia floral defensins with unique prodomains as novel candidates for development of Fusarium wilt resistance in transgenic banana plants, *PLoS ONE* 7(6), e39557.

Ghag, S. B., Shekhawat, U. K. S. and Ganapathi, T. R. (2014a). Host-induced post-transcriptional hairpin RNA-mediated gene silencing of vital fungal genes confers efficient resistance against Fusarium wilt in banana, *Plant Biotechnol. J.* 12(5), 541–553.

Ghag, S. B., Shekhawat, U. K. S. and Ganapathi, T. R. (2014b). Transgenic banana plants expressing a Stellaria media defensin gene (Sm-AMP-D1) demonstrate improved resistance to *Fusarium oxysporum*, *Plant Cell Tissue Organ Cult.* 119(2), 247–255.

Ghag, S. B., Shekhawat, U. K. S. and Ganapathi, T. R. (2014c). Native cell-death genes as candidates for developing wilt resistance in transgenic banana plants, *AoB Plants* 6, plu037.

Ghosh, A., Ganapathi, T. R., Nath, P. and Bapat, V. A. (2009). Establishment of embryogenic cell suspension cultures and Agrobacterium-mediated transformation in an important Cavendish banana cv. Robusta (AAA), *Plant Cell. Tiss. Org. Cult.* 97(2), 131–139.

Heslop-Harrison, J. S. and Schwarzacher, T. (2007). Domestication, genomics and the future for banana, *Ann. Bot.* 100(5), 1073–1084.

Hu, C., Wei, Y., Huang, Y. and Yi, G. (2013). An efficient protocol for the production of chit42 transgenic Furenzhi banana (*Musa* spp. AA group) resistant to *Fusarium oxysporum*, *In Vitro Cell. Dev. Biol.-Pl.* 49(5), 584–592.

Huang, H.-E., Ger, M.-J., Yip, M.-K., Chen, C., Pandey, A. and Feng, T.-Y. (2004). A hypersensitive response was induced by virulent bacteria in transgenic tobacco plants overexpressing a plant ferredoxin-like protein (pflp), *Physiol. Mol. Plant Pathol.* 64(2), 103–110.

Huang, H. E., Liu, C. A., Lee, M. J., Kuo, C. G., Chen, H. M., Ger, M. J., Tsai, Y. C., Chen, Y. R., Lin, M. K. and Feng, T. Y. (2007). Resistance enhancement of transgenic tomato to bacterial pathogens by the heterologous expression of sweet pepper ferredoxin-I protein, *Phytopathology* 97(8), 900–906.

Ismail, R. M., El-Domyati, F. M., Wagih, E. E., Sadik, A. S. and Abdelsalam, A. Z. E. (2011). Construction of banana bunchy top nanovirus-DNA-3 encoding the coat protein gene and its introducing into banana plants cv. Williams, *J. Genet. Eng. Biotechnol.* 9(1), 35–41.

Kaur, N., Alok, A., Shivani, Kaur, N., Pandey, P., Awasthi, P. and Tiwari, S. (2018). CRISPR/Cas9-mediated efficient editing in phytoene desaturase (PDS) demonstrates precise manipulation in banana cv. Rasthali genome, *Funct. Integr. Genomic.* 18(1), 89–99.

Khanna, H., Becker, D., Kleidon, J. and Dale, J. L. (2004). Centrifugation assisted Agrobacterium tumefaciens mediated transformation (CAA) of embryogenic cell suspensions of banana (*Musa* spp.) Cavendish AAA and Lady Finger AAB, *Mol. Breed.* 14(3), 239–252.

Kosky, R. G., Chong-Pérez, B., López-Torres, J., Reyes, M., Bermúdez-Caraballoso, I., Montalvo Martín, N., Machado-Rodriguez, J. M., Portal, O., Ocaña, B., Alvarado-Capó, Y., Leiva-Mora, M., Acosta-Suárez, M., Cruz-Martin, M., Roque, B. and Hernández, L. (2010). Plantain (*Musa* spp. cv. Navolean AAB) transgenic plants from Agrobacterium tumefaciens-mediated transformation of embryogenic cell suspensions, *Biotechnol. Veget.* 10, 209–218.

Kovács, G., Sági, L., Jacon, G., Arinaitwe, G., Busogoro, J. P., Thiry, E., Strosse, H., Swennen, R. and Remy, S. (2013). Expression of a rice chitinase gene in transgenic banana ('Gros Michel', AAA genome group) confers resistance to black leaf streak disease, *Transgenic Res.* 22(1), 117–130.

Kumar, G. B. S., Ganapathi, T. R., Revathi, C. J., Srinivas, L. and Bapat, V. A. (2005). Expression of hepatitis B surface antigen in transgenic banana plants, *Planta* 222(3), 484–493.

Kumar, G. B. S., Srinivas, L. and Ganapathi, T. R. (2011). Iron fortification of banana by the expression of soybean ferritin, *Biol. Trace Elem. Res.* 142(2), 232–241.

Kusumastuti, T. R., Esyantia, R. R. and Dwivany, F. M. (2017). Agrobacterium tumefaciens-mediated transformation of embryogenic callus and somatic embryos of the banana cv "Ambon Lumut" (*Musa acuminata*), *Life Sci.* 2(6), 599–616.

Liau, C. H., Lu, J. C., Prasad, V., Hsiao, H. H., You, S. J., Lee, J. T., Yang, N. S., Huang, H. E., Feng, T. Y., Chen, W. H. and Chan, M. T. (2003). The sweet pepper ferredoxin-like protein (pflp) conferred resistance against soft rot disease in Oncidium orchid, *Transgenic Res.* 12(3), 329–336.

Liu, J., Gao, P., Sun, X., Zhang, J., Sun, P., Wang, J., Jia, C., Zhang, J., Hu, W., Xu, B. and Jin, Z. (2017). Efficient regeneration and genetic transformation platform applicable to five Musa varieties, *Electron. J. Biotechnol.* 25, 33–38.

Magambo, B. (2012). Generating transgenic banana (cv. Sukali Ndizi) resistant to Fusarium Wilt, Masters by Research thesis, Queensland University of Technology.

Mahdavi-Darvari, F., Meon, S. and Maziah, M. (2012). Expression of Rice Thaumatin-Like Protein Gene in Transgenic Banana Plants Enhances Resistance to Fusarium Wilt. *Appl. Biochem. Biotechnol.*. 166, 1008–19.

Maryani, N., Lombard, L., Poerba, Y. S., Subandiyah, S., Crous, P. W. and Kema, G. H. J. (2019). Phylogeny and genetic diversity of the banana Fusarium wilt pathogen *Fusarium oxysporum* f. sp. cubense in the Indonesian centre of origin, *Stud. Mycol.* 92, 155–194.

Mason, H. S., Lam, D. M. and Arntzen, C. J. (1992). Expression of hepatitis B surface antigen in transgenic plants, *Proc. Natl Acad. Sci. U. S. A.* 89(24), 11745–11749.

Matovu, M. (2016). Molecular and biochemical characterisation of transgenic banana lines containing iron uptake and storage enhancing genes, PhD thesis, Queensland University of Technology.

May, G. D., Afza, R., Mason, H. S., Wiecko, A., Novak, F. J. and Arntzen, C. J. (1995). Generation of transgenic banana (*Musa acuminata*) plants via Agrobacterium-mediated transformation, *Biotechnology* 13(5), 486–492.

Maziah, M., Sariah, M. and Sreeramanan, S. (2007). Transgenic banana Rasthali (AAB) with β-1, 3-glucanase gene for tolerance to Fusarium wilt race 1 disease via Agrobacterium-mediated transformation system, *Plant Pathol. J.* 6, 271–282.

Metje-Sprink, J., Menz, J., Modrzwjewski, D. and Sprink, T. (2019). DNA-free genome editing: past, present and future, *Front. Plant Sci.* 9, 1957.

Mlalazi, B., Welsch, R., Namanya, P., Khanna, H., Geijskes, R. J., Harrison, M. D., Harding, R., Dale, J. L. and Bateson, M. (2012). Isolation and functional characterisation of banana phytoene synthase genes as potential cisgenes, *Planta* 236(5), 1585–1598.

Mohandas, S., Sowmya, H. D., Saxena, A. K., Meenakshi, S., Rani, R. T. and Mahmood, R. (2013). Transgenic banana cv Rasthali (AAB, Silk gp) harboring Ace-AMP1 gene imparts enhanced resistance to *Fusarium oxysporum* f. sp cubense race 1, *Sci. Hortic.* 164, 392–399.

Muwonge, A., Tripathi, J., Kunert, K. and Tripathi, L. (2016). Expressing stacked HRAP and PFLP genes in transgenic banana has no synergistic effect on resistance to Xanthomonas wilt disease. *S. Afr. J. Bot.* 104, 125–133.

Mware, B. O. (2016). Development of banana bunchy top virus resistance in bananas: RNAi approach, PhD thesis, Queensland University of Technology.

Naim, F., Dugdale, B., Kleidon, J., Brinin, A., Shand, K., Waterhouse, P. and Dale, J. (2018). Gene editing the phytoene desaturase alleles of Cavendish banana using CRISPR/Cas9, *Transgen. Res.* 27, 451–460.

Namukwaya, B., Tripathi, L., Tripathi, J. N., Arinaitwe, G., Mukasa, S. B. and Tushemereirwe, W. K. (2012). Transgenic banana expressing pflp gene confers enhanced resistance to Xanthomonas wilt disease, *Transgenic Res.* 21(4), 855–865.

Negi, S., Tak, H. and Ganapathi, T. R. (2015a). Cloning and functional characterization of MusaVND1 using transgenic banana plants, *Transgenic Res.* 24(3), 571–585.

Negi, S., Tak, H. and Ganapathi, T. R. (2015b). Functional characterization of secondary wall deposition regulating transcription factors MusaVND2 and MusaVND3 in transgenic banana plants, *Protoplasma* 253(2), 431–446.

Negi, S., Tak, H. and Ganapathi, T. R. (2018). A banana NAC transcription factor (MusaSNAC1) impart drought tolerance by modulating stomatal closure and H_2O_2 content, *Plant Mol. Biol.* 96(4–5), 457–471.

Negi, S., Tak, H. and Ganapathi, T. R. (2019). Overexpression of MusaNAC68 reduces secondary wall thickness of xylem tissue in banana, *Plant Biotechnol. Rep.* 13(2), 151–160.

OECD (2018). Conference on Genome Editing: Applications in Agriculture. Paris: Organization for Economic Cooperation and Development (OECD), Available at: http://www.oecd.org/environment/genome-editing-agriculture/ (Accessed May 31, 2019).

Paine, J. A., Shipton, C. A., Chaggar, S., Howells, R. M., Kennedy, M. J ., Vernon, G., Wright, S. Y., Hinchliffe, E., Adams, J. L., Silverstone, A. L. and Drake, R. (2005). Improving the nutritional value of Golden Rice through increased pro-vitamin A content, *Nat. Biotechnol.* 23(4), 482–487.

Pandey, A. K., Ger, M. J., Huang, H. E., Yip, M. K., Zeng, J.. and Feng, T. Y. (2005). Expression of the hypersensitive response-assisting protein in Arabidopsis results in harpin-dependent hypersensitive cell death in response to Erwinia carotovora, *Plant Mol. Biol.* 59(5), 771–780.

Paul, J. Y., Becker, D. K., Dickman, M. B., Harding, R. M., Khanna, H. K. and Dale, J. L. (2011). Apoptosis-related genes confer resistance to Fusarium wilt in transgenic 'Lady Finger' bananas, *Plant Biotechnol. J.* 9(9), 1141–1148.

Paul, J. Y., Khanna, H. K., Kleidon, J., Hoang, P., Geijskes, J., Daniells, J., Zaplin, E., Rosenberg, Y., James, A., Mlalazi, B., Deo, P., Arinaitwe, G., Namanya, P., Becker, D., Tindamanyire, J., Tushemereirwe, W., Harding, R. and Dale, J. (2017). Golden bananas in the field: elevated fruit pro-vitamin A from the expression of a single banana transgene, *Plant Biotechnol. J.* 15(4), 520–532.

Pei, X. W., Chen, S. K., Wen, R. M., Ye, S., Huang, J. Q., Zhang, Y., Wang, B., Wang, Z. and Jia, S. (2005). Creation of transgenic bananas expressing human lysozyme gene for Panama wilt resistance, *J. Integr. Plant Biol.* 47(8), 971–977.

Peraza-Echeverria, S., Dale, J. L., Harding, R. M., Smith, M. K. and Collet, C. (2008). Characterization of disease resistance gene candidates of the nucleotide binding site (NBS) type from banana and correlation of a transcriptional polymorphism with resistance to *Fusarium oxysporum* f.sp. cubense race 4, *Mol. Breed.* 22(4), 565–579.

Pillay, M. and Tenkouano, A. (2011). *Banana Breeding: Progress and Challenges*, CRC Press, Boca Raton, FL.

Ploetz, R. C. (2005). Panama disease: an old Nemesis rears its ugly head. Part 1: the beginnings of the banana export trades, *Plant Health Prog.* 6(1), 18.

Rai, M. K. and Shekhawat, N. S. (2014). Recent advances in genetic engineering for improvement of fruit crops, *Plant Cell. Tiss. Org. Cult.* 116(1), 1–15.

Roderick, H., Tripathi, L., Babirye, A., Wang, D., Tripathi, J., Urwin, P. E. and Atkinson, H. J. (2012). Generation of transgenic plantain (*Musa* spp.) with resistance to plant pathogenic nematodes, *Mol. Plant Pathol.* 13(8), 842–851.

Rustagi, A., Jain, S., Kumar, D., Shekhar, S., Jain, M., Bhat, V. and Sarin, N. B. (2015). High efficiency transformation of banana [Musa acuminata L. cv. Matti (AA)] for enhanced tolerance to salt and drought stress through overexpression of a peanut salinity-induced pathogenesis-related class 10 protein, *Mol. Biotechnol.* 57(1), 27–35.

Sagi, L., Panis, B., Remy, S., Schoofs, H., De Smet, K., Swennen, R. and Cammue, B. P. (1995). Genetic transformation of banana and plantain (*Musa* spp.) via particle bombardment, *Biotechnology* 13(5), 481–485.

Santos, E., Remy, S., Thiry, E., Windelinckx, S., Swennen, R. and Sági, L. (2009). Characterization and isolation of a T-DNA tagged banana promoter active during in vitro culture and low temperature stress, *BMC Plant Biol.* 9, 77.

Shekhawat, U. K. S. and Ganapathi, T. R. (2013). MusaWRKY71 overexpression in banana plants leads to altered abiotic and biotic stress responses, *PLoS ONE* 8(10), e75506.

Shekhawat, U. K. S. and Ganapathi, T. R. (2014). Transgenic banana plants overexpressing Musa bZIP53 display severe growth retardation with enhanced sucrose and polyphenol oxidase activity, *Plant Cell Tiss. Org. Cult.* 116(3), 387–402.

Shekhawat, U. K. S., Ganapathi, T. R. and Hadapad, A. B. (2012). Transgenic banana plants expressing small interfering RNAs targeted against viral replication initiation gene display high-level resistance to banana bunchy top virus infection, *J. Gen. Virol.* 93(8), 1804–1813.

Shekhawat, U. K. S., Srinivas, L. and Ganapathi, T. R. (2011). MusaDHN-1, a novel multiple stress-inducible SK3-type dehydrin gene, contributes affirmatively to drought-and salt-stress tolerance in banana, *Planta* 234(5), 915–932.
Sreedharan, S., Shekhawat, U. K. S. and Ganapathi, T. R. (2012). MusaSAP1, a A20/AN1 zinc finger gene from banana functions as a positive regulator in different stress responses, *Plant Mol. Biol.* 80(4–5), 503–517.
Sreedharan, S., Shekhawat, U. K. S. and Ganapathi, T. R. (2013). Transgenic banana plants overexpressing a native plasma membrane aquaporin MusaPIP1; 2 display high tolerance levels to different abiotic stresses, *Plant Biotechnol. J.* 11(8), 942–952.
Sreedharan, S., Shekhawat, U. K. S. and Ganapathi, T. R. (2015). Constitutive and stress-inducible overexpression of a native aquaporin gene (MusaPIP2; 6) in transgenic banana plants signals its pivotal role in salt tolerance, *Plant Mol. Biol.* 88(1–2), 41–52.
Tak, H., Negi, S. and Ganapathi, T. R. (2017a). Overexpression of MusaMYB31, a R2R3 type MYB transcription factor gene indicate its role as a negative regulator of lignin biosynthesis in banana, *PLoS ONE* 12(2), e0172695.
Tak, H., Negi, S. and Ganapathi, T. R. (2017b). Banana NAC transcription factor Musa NAC042 is positively associated with drought and salinity tolerance, *Protoplasma* 254(2), 803–816.
Tak, H., Negi, S., Gupta, A. and Ganapathi, T. R. (2018). A stress associated NAC transcription factor MpSNAC67 from banana (Musa × paradisiaca) is involved in regulation of chlorophyll catabolic pathway, *Plant Physiol. Biochem.* 132, 61–71.
Tang, K., Sun, X., Hu, Q., Wu, A., Lin, C.-H., Lin, H., Twyman, R. M., Christou, P. and Feng, T. (2001). Transgenic rice plants expressing the ferredoxin-like protein (AP1) from sweet pepper show enhanced resistance to Xanthomonas oryzae pv. oryzae, *Plant Sci.* 160(5), 1035–1042.
Thangavelu, R., Mostert, D., Gopi, M., Devi, P. G., Padmanaban, B., Molina, A. B. and Viljoen, A. (2019). First detection of *Fusarium oxysporum* f. sp. Cubense tropical race 4 (TR4) on Cavendish banana in India, *Eur. J. Plant Pathol.* 154(3), 777–786.
Tripathi, J. N., Lorenzen, J., Bahar, O., Ronald, P. and Tripathi, L. (2014a). Transgenic expression of the rice Xa21 pattern-recognition receptor in banana (*Musa* sp.) confers resistance to *Xanthomonas campestris* pv. Musacearum, *Plant Biotechnol. J.* 12(6), 663–673.
Tripathi, J. N., Muwonge, A. and Tripathi, L. (2012). Highly efficient regeneration and transformation protocol for plantain cv. Gonja manjaya (*Musa* spp. AAB) using embryogenic cell suspension, *In Vitro Cell. Dev. Biol.-Pl.* 48(2), 216–224.
Tripathi, J. N., Ntui, V. O., Ron, M., Muiruri, S. K., Britt, A. and Tripathi, L. (2019). CRISPR/Cas9 editing of endogenous banana streak virus in the B genome of *Musa* spp. overcomes a major challenge in banana breeding, *Commun. Biol.* 2, Article number: 46.
Tripathi, J. N., Oduor, R. O. and Tripathi, L. (2015a). A high-throughput regeneration and transformation platform for production of genetically modified banana, *Front. Plant Sci.* 6, 1025.
Tripathi, L., Mwaka, H., Tripathi, J. N. and Tushemereirwe, W. K. (2010). Expression of sweet pepper Hrap gene in banana enhances resistance to Xanthomonas campestris pv. Musacearum, *Mol. Plant Pathol.* 11(6), 721–731.
Tripathi, L., Tripathi, J. N. and Hughes, J. A. (2005). Agrobacterium-mediated transformation of plantain (*Musa* spp.) cultivar Agbagba, *Afr. J. Biotechnol.* 4, 1378–1383.

Tripathi, L., Tripathi, J. N. and Tushemereirwe, W. K. (2008). Rapid and efficient production of transgenic East African Highland Banana (*Musa* spp.) using intercalary meristematic tissues, *Afr. J. Biotechnol.* 7, 1438–1445.

Tripathi, L., Tripathi, J. N., Kiggundu, A., Korie, S., Shotkoski, F. and Tushemereirwe, W. K. (2014b). Field trial of Xanthomonas wilt disease-resistant bananas in East Africa, *Nat. Biotechnol.* 32(9), 868–870.

Tripathi, L., Babirye, A., Roderick, H., Tripathi, J. N., Changa, C., Urwin, P. E., Tushemereirwe, W. K., Coyne, D. and Atkinson, H. J. (2015b). Field resistance of transgenic plantain to nematodes has potential for future African food security, *Sci. Rep.* 5, 8127.

USDA (2018). Secretary perdue issues USDA statement on plant breeding innovation, Available at: https://www.usda.gov/media/press-releases/2018/03/28/secretary-perdue-issues-usda-statement-plant-breeding-innovation.

Vishnevetsky, J., White, T. L. Jr., Palmateer, A. J., Flaishman, M., Cohen, Y., Elad, Y., Velcheva, M., Hanania, U., Sahar, N., Dgani, O. and Perl, A. (2011). Improved tolerance toward fungal diseases in transgenic Cavendish banana (Musa spp: AAA group) cv. Grand Nain, *Transgenic Res.* 20(1), 61–72.

Waltz, E. (2016). CRISPR-edited crops free to enter market, skip regulation, *Nat. Biotechnol.* 34(6), 582.

Waltz, E. (2018). With a free pass, CRISPR-edited plants reach market in record time, *Nat. Biotechnol.* 36(1), 6–7.

Whelan, A. I. and Lema, M. A. (2015). Regulatory framework for gene editing and other new breeding techniques (NBTs) in Argentina, *GM Crops Food* 6(4), 253–265.

Yadav, K., Patel, P., Srivastava, A. K. and Ganapathi, T. R. (2017). Overexpression of native ferritin gene MusaFer1 enhances iron content and oxidative stress tolerance in transgenic banana plants, *PLoS ONE* 12(11), e0188933.

Ye, X., Al-Babili, S., Kloti, A., Zhang, J., Lucca, P., Beyer, P. and Potrykus, I. (2000). Engineering the provitamin A (β-carotene) biosynthetic pathway into (carotenoid-free) rice endosperm, *Science* 287(5451), 303–305.

Yip, M. K., Huang, H. E., Ger, M. J., Chiu, S. H., Tsai, Y. C., Lin, C. I. and Feng, T. Y. (2007). Production of soft rot resistant calla lily by expressing a ferredoxin-like protein gene (pflp) in transgenic plants, *Plant Cell Rep.* 26(4), 449–457.

Yip, M.-K., Lee, S.-W., Su, K.-C., Lin, Y.-H., Chen, T.-Y. and Feng, T. (2011). An easy and efficient protocol in the production of pflp transgenic banana against Fusarium wilt, *Plant Biotechnol. Rep.* 5(3), 245–254.

Zheng, S. J., García-Bastidas, F. A., Li, X., Zeng, L., Bai, T., Xu, S., Yin, K., Li, H., Fu, G., Yu, Y., Yang, L., Nguyen, H. C., Douangboupha, B., Khaing, A. A., Drenth, A., Seidl, M. F., Meijer, H. J. G. and Kema, G. H. J. (2018). New geographical insights of the latest expansion of Fusarium oxysporum f. sp. cubense tropical race 4 into the greater Mekong subregion, *Front. Plant Sci.* 9, 457.

Chapter 15

The usage of phenotyping, genetics and functional genomics approaches to improve environmental stress factors in banana

Sebastien Christian Carpentier, Bioversity International and Katholieke University of Leuven (KUL), Belgium; and David Eyland, Katholieke University of Leuven (KUL), Belgium

1 Introduction

In an agricultural approach, an abiotic 'stress' is a disadvantageous influence coming from the environment threatening yield. Hence, growers focus on reducing abiotic stress to safeguard their production and profit. Plant growth is determined by the interaction of the genotype with the environment, while crop growth is determined by the interaction of the genotype with the environment and the farm management (G×E×M). Absolute classifications of genotypes towards tolerance are therefore not possible. The terms 'susceptible' and 'tolerant' are relative terms which depend on the given set of genotypes evaluated under a given set of environmental conditions. Ideally, all genotypes need to be screened in the field for at least two crop cycles and within the area of interest. However, screening numerous banana genotypes from start to harvest would be extremely labour, time and cost-intensive. Therefore, a workflow is required where the existing biodiversity is first screened through an early screening process in a controlled environment, especially if a link is to be made to molecular analysis. A controlled setup ensures a known and acceptable variability in soil water potential, vapour pressure deficit (VPD) and light. The

http://dx.doi.org/10.19103/AS.2020.0070.22

focus of early screening is high throughput, which means the screening of many plants/genotypes as thorough as possible within a limited time. The more variation that can be attributed to a difference in genetic makeup without being overshadowed by noise due to the environment, the more precise the differential reactions between genotypes can be distinguished. Therefore, the difference between genotypes can be detected much more accurately in a controlled setup, while the time needed for screening is reduced. To be relevant for banana breeding, we need efficient variables that are a good proxy for yield in the specified target region of interest and that can be correlated to genetic markers.

2 Crop physiology and phenotyping

Screening bananas in the field is challenging but absolutely necessary. It involves detailed and careful planning to accommodate different genotypes with variations in crop duration. Therefore, it is recommended to plan the experiment carefully and to evaluate genotypes with a comparable crop cycle. Special attention needs to be given to tolerance traits during the critical stages in the ontogeny (Gibbs and Turner, 2018). The life cycle of bananas can be divided into three phases: the vegetative, floral and fruit phase (Fig. 1).

The vegetative phase begins after planting or when the dominant sucker starts growing and ends when the apex becomes reproductive. During the vegetative phase, any disturbing environmental factor like water deficit will delay the transition to floral phase (Fortescue et al., 2011, Kissel et al., 2015). The interaction between seasonal changes in temperature, photoperiod and soil water balance influences the timing of the transition from vegetative to floral phase in bananas, contributing to seasonal variation in flowering and bunch harvest (Gibbs and Turner, 2018). At the start of the floral phase, the apex changes from leaf formation towards the formation of the inflorescence. During this crucial floral phase, the development of the female flowers inside the pseudostem spans 12–13 weeks in the tropics and is up to twice as long in the subtropics. So the sequence of floral initiation, ovule primordium formation and megaspore mother-cell formation occurs while the inflorescence is inside the pseudostem. Any effect of environmental conditions during this time will reflect in the shape and anatomy of the fruit. For example, low temperatures (< 18°C) during ovary differentiation are associated with deformed fruits. The floral phase ends when the inflorescence is extruded from the top of the pseudostem. So when the inflorescence emerges from the pseudostem it is almost at anthesis. Thereafter the fruit phase starts and lasts until fruit maturity. The sequence of events during bunch formation takes place in the pseudostem and is best observed with reference to retrospective counting of leaves, starting at bunch emergence. The start of the floral phase (bunch formation)

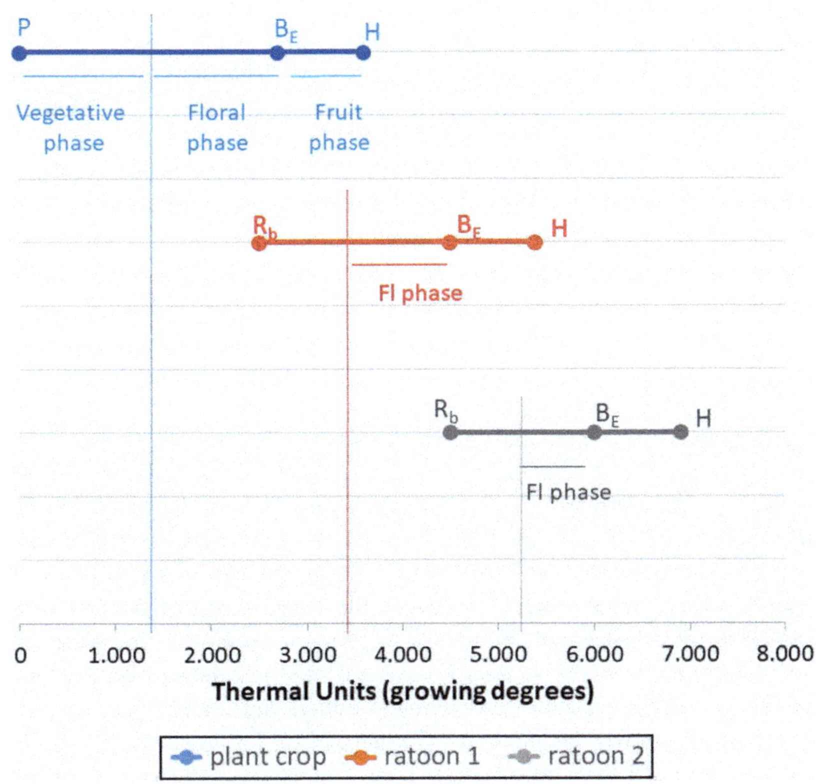

Figure 1 Crop cycle and distribution of harvest of the plant and two ratoon crops. P: Planting date; Fl: Floral phase; Rb: Time at which the ratoon begins (selection of suckers and elimination of the others); B_E: time of bunch emergence; H: Harvest. Figure adapted from Turner (1995) and Robinson (1996a). The time frame in thermal units is indicative. The real thermal units depend on the Environment × Genotype × Management interaction (G × E × M). A water deficit during the vegetative phase will delay the transition from vegetative to floral phases, prolonging the crop cycle. A water deficit during the floral phase will prolong the crop cycle and reduce the final yield. The number of fruits per hand is especially affected (Gibbs and Turner, 2018). Water deficit during the fruit phase affects fruit filling, reducing both the bunch and hands in size and weight. Flower abortion might also influence the number of fruits per hand.

begins at leaf −11 (11 leaves before bunch emergence, range 8–14) (Gibbs and Turner, 2018). Lassoudière refers to the end of the vegetative phase in a ratoon practically as F10 + 12, which means that 12 leaves after the first leaf reaching a width of 10 cm (F10) the vegetative phase ends (Lassoudière, 2012). The floral phase is the most vulnerable as there is a competition for resources between the developing bunch, growth of emerging leaves and elongation of the true aerial stem. A disturbing environmental factor during the floral phase will prolong the crop cycle and will reduce yield. Especially the number of fruits per hand is affected (Gibbs and Turner, 2018). A disturbing environmental factor during the fruit phase affects fruit filling and reduces hand size and weight. Late flower abortion might also influence the number of fruits per hand. Machovina and Feeley have modelled Cavendish production and ranked economical and abiotic factors (Machovina and Feeley, 2013). The mean annual temperature, the minimum temperature of the coldest month, the maximum temperature of the warmest month, the precipitation of the driest quarter and the annual precipitation were the abiotic factors having the largest impact. Temperature variables contributed 10.4% and precipitation variables contributed 2.6% to the model fit.

2.1 Water

A plant's reaction towards water deficit (phenotype) is not solely determined by the genotype, but also by the developmental stage and environment (relationship between the soil-to-canopy-hydraulic conductance (k), soil water potential (Ψ_{soil}), leaf water potential (Ψ_l) and the maximal transpiration (E_{max})). One assumes that the transpiration (E) is regulated by the root/vascular conductivity and the stomata to prevent the leaf water potential (Ψ_l) from falling below a critical water potential (Ψ_{l_crit}). The hormone abscisic acid (ABA) is sent by the roots when Ψ_{soil} is too low and can be released by the leaves when Ψ_{l_crit} is reached. Also other chemical signals such as pH, cytokinins and strigolactones can play the same role (Tardieu, 2016, Tardieu et al., 2017). Note that a reduction of E can also be driven by a reduction in radiation, as this is the primary stomatal opening impulse.

Field capacity quantifies the soil water retained after the soil is flooded and allowed to drain freely. When drainage ceases, the water content is defined as field capacity. The matrix suction (tension) of water held in the micropores at field capacity is not the same for all soils and depends on the soil texture, structure and porosity. Brady mentions values ranging from −10 kPa to −30 kPa (pF between 2 and 2.47) (Brady, 1984). The actual water content and potential at field capacity need to be determined experimentally. Robinson and Alberts are stating that for bananas, a mean soil moisture content of −55 kPa (pF 2.74) in the root zone was sufficient to cause a water deficit, reducing growth speed and final yield. However, they also reported that although not causing a clear growth reduction, a soil moisture content of −23 kPa (pF 2.36) resulted in a lower bunch mass (Robinson and Alberts, 1986). Turner and Thomas found that a soil water potential of −62 kPa (pF 2.78) stopped gas exchange and mentioned −30 kPa as inducing water deficit (Thomas and Turner, 1998). Ekanayake and co-workers mentioned practices in commercial plantations in which irrigation is applied when the soil water potential exceeds −20 kPa (pF 2.3) (Ekanayake et al., 1994). Carr also mentioned the limit of −15 kPa to−-20 kPa at 20 cm depth (Carr, 2009). Robinson and Bower mentioned a build-up of water deficit at soil water potentials of −20 kPa to −25 kPa (pF 2.3–2.4) (Robinson and Bower, 1987). To maintain gas exchange rates the soil water potential should not fall below −15 kPa to −33 kPa at 0.2–0.3 m depth. In practice for the commercial Cavendish bananas, it is said that 25 mm per week or 6 mm per day is needed to achieve maximum productivity (Robinson, 1996b). In commercial plantations the water requirements are met by rainfall and supplementary irrigation. The proportion of irrigation varies widely over the world and the actual requirement depends on the developmental stage and the environment (mainly radiation and VPD). The average annual precipitation in areas economically viable for banana culture has been modelled to be 2167.5 mm while the modelled average precipitation during the driest quarter was 154.3 mm (Machovina and Feeley, 2013). It is estimated that by 2070 this water demand will increase by 12–15%, mainly due to rising temperature (Calberto et al., 2015).

Moreover, radiation, VPD and wind are interacting with the effect of soil water potential (Robinson and Bower, 1988). This complex interaction makes it impossible to get one specific value where a reduction or complete stop in transpiration takes place. Whatever the absolute value and the conditions are, it is clear that bananas respond very early to reduced soil water content. Even when the water potential of the soil is sufficiently high, transpiration rates can be limited by dry air. VPD values of 2.0–2.3 kPa have been mentioned to limit transpiration (Carr, 2009). But similarly to the soil water potential, it is difficult to specify one VPD limit of response due to the influence of solar radiation and wind. The reactive behaviour of banana towards the soil water content and VPD can be considered as isohydric behaviour. This terminology was introduced to describe the daily patterns of water relations (Berger-Landefeldt, 1936). Isohydric behaviour is also considered as 'conservative behaviour': there is a strong regulation of the leaf water potential and stomatal aperture is reduced to ensure that a high leaf water potential is maintained. This apparent 'feedforward' mechanism was shown to be genotype-specific in bananas (Thomas et al., 1998) and might be a very early and sensitive feedback mechanism where a low plant water status triggers a lower stomatal conductance (Buckley, 2005).

But what is the cause of this sensitivity to water deficit and high VPDs in bananas? Presumably not the ability of the roots to take up water from the soil. Turner and Thomas determined the root water potential and concluded that the roots were able to establish a lower water potential than the soil at the stage where the plant had no more leaf emergence (Turner and Thomas, 1998). It is therefore unlikely that the capacity of banana roots to withdraw water from the soil is the reason for the high sensitivity to soil water deficit (Turner et al., 2007). The root system of banana plays an important role in maintaining the plants hydrated during water deficit. Once the stomata are closed, root pressure becomes the dominant force supplying water and nutrients to the shoot. With this reaction bananas can survive long periods of soil water deficit but the disadvantage of this conservative behaviour is that photosynthesis and thereby production are rapidly affected even under mild soil drying. Root anatomy plays a major role in root hydraulics, influencing axial conductance and the distribution of water uptake along the root, with a key role for water channels called aquaporins (Bramley et al., 2009). Aquaporins contribute to the radial water transport because they allow water to be transported across cell membranes where transport can continue through the cytoplasm via plasmodesmata (symplastic transport) or across cell membranes (transcellular transport). This symplastic transport together with water flow through cell-wall structures (apoplast) and the transcellular transport determine the complete radial transport (Maurel et al., 2015).

It is known that both root and leaf hydraulic conductance vary across species (Franks, 2006). The hydraulic conductivity of banana roots, when well supplied with water, is somewhat higher but of the same order of magnitude as that of maize (Aguilar et al., 2003). Root conductivity may fall as soil water content decreases. This is the case for many plant species, but this situation is not universal (Bramley et al., 2007). The magnitude and significance of changes in root-hydraulic conductivity in bananas is yet to be established. Embolisms, or cavitation of xylem vessels, cause malfunction of the conducting tissues reducing the supply of water to the shoot. Embolisms can occur in the roots where up to 70% of the conducting vessels may be affected (Turner, 2003). Reduction of stomatal aperture and associated transpiration prevents embolisms to occur and allows those present to be refilled. When the plants' hydraulic system is operating close to its maximum capacity, the transpiration is maximal and the leaf water potential is in the vicinity of the cavitation threshold. Under these conditions a hydro-active negative feedback regulation of stomatal conductance produces a homeostasis in the leaf water potential (Buckley, 2005). This could explain the isohydric behaviour of bananas. However, it has been shown in sunflower that stomatal closure can occur at water potentials well above xylem cavitation thresholds (Huber et al., 2019). What is more likely is that in bananas an increased water loss causes a hydraulic mismatch between leaf water supply and transpiration rate. These hydraulic mismatches have been linked to stomatal oscillations (Barrs and Klepper, 1968, Farquhar and Cowan, 1974) and have been observed in bananas (Zait et al., 2017).

Fluctuations of leaf turgor pressure induced by an imbalance between the evaporative demand and the provision of water by stem flux may be a direct cause of stomatal oscillation under conditions of high transpiration demand (Wang et al., 2001, Yang et al., 2005). Using a new pressure probe on bananas, Zait and colleagues showed that oscillations in leaf turgor pressure were synchronized with stomatal oscillations and balanced only when transpiration equalled water supply (Zait et al., 2017). Zait and colleagues observed turgor oscillations under high transpiration rates and whether cavitation plays any role in oscillations remains speculation (Buckley, 2005). We have also observed those oscillations and we see that they are genotype-specific and that they are triggered when the VPD or light is increased (unpublished results). In bananas, oscillations and stomatal anatomy seem to be genotype-specific (see also Section 2.3 'Light'). The stomatal conductance is, besides a behavioural characteristic, dependent on the stomatal density and the stomatal size, two genetically determined components. The number of stomata has been reported to vary according to the banana genotype (Vandenhout et al., 1995, Craenen et al., 1997) and also the stomatal size differs between genotypes (unpublished results). Although in bananas stomata are present on both abaxial and adaxial sides, the abaxial density is up to 4 times higher (Carr, 2009). In other crops, a higher density of smaller stomata has been reported to result in higher stomatal conductance (Muchow and Sinclair, 1989, Ohsumi et al., 2007), which often is translated in higher photosynthesis rates. In other crops, traditional breeding has inadvertently or actively selected for increases in stomatal conductance as a means to improve CO_2 diffusion and hence photosynthesis and yield (Fischer et al., 1998). However, the downside of high stomatal conductance is a lower water-use efficiency. *Musa* genotypes have 4–6 large subsidiary cells surrounding the stomata. These cells regulate solute exchange in and out of the guard cells, and their presence in grasses has been linked with improved transpiration-efficient stomata (Franks and Farquhar, 2007). Therefore, it needs to be investigated further how influential the subsidiary cells are on the actual phenotype and whether breeding for this trait is needed. Not only the density and size, but also the stomatal responsivity has been observed to vary at interspecific and intra-specific levels (Drake et al., 2013, McAusland et al., 2016, Faralli et al., 2019). McAusland and co-workers also report that there is diversity in stomatal opening time towards a radiation signal (McAusland et al., 2016). The best way to phenotype transpiration and stomatal behaviour is by gas exchange measurements and gravimetric phenotyping (Fig. 2).

Most research focusses on studying the differences between well-watered and water-deficit conditions (Santos et al., 2018). However, constitutive morphological differences are also very important and do contribute towards drought tolerance. It is known that there is quite some variability in terms of the root architecture and root:shoot ratio (Swennen et al., 1986, Blomme et al., 1999, Draye, 2002, van Wesemael et al., 2019). The total root length of bananas seems to be twice the length of plantains (Swennen et al., 1986). It is also known that under a water deficit the growth of the roots is less inhibited than that of the growth of the shoot. van Wesemael and colleagues (2019) showed that there were significant genetic differences in the investment in root versus shoot growth. Eight genotypes (five ABB, one AAB and two AAA) were identified

Figure 2 High-throughput phenotyping at the Phenodyn platform (Berthezene et al., 2015). About 168 balances are available for online transpiration monitoring. The weight of every pot is registered every 5 min. Hence the transpiration response of the plant to changing radiation, vapour pressure deficit and soil water potential is registered. This experiment took place in the framework of the EPPN^{2020} European Plant Phenotyping Network project in collaboration with the French National Research Institute for Agriculture, Food and Environment (INRAE).

which invested under osmotic stress significantly more in root growth than in leaf growth (van Wesemael et al., 2019).

Investment in root biomass is a strategy which can lead to drought avoidance and supports further transpiration and growth. It enables the exploration of a larger soil volume for water uptake, but it has a significant respiration cost, an investment that will be lost if no extra water is found or sufficient water remains available. Root traits such as length, branching or depth interact with shoot traits such as leaf area or stomatal control, thereby generating feedback. As a consequence, optimum root systems for water uptake at a given time are not always those associated with the highest yields (Tardieu et al., 2017). The success of this drought-avoidance phenotype depends on the actual climate conditions throughout the growth period. During a relatively short dry season, followed by a rainy season with excess water supply, the deep-rooting phenotype is favoured as it largely avoids drought and supports continued transpiration and growth during the adverse conditions. This phenotype reduces the impact of stress on the yield compared to shallow-rooting phenotypes which show growth cessation.

Leaves are important in determining plant-hydraulic capacity because they constitute a bottleneck in the path from roots to stomata and represent at least one-quarter of the whole plant resistance to water movement. Leaf-hydraulic conductance varies more than 65-fold across species, reflecting differences in the anatomy of the petiole and the venation architecture, as well as pathways beyond the xylem through living tissues to the sites of evaporation (Sack and Holbrook, 2006). Leaf venation architecture is tremendously variable across species and is thus a striking example of diversity in a complex and critical plant feature (McKown et al., 2010). Vascular traits can impact whole-plant performance as hydraulic supply influences stomatal aperture and thus transpiration and leaf photosynthetic rate. This in combination with the leaf area, stomatal density and stomatal size determine the maximal transpiration rate. In bananas, not much is done to investigate the diversity in drought-tolerance/avoidance (sub)traits. An often unconsidered drought-avoidance trait is the leaf wax content and quantity. Wax has been hypothesized to reduce epidermal water losses, which are especially important during the night or under conditions where stomata remain closed. Leaf wax production commences at the time of leaf emergence and reaches a maximum at leaf position 3 or 4, some 30–40 days after emergence (Freeman and Turner, 1985). The predominant chemical groups in the leaf waxes are paraffins, primary alcohols and fatty acids, irrespective of the genome (Freeman and Turner, 1985). Large differences in form and concentration were observed between varieties containing different proportions of the *M. acuminata* and *M. balbisiana* genome. These differences in waxes can definitely play a key role in the water usage of a plant.

Next to a water deficit, the opposite can also happen. As Machovina and Feeley illustrated, periods of excessive rainfall can occur in banana-producing areas (Machovina and Feeley, 2013). In any case, if the soil is not well-drained, the field becomes flooded and anoxic. The problem of excess water in banana plantations is normally managed using drainage. Although banana has aerenchyma, flooding has a large influence on plant performance (Aguilar et al., 1999, Aguilar et al., 2003, Bramley et al., 2007). Bananas subjected to flooding for up to 72 hours are not seriously damaged, provided the water table decreases rapidly (Stover, 1972). Though not causing serious yield decline, short-term O_2 deficiency in the root zone adversely affects root aeration. Within minutes to hours, root functions such as net solute transport, conductivity and elongation are affected. Genotype-specific responses towards oxygen deficiency within the *Musa* diversity have been suggested (Aguilar et al., 2003).

2.2 Temperature

The rate of development of a banana plant is influenced by temperature and since in a certain environment temperature is more stable, the short-term growth is predicted to be more a function of water supply (Turner, 1990). However, in contrast to water, the temperature cannot be controlled by farm management. The importance of temperature is illustrated in the use of thermal time rather than absolute time to indicate the duration between planting and harvest. Thermal time is often expressed as growing degree days (GDD). To calculate GDD one needs to take the integral of warmth above a certain base temperature. The base temperature where growth stops is genotype-dependent. Turner and Hunt investigated 30 genotypes and determined the base temperature via least squares extrapolation. The estimated base temperature varied between 8°C and 14°C (Turner and Hunt, 1984, Turner, 1995). The base temperature for Cavendish was determined at 14°C (Ganry and Chillet, 2008). It is estimated that depending on the *Musa* species and on the other environmental-limiting factors (water deficit, photoperiod), bananas need between 2400 GDD and 5000 GDD to complete their cycle (Turner et al., 2009, Turner and Fortescue, 2010, Mboula, 2014, Turner et al., 2016). Plants of the subgroup Cavendish are believed to go from bunch emergence to harvest in 900 GDD (Ganry and Chillet, 2008) (see also Fig. 1). The wild bananas, particularly *M. acuminata* (A) and its subspecies, and *M. balbisiana* (B), are distributed across 30° of latitude from the equator including climates that vary in temperature, amount and distribution of rainfall and day length (Simmonds, 1962). It is clear that lower temperatures (higher altitudes) will delay the flowering time, but that the impact depends on the genotype (Turner et al., 2016). Since it is not possible to control temperature in the field and since more time is needed to reach the critical GDD in a milder climate, it is economically perfectly understandable that temperature is the most important variable determining banana yield. Apart from the developmental delay when growing under suboptimal temperatures, bananas can be damaged by exposure to chilling temperatures. Leaves of fruit-bearing banana plants begin to die around 10–20 days after exposure to damaging low temperatures. Apart from temperatures that are too low, also too high temperatures can be detrimental, especially under low VPD when transpiration is not efficient or when under water deficit when not enough water is available for transpirational cooling. Values of 10 min at 47°C are mentioned to be lethal to leaves (Turner, 1995). But also temperatures lower than 47°C cause damage. Thomas and co-workers examined the photosynthesis rate in three different banana cultivars (Bluggoe (ABB), Lady Finger (AAB) and Williams (AAA Cavendish type)) and extrapolated that net photosynthesis would become zero if leaf temperature would reach 43°C (Thomas et al., 1998). Further research seems necessary to link the temperature damage to the exposure time and to understand the combining effect of temperature with other environmental conditions.

The optimal annual mean temperature for commercial banana culture has been modelled to be 26.2°C (with a minimum of 22.6°C and maximum of 29.1°C), the average minimum temperature of the coldest month 20.3°C (with a minimum of 12.7°C and maximum of 23.9°C) and the average maximum temperature of the warmest month 32.4°C (with a minimum of 27.4°C and maximum of 38.1°C) (Machovina and Feeley, 2013). A change in the areas suitable to grow bananas in a profitable way has been predicted (Machovina and Feeley, 2013, Varma and Bebber, 2019). Areas that are currently too cold will become more suitable and several of the current banana-producing areas will become too warm. It is however estimated that by 2070 the total land area suitable for bananas will increase by 50% (Calberto et al., 2015).

To evaluate the influence of temperature on growth, we have designed the bananaTainer, where 504 plants can be grown simultaneously under controlled environmental conditions. The container has control over temperature, relative humidity, light and water/nutrients. About 168 plants of 16 different genotypes are grown at three slightly different temperatures for 6 weeks (Figs 3 and 4). We simulate in the container the climate of the East African Highlands that have a mild climate (Turner et al., 2016) and we assess the impact on growth at different temperatures. Significant temperature-dependent growth differences have been observed between genotypes (van Wesemael, 2019).

2.3 Light

The natural general habitat of wild banana species is a forest environment. Wild bananas are opportunistic pioneer plants that occur in disturbed areas of forest and thrive at places in the lowland forests where light is abundant. Many clumps of wild bananas survive only for a few generations because they are seral plants, which do not survive when the climax vegetation becomes established (Simmonds, 1962, Gibbs and Turner, 2018). Wild banana species require high temperatures, high humidity and high light intensities. In the

Figure 3 The bananaTainer in which 504 plants can be grown simultaneously. The container has control over temperature, humidity, light and water/nutrients. About 168 plants of 16 different genotypes are grown at three slightly different temperatures for 6 weeks.

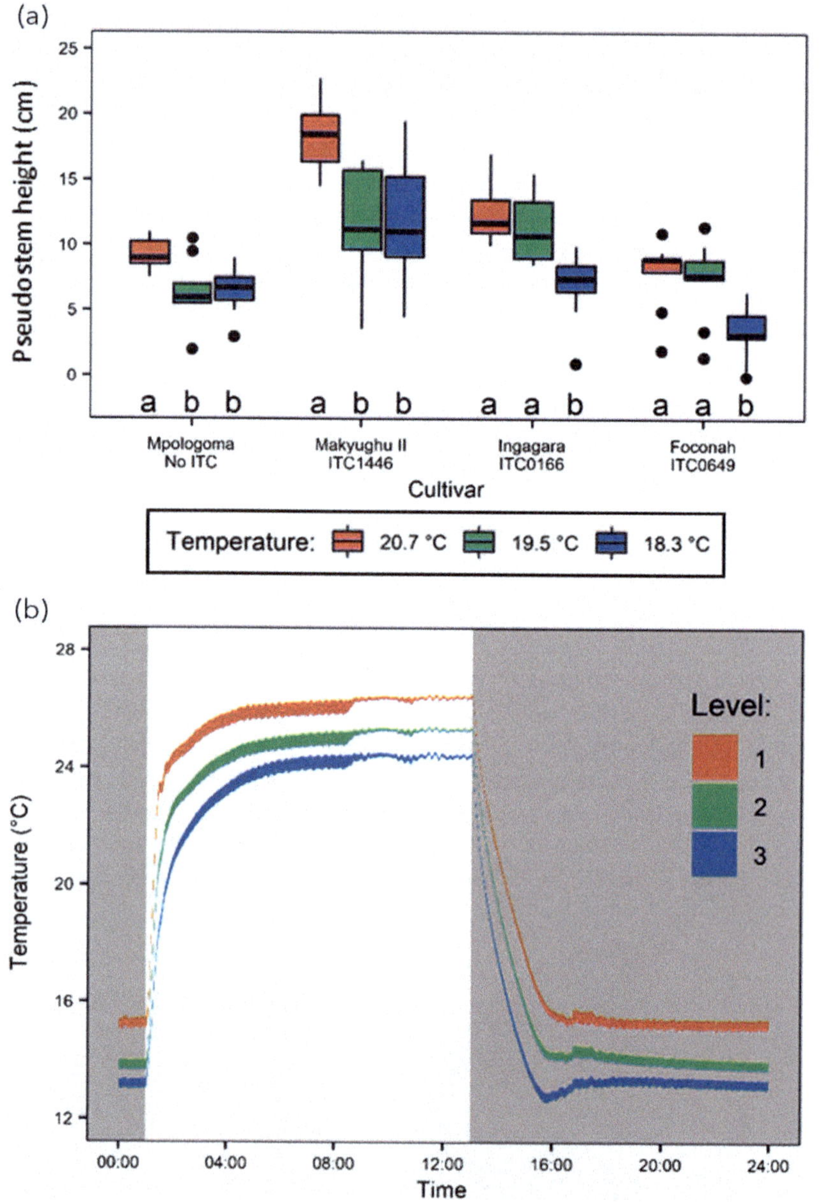

Figure 4 (a) Impact of the three different temperature regimes on pseudostem height of four different banana cultivars. (b) We established growth trials with three relevant temperature regimes (24 h average temperature: 20.7°C, 19.5°C and 18.3°C). These regimes represent a span of the existing temperatures in the East African highlands (Fick and Hijmans, 2017). Significant growth differences between temperature treatments are indicated by different letters (Tukey HSD, following cultivar-specific 1-way ANOVA (growth ~ treatment), alpha = 0.05). Grey and white areas indicate the night and day respectively (van Wesemael, 2019).

process of domestication humans have brought the crop into open fields, where the protective surrounding canopy of the forest is no longer present. An open field is subjected to higher levels of solar radiation and higher VPDs, resulting in higher evaporative demands than the transient habitats in the forest. Bananas are facultative long-day plants, meaning that long photoperiods (>12 h) promote flowering but are not essential. There is evidence that there is genetic variation within *Musa* spp. in their response to photoperiod (Turner and Fortescue, 2010, Fortescue et al., 2011).

Light also plays a crucial role in determining the transpiration rate. Photosynthesis light-response curves are ideal to predict what light intensities are ideal for a species. Turner and colleagues have made light-response curves of three different banana cultivars (Bluggoe (ABB), Lady Finger (AAB) and Williams (AAA Cavendish type)). Photosynthesis seems to saturate around 1400 µmol/m^2 s (Thomas et al., 1998). We confirmed this by analysing six different genotypes (Fig. 5a). We observed a genotype-specific maximum level and noted that the higher the light intensity, the less efficient the photosynthesis becomes (Fig. 5b). Light intensities above 1400 µmol/m^2 s were not increasing CO_2 uptake and might even be undesirable. Note that plants' acclimatization to higher or lower light environment influence light sensitivity and thereby plant performance. Siles and colleagues studied the light response of banana under three levels of natural shade (25%, 50% and 75% of full solar irradiance) and full sun control. Plants which have been grown under lower light intensities

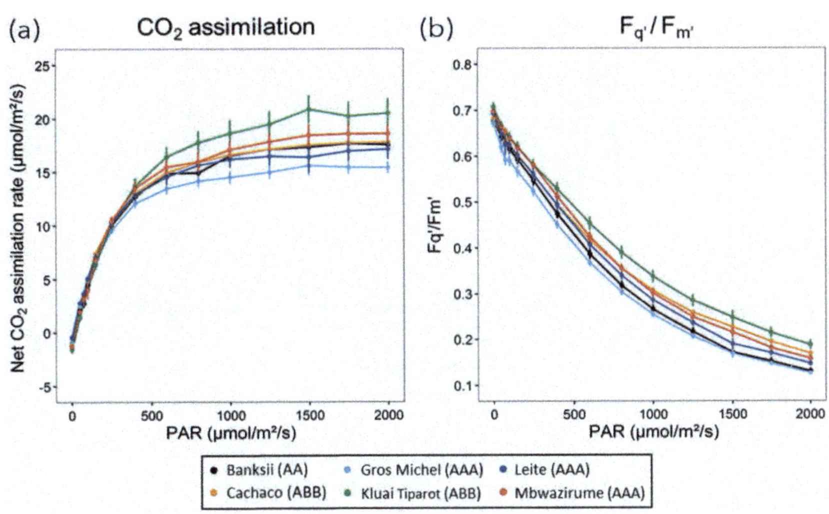

Figure 5 Light response curve of six banana genotypes. (a) Net CO_2 assimilation in function of light (b) PSII operating efficiency, a proxy of photosynthesis light-use efficiency. Results are the average of six biological replicates, error bars indicate SE. PAR: photosynthetic active radiation.

showed lower light-saturation points (Siles et al., 2011). Eckstein and colleagues investigated the effect of shading via windbreaker plants. In their experiment the shading reduced the light intensity by 69%. This resulted in a reduction of photosynthesis with 27% and transpiration with 38% compared with unshaded plants (average 1583 µmol/m^2) (Eckstein et al., 1997). The yield was consequently reduced from 47.4 tonnes to 41.2 tonnes per hectare per year. So if there is sufficient water available to meet the evaporative demand, banana yields more in full sunlight. However, a 70% reduction of light resulted only in a 13% yield reduction. So one needs to find the right balance between light and water availability as high light intensities drive transpiration and will only be beneficial when sufficient water is available.

Not only the light intensity but also the spectrum plays a crucial role. It is known that blue light drives stomatal opening (Zeiger and Field, 1982). It has also been suggested that blue light affects the hydraulic conductance (Savvides et al., 2011). Also banana reacts to blue light by opening its stomata and the sensitivity to opening seems to be genotype-dependent. Some genotypes are very efficient to react and increase their stomatal conductance while others react more moderately. We have been experimenting with different proportions of red and blue light (Fig. 6). We show that the Bluggoe type Cachaco has a more 'conservative' response to the opening impulse received by light (Fig. 7). A sudden switch in light triggers a fast opening of the stomata, where the stomata seem to open more than needed. This causes in many 'risk taking' genotypes

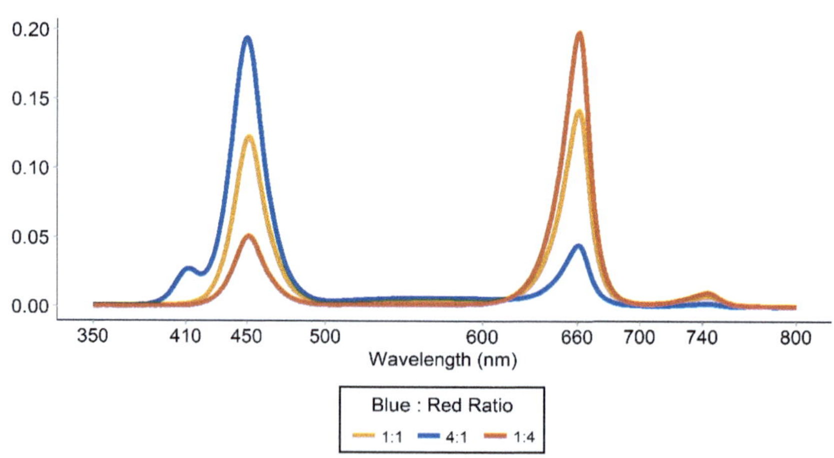

Figure 6 Three different spectra with different blue- and red-light intensities. The spectrum represented in orange had an intermediate amount of blue light and the overall blue- and red-light intensity were equal (1:1). The spectrum represented in blue had the highest amount of blue light, which was four times more intense than its red-light component (4:1). The spectrum represented in red has the lowest amount of blue light and had a four times higher red-light intensity (1:4).

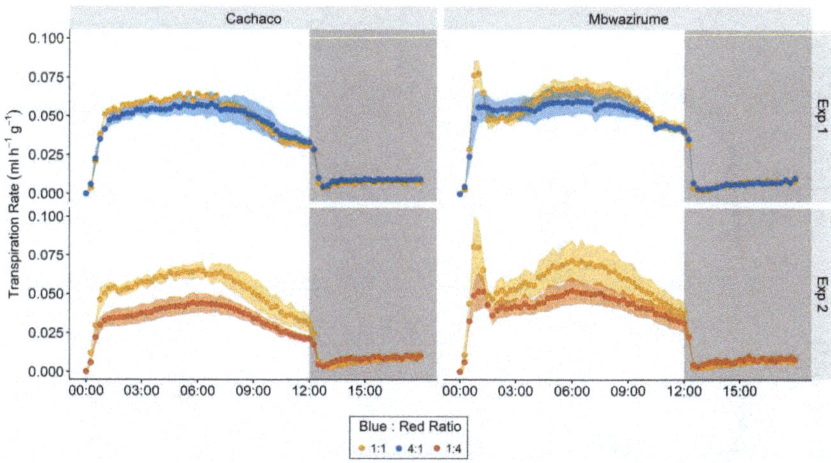

Figure 7 Mean transpiration rates of two genotypes under three different spectra (see Fig. 6 for the colour legend). Both genotypes react to the three different intensities of blue light. In the spectrum where the intensity of red and blue light are equal (orange), the highland banana Mbwazirume has a strong reaction where the stomata seem to open more than needed in both experiments. This increased water loss via increased transpiration might cause a hydraulic mismatch between leaf water supply and transpiration rate, which leads to a wrong way hydropassive response. This response opens the stomata even more and is caused by rapid epidermal water loss upon increasing the blue light, resulting in turgor loss and reduced backpressure towards guard cells and stomatal opening. A higher intensity of blue light seems to alleviate this phenomenon (blue lines). The hypothesis is that upon even higher blue-light intensities also the hydraulic conductance is increased resulting in a smaller mismatch (Savvides et al., 2011, Zait et al., 2017). The Bluggoe type Cachaco (ABB) did not display this phenomenon or at least much less, meaning that the stomata are not so sensitive to blue light and/or the hydraulic conductance is much better leading to no mismatch. The increase in blue-light intensity did not increase the transpiration. The lowest blue-light intensity lowered the transpiration and does not provoke the hydraulic mismatch (red lines). The transpiration rate was recorded via online gravimetry and is normalized for the biomass, with standard deviation of three individuals (shaded). Grey and white areas indicate the night and day respectively.

a hydraulic mismatch between leaf water supply and transpiration rate, which leads to wrong-way hydropassive response and oscillations (Raschke, 1970, McAdam and Brodribb, 2015, Zait et al., 2017). A fast response to light enables an efficient uptake of CO_2 but contains a risk since it lowers the water-use efficiency as relatively more water is lost than the amount of CO_2 gained (Fig. 8).

We have also observed a diurnal pattern where stomata respond less to the constant light impulse and reduce their stomatal opening after a few hours of light exposure (Fig. 8) (van Wesemael et al., 2019). Also Mencuccini and colleagues have described a similar phenomenon in common bean (*Phaseolus vulgaris* L.) where the stomata respond less to a soil signal in the afternoon (Mencuccini et al., 2000). This phenomenon leads to a higher water-use efficiency and the timing of reduction is genotype-dependent in banana (Fig. 8).

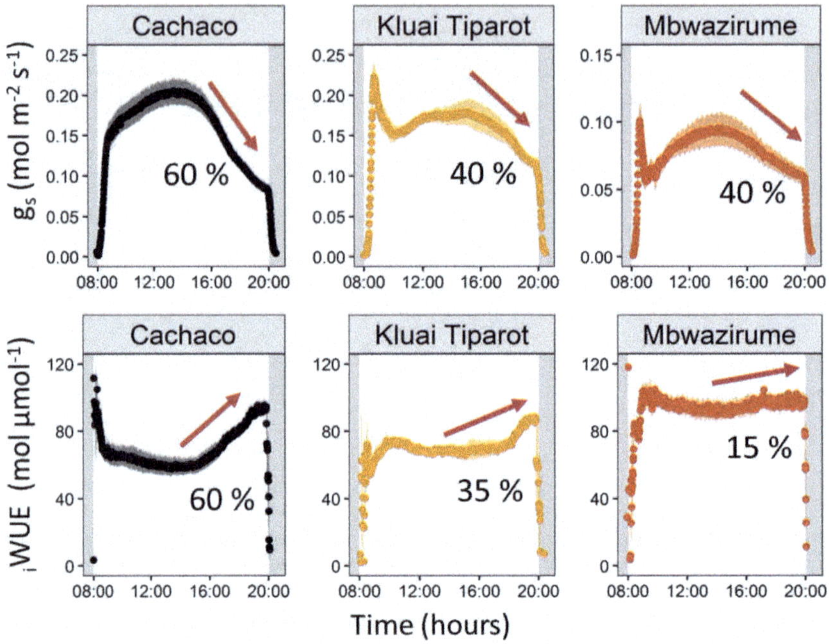

Figure 8 The reduction of stomatal conductivity (g_s) in the afternoon increased the intrinsic water-use efficiency ($_i$WUE). All plants are exposed to a constant light intensity of 350 µmol m^{-2}s^{-1}, a temperature of 25°C and a relative humidity of 75% in the growth chamber and leaf cuvette. All genotypes showed a g_s reduction in the afternoon, which led to an increase in $_i$WUE. The Bluggoe type Cachaco (ABB) displayed this phenomenon the most, resulting in a $_i$WUE increase of 60%. Note also here the absence of the strong g_s increase at the onset of light in Cachaco, while a g_s peak was observed in Kluai Tiparot and Mbwazirume. Data are the mean ± SE (n = 5).

Some genotypes are 'risk takers' and send the signal to reduce the stomatal opening relatively late in the day, while others are 'conservative' and reduce the stomatal conductivity sooner (van Wesemael et al., 2019). We hypothesize that this phenomenon might be the result of a feedback mechanism (Delorge et al., 2014) or the effect of a diurnal response (Mencuccini et al., 2000). Yang and colleagues mention that at the end of the day, it seems that the decrease in photosynthesis and energy supply initiates a loss of water and reduces turgor pressure, which causes stomatal closing (Yang et al., 2005).

3 Biotechnology to assist breeding

To obtain sustainable agriculture, the right genotype should be grown in the right environment. The flexibility of genotypes towards the environment is determined by the genetic diversity (G). Identifying sources of natural variation with tolerance to unfavourable abiotic conditions while minimizing the yield penalty is an absolute priority in current banana-breeding programmes. Breeding efforts need to address disruptions to production and food security by placing a greater emphasis on identifying sources of germplasm material adapted to the forecasted, more variable warmer and drier climates. Germplasm collections provide valuable genetic resources with potential to augment the resilience of agroecosystems towards adverse conditions. An example of such a collection is the Bioversity International banana collection holding over 1500 accessions (Vandenhouwe et al., 2003). Understanding the physiology and applying phenotyping to screen for interesting genotypes is the necessary first step. Secondly, correlations between phenotype and genetic markers can be identified, which will speed up breeding through marker-assisted selection. The identification of genetic markers and the complexity within a non-model crop as banana are discussed below.

3.1 Genomics, SNP technology and genome-wide association studies

DNA markers proved to be useful for characterizing germplasm in various plant species and microsatellite markers (simple sequence repeats (SSR)) gained special popularity, thanks to their co-dominant inheritance, good reproducibility and high abundance in plant genomes. SSR and other DNA markers are still used as an effective tool in elucidating the genetic diversity among banana species, subspecies and landraces (Christelová et al., 2011a,b, Christelova et al., 2017, Janssens et al., 2016, Perrier et al., 2018). To make a link to a specific phenotype/trait, a high level of marker saturation is needed where a multitude of markers is found evenly spread across all the chromosomes. In the previous section we mentioned the phenotyping/physiology part of the germplasm as a bottleneck. While genetics has been a bottleneck for a long time in bananas rapid development of genotyping technology, including high throughput methods such as diversity array technology (DArT) or next-generation sequencing systems, at low price points has now removed this limiting factor. Diversity array technology is a DNA hybridisation-based molecular marker technique that can detect simultaneously variation at numerous genomic loci without sequence information and has been applied in banana as an alternative for SSR (Risterucci et al., 2009). Soon other molecular and analytical tools were developed to support the breeding programmes such as single nucleotide polymorphism (SNP)-chip, genotyping by sequencing (GBS), restriction site-associated DNA RAD-seq, and diversity arrays technology (DArTseq). Plant genome-wide association studies (GWAS) are currently being applied with some success. The cost-effective next-generation sequencing technologies (usually GBS or RADseq technology) have opened up the technologies for many crops. Both GWAS and genomic selection ((GS), a method for predicting phenotypes by the use of genome-wide marker data) are receiving considerable attention among plant breeders. Association genetics can quickly and efficiently delineate regions of the genome that control traits and provide markers to accelerate breeding by marker-assisted selection (Hamblin et al., 2011).

Many crops, including banana, are polyploid, complicating the identification of required markers. Applying GWAS for banana is challenging due to many deviations from the assumption of Mendelian genetics. Many cultivated varieties are allopolyploid hybrids, which complicates SNP mapping and increases genetic heterogeneity. Cultivated bananas are moreover sterile, which drastically reduces the occurrence of sexual reproduction. Edible bananas are clonally propagated giving rise to high rates of clonal diversification, which lead to confounding effects and tricky associations due to the clonal population structure (Sardos et al., 2016). Despite the challenges, GBS-based GWAS has already been applied in banana to associate to parthenocarpy and bunch weight (Sardos et al., 2016, Nyine et al., 2019). The publication of the *Musa acuminata* ssp. *malaccensis* genome and later the *M. balbisiana*, *M. acuminata* ssp. *banksii*, *M. acuminata* ssp. *zebrina* and *M. acuminata* ssp. *burmannica* genomes were major breakthroughs (D'hont et al., 2012, Davey et al., 2013, Martin et al., 2016, Rouard et al., 2018, Wang et al., 2019). The publication and further annotation of those genomes gave a boost to functional genomics approaches such as transcriptomics and proteomics that are a good alternative for GWAS or GS. Associative transcriptomics and proteomics use respectively transcriptome sequencing (mRNAseq) and mass spectrometry to identify and score molecular markers representing variation in both gene/allele sequences and gene expression/protein abundance, and correlate this with trait variation (Harper et al., 2012, van Wesemael et al., 2018, Cenci et al., 2019). However, as indicated above, a huge task in all the approaches is a deep understanding and accurate quantification of the traits. The published GWAS cases so far have focussed on 'simple' traits. Tolerance towards abiotic stress such as water deficit, flooding, heat stress, chilling stress, wind stress, light stress are all complex multi-gene traits that are very challenging to phenotype accurately (see above). Phenotyping should thereby focus on crucial subtraits and look for correlation with genetic markers.

Polyploids and especially allopolyploids likely have an evolutionary advantage due to their high level of heterozygosity (Barker et al., 2016). Genome reorganization and the associated greater flexibility in gene expression allow these plants to cope with immediate stresses which may be beyond the tolerances of their progenitors/parents. Gene loss or silencing, neo- and/or sub-functionalization, intergenomic transfer, allele dominance/co-dominance, differences in transcription/translation efficiency and post-translational modifications, exemplify how the genome, transcriptome, and proteome are regulated following polyploidization events. A polyploid genome is thus a patchwork of gene variants enabling many genotype × environment interactions. Gene variants unique for a preferential phenotype are most probable candidate markers controlling the observed phenotype. Organ or tissue-specific silencing or overexpression of one parental homologue is quite common (Adams et al., 2003, Carpentier et al., 2011, Vanhove et al., 2015, van Wesemael et al., 2018). The link between gene or protein abundance and an SNP or single amino acid polymorphisms (SAAP) and a preferential phenotype is more stringent with knowledge of the tissue, time point and environmental condition of the gene/protein activity (Zivy et al., 2015). RNA sequencing effectively combines gene expression quantification with gene sequencing and allows also SNP calling (van Wesemael et al., 2018, Cenci et al., 2019). Yet, most read-mapping software tools have difficulties to process complex (polyploid) genomes. The read-mapping efficiency to the reference genome might be biased, and the degree of heterozygosity greatly increases computational effort, hampering quantitative results per genome (Page et al., 2013). Consequently the RNA reads are not separated and traced back to their (sub-)genome. Algorithms like PolyCat and HANDS2 process read based on classification towards their genome of origin, but depend heavily on the presence and the quality of reference genomes (Page et al., 2013, Khan et al., 2016). Since not all cultivars of interest carry one of the reference sequences, mapping efficiency biases can still occur when one reference genome is more closely linked to a constituting genome than the other (Page et al., 2013).

Proteomics allows to pick up and to quantify the actual differential products without previous genome knowledge (Samyn et al., 2007, Carpentier et al., 2011). Two-dimensional electrophoresis-based (2DE) proteomics is very useful to identify allele-specific protein isoforms in complex protein families (Carpentier, 2016). This is exemplified in banana for the HSP70 protein family (Vanhove et al., 2015). However, 2DE is no longer the tool of choice in high-throughput differential proteomics because of the labour and time involved in producing 2DE gels. Via liquid chromatography with tandem mass spectrometry (LC-MSMS) multiple allele-specific products (tryptic-specific peptides) can be quantified in a high-throughput manner without required prior knowledge (Carpentier and America, 2014). Proteomes provide insights into the consequences of genomic merging and reorganization (Carpentier et al., 2011, Koh et al., 2012, Hu et al., 2014, Soltis et al., 2016, Campos et al., 2018).

3.2 Genetic modification and genome editing

Conventional banana breeding is complex and time-consuming. Specific challenges in breeding bananas for drought stress include the plants' perennial nature, non-seasonal flowering, physical size and reproductive barriers to hybridization that complicates the selection of superior genotypes in improvement schemes (Brown et al., 2020). Some of these intercrossing constraints might be alleviated by applying genetic modifications and genome editing in banana breeding (Tripathi et al., 2019). During the last decade, studies of banana genetic modification have focussed on *Agrobacterium*-mediated transformation or microprojectile bombardment (Ságí et al., 1995, Ganapathi et al., 2001, Khanna et al., 2004, Tripathi et al., 2015, Becker et al., 2000). However, novel genome-editing techniques offer improved efficiency and precision. Among these recently developed tools, CRISPR/Cas9 is feasible in plants and has received the most attention because of its lower cost, higher precision and the possibility to edit multiple sites across the genome (Woo et al., 2015, Bortesi and Fischer, 2015). The CRISPR/Cas9 technique has shown the potential to improve crop plants and has been developed for many plant species. Most progress has been made in model crops, but CRISPR/Cas9 has also been applied to many non-model crops including banana. As a proof of principle, Kaur et al. (2018) edited the *PDS* gene, resulting in an albino phenotype (Kaur et al., 2018). Tripathi and colleagues applied CRISPR/Cas9 to inactivate the endogenous banana streak virus in the B genome (Tripathi et al., 2019) and the group of J. Dale engineered the phytoene synthase to create golden bananas (Naim et al., 2018). Abiotic tolerances (i.e. water deficit, flooding, heat stress, chilling stress, wind stress and light stress) are all complex multi-gene traits and the identification of targets genes prior to be able to conduct genome editing is critical to alter the desired tolerance trait. The focus of phenotyping should therefore be on identifying and quantifying specific subtraits of abiotic tolerance. Through integration of phenotyping with genomics and proteomics techniques, genomic selection targets may be more efficiently identified based on a detailed physiological understanding of the desired trait (Rengel et al., 2012).

4 Conclusions

Temperature and precipitation are the abiotic factors having the biggest impact on banana production. Banana generally responds very sensitively to soil water deficit and high VPDs. Even under relatively high soil water potentials the yield is reduced and harvest postponed. Similarly, stomatal closure can occur under high VPDs even when sufficient water is available in the soil. Temperature strongly determines banana production and the use of a genotype-specific thermal time is recommended rather than absolute time to indicate development. Tolerances towards abiotic stresses such as water deficit and exposure to suboptimal temperatures are complex multi-gene traits and phenotyping should thereby focus on identifying and quantifying crucial traits. We propose a high-throughput phenotyping workflow under controlled environmental conditions as a first step in screening the existing biodiversity for relevant traits. Secondly, correlations between phenotypic traits and genetic markers need to be identified. Techniques such as GWAS, GS and associative transcriptomics and proteomics have been applied in banana and offer great potential. The identification of these markers will speed up breeding through marker-assisted selection and can be used for genome editing.

5 Where to look for further information

- https://www.crop-diversity.org/mgis/
  - Information on the *Musa* accessions present in 30 collections around the world.
- http://www.musanet.org/
  - The global collaborative framework for Musa genetic resources with general information, publications and projects.
- http://www.promusa.org/
  - A network of people promoting scientific discussions on bananas with general information and scientific symposia.
- http://www.egc.com/useful_info_lighting.php
  - Information about light conditions in growth chambers.
- https://banana-genome-hub.southgreen.fr/
  - The website of the banana genomics consortium.
- https://eppn2020.plant-phenotyping.eu/
  - Website on phenotyping tools and collaborations.
- https://www.cost.eu/actions/FA1306/
  - Website on the integration of phenotyping and biotechnology collaboration.

6 Acknowledgements

This work is mainly based on the results of PhD projects of Ewaut Kissel, Jelle van Wesemael and David Eyland. They are gratefully acknowledged for their hard work during their project. This work was supported by the Bioversity International projects 'Adding value to the ITC collection through molecular and phenotypic characterization', 'Phensee data' and 'More fruit for food security: developing climate-smart bananas for the African Great Lakes region', financed by the Belgian Development Cooperation; the CGIAR Research Program on Roots, Tubers and Bananas (RTB) and the KULeuven project LEAPAGRI. The Global Trust Foundation is gratefully acknowledged for the financing of the crop wild relative project and the PhD of David Eyland. COST Action Phenomenal FA 1306 is gratefully acknowledged for financing the participation at numerous meetings on phenotyping and omics integration and

Tracy Lawson is gratefully acknowledged for hosting us. EPPN^{2020} is gratefully acknowledged for financing the BananaDyn project in collaboration with INRAE Montpellier and François Tardieu and Claude Welcker are gratefully acknowledged for hosting us.

7 References

Adams, K. L., Cronn, R., Percifield, R. and Wendel, J. F. 2003. Genes duplicated by polyploidy show unequal contributions to the transcriptome and organ-specific reciprocal silencing. *Proceedings of the National Academy of Sciences of the United States of America* 100(8), 4649–4654.

Aguilar, E. A., Turner, D. W., Gibbs, D. J., Armstrong, W. and Sivasithamparam, K. 2003. Oxygen distribution and movement, respiration and nutrient loading in banana roots (Musa spp. L.) subjected to aerated and oxygen-depleted environments. *Plant and Soil* 253(1), 91–102.

Aguilar, E. A., Turner, D. W. and Sivasithamparam, K. 1999. Aerenchyma formation in roots of four banana (Musa spp.) cultivars. *Scientia Horticulturae* 80(1–2), 57–72.

Barker, M. S., Arrigo, N., Baniaga, A. E., Li, Z. and Levin, D. A. 2016. On the relative abundance of autopolyploids and allopolyploids. *New Phytologist* 210(2), 391–398.

Barrs, H. D. and Klepper, B. 1968. Cyclic variations in plant properties under constant environmental conditions. *Physiologia Plantarum* 21(4), 711–730.

Becker, D. K., Dugdale, B., Smith, M. K., Harding, R. M. and Dale, J. L. 2000. Genetic transformation of Cavendish banana (Musa spp. AAA group) cv 'Grand Nain' via microprojectile bombardment. *Plant Cell Reports* 19(3), 229–234.

Berger-Landefeldt, U. 1936. *Das Wasserhaushalt der Alpenpflanzen (Bibliotheca Botanica, Vol. 115)*. Stuttgart: Schweizerbartsche Verlagsbuchhandlung.

Berthezene, S., Brichet, N., Negre, V., Parent, B., Suard, B., Tireau, A., Turc, O., Tardieu, F. and Welcker, C. 2015. PHENODYN: a high throughput platform for measurement of organ elongation rate and plant transpiration with high temporal resolution. 2015; Recent progress in drought tolerance from genetics to modelling, Conference Handbook, Montpellier, France, 2015-06-08-2015-06-09, 80.

Blomme, G., Draye, X., Rufyikiri, G., Declerck, S., De Waele, D., Tenkouano, A. and Swennen, R. 1999. Progress in understanding the roots of Musa spp. INIBAP annual report, 14–19.

Bortesi, L. and Fischer, R. 2015. The CRISPR/Cas9 system for plant genome editing and beyond. *Biotechnology Advances* 33(1), 41–52.

Brady, N. C. 1984. *The Nature and Properties of Soils*. New York: Macmillian.

Bramley, H., Turner, D. W., Tyerman, S. D. and Turner, N. C. 2007. Water flow in the roots of crop species: the influence of root structure, aquaporin activity, and waterlogging. *Advances in Agronomy* 96, 133–196.

Bramley, H., Turner, N. C., Turner, D. W. and Tyerman, S. D. 2009. Roles of morphology, anatomy, and aquaporins in determining contrasting hydraulic behavior of roots. *Plant Physiology* 150(1), 348–364.

Brown, A., Carpentier, S. C. and Swennen, R. 2020. Breeding climate-resilient bananas. In: Kole, C. (Ed.) *Genomic Designing of Climate-Smart Fruit Crops*. Cham: Springer, pp. 91–115.

Buckley, T. N. 2005. The control of stomata by water balance. *New Phytologist* 168(2), 275–292.

Calberto, G., Staver, C. and Siles, P. 2015. An assessment of global banana production and suitability under climate change scenarios. *Climate Change and Food Systems: Global Assessments and Implications for Food Security and Trade*, Rome: Food Agriculture Organization of the United Nations (FAO).

Campos, N. A., Swennen, R. and Carpentier, S. C. 2018. The plantain proteome, a focus on allele specific proteins obtained from plantain fruits. *Proteomics* 18(3–4), 29333696.

Carpentier, S. C., Panis, B., Renaut, J., Samyn, B., Vertommen, A., Vanhove, A. C., Swennen, R. and Sergeant, K. 2011. The use of 2D-electrophoresis and de novo sequencing to characterize inter- and intra-cultivar protein polymorphisms in an allopolyploid crop. *Phytochemistry* 72(10), 1243–1250.

Carpentier, S. C. 2016. Multiple testing and pattern recognition in 2-DE proteomics. In: Marengo, E. and Robotti, E. (Eds) *2-D PAGE Map Analysis*. New York: Springer.

Carpentier, S. C. and America, T. 2014. Proteome analysis of orphan plant species, fact or fiction? In: Jorrin-Novo, J., Komatsu, S., Weckwerth, W. and Wienkoop, S. (Eds) *Plant Proteomics*. New York: Springer.

Carr, M. K. V. 2009. The water relations and irrigation requirements of banana (Musa spp.). *Experimental Agriculture* 45(3), 333–371.

Cenci, A., Hueber, Y., Zorrilla-Fontanesi, Y., van Wesemael, J., Kissel, E., Gislard, M., Sardos, J., Swennen, R., Roux, N., Carpentier, S. C. and Rouard, M. 2019. Effect of paleopolyploidy and allopolyploidy on gene expression in banana. *BMC Genomics* 20(1), 244.

Christelova, P., De Langhe, E., Hribova, E., Cizkova, J., Sardos, J., Husakova, M., Van Den Houwe, I., Sutanto, A., Kepler, A. K., Swennen, R., Roux, N. and Dolezel, J. 2017. Molecular and cytological characterization of the global Musa germplasm collection provides insights into the treasure of banana diversity. *Biodiversity and Conservation* 26(4), 801–824.

Christelová, P., Valárik, M., Hřibová, E., Van Den Houwe, I., Channelière, S., Roux, N. and Doležel, J. 2011a. A platform for efficient genotyping in Musa using microsatellite markers. *AoB Plants* 2011, plr024.

Christelová, P., Valárik, M., Hřibová, E., De Langhe, E. and Doležel, J. 2011b. A multi gene sequence-based phylogeny of the Musaceae (banana) family. *BMC Evolutionary Biology* 11, 103.

Craenen, K., Coosemans, J. and Ortiz, R. 1997. The role of stomatal traits and epicuticular wax in resistance to Mycosphaerella fijiensis in Banana and Plantain (Musa spp.). *Tropicultura* 15, 136–140.

D'hont, A., Denoeud, F., Aury, J. M., Baurens, F. C., Carreel, F., Garsmeur, O., Noel, B., Bocs, S., Droc, G., Rouard, M., Da Silva, C., Jabbari, K., Cardi, C., Poulain, J., Souquet, M., Labadie, K., Jourda, C., Lengellé, J., Rodier-Goud, M., Alberti, A., Bernard, M., Correa, M., Ayyampalayam, S., Mckain, M. R., Leebens-Mack, J., Burgess, D., Freeling, M., Mbéguié-A-Mbéguié, D., Chabannes, M., Wicker, T., Panaud, O., Barbosa, J., Hribova, E., Heslop-Harrison, P., Habas, R., Rivallan, R., Francois, P., Poiron, C., Kilian, A., Burthia, D., Jenny, C., Bakry, F., Brown, S., Guignon, V., Kema, G., Dita, M., Waalwijk, C., Joseph, S., Dievart, A., Jaillon, O., Leclercq, J., Argout, X., Lyons, E., Almeida, A., Jeridi, M., Dolezel, J., Roux, N., Risterucci, A. M., Weissenbach, J., Ruiz, M., Glaszmann, J. C., Quétier, F., Yahiaoui, N. and Wincker, P. 2012. The banana

(Musa acuminata) genome and the evolution of monocotyledonous plants. *Nature* 488(7410), 213–217.
Davey, M. W., Gudimella, R., Harikrishna, J. A., Sin, L. W., Khalid, N. and Keulemans, J. 2013. A draft Musa balbisiana genome sequence for molecular genetics in polyploid, inter-and intra-specific Musa hybrids. *BMC Genomics* 14, 683.
Delorge, I., Janiak, M., Carpentier, S. and Van Dijck, P. 2014. Fine tuning of trehalose biosynthesis and hydrolysis as novel tools for the generation of abiotic stress tolerant plants. *Frontiers in Plant Science* 5, 147.
Drake, P. L., Froend, R. H. and Franks, P. J. 2013. Smaller, faster stomata: scaling of stomatal size, rate of response, and stomatal conductance. *Journal of Experimental Botany* 64(2), 495–505.
Draye, X. 2002. Banana roots: architecture and genetics. *Plant Roots: the Hidden Half* 3, 261–277.
Eckstein, K., Robinson, J. C. and Fraser, C. 1997. Physiological responses of banana (Musa AAA; Cavendish sub-group) in the subtropics. VII. Effects of windbreak shading on phenology, physiology and yield. *Journal of Horticultural Science* 72(3), 389–396.
Ekanayake, I., Ortiz, R. and Vuylsteke, D. 1994. Influence of leaf age, soil moisture, VPD and time of day on leaf conductance of various Musa genotypes in a humid forest-moist savanna transition site. *Annals of Botany* 74(2), 173–178.
Faralli, M., Cockram, J., Ober, E., Wall, S., Galle, A., Van Rie, J., Raines, C. A. and Lawson, T. 2019. Genotypic, developmental and environmental effects on the rapidity of gs in wheat: impacts on carbon gain and water-use efficiency. *Frontiers in Plant Science* 10, 492.
Farquhar, G. D. and Cowan, I. R. 1974. Oscillations in stomatal conductance: the influence of environmental gain. *Plant Physiology* 54(5), 769–772.
Fick, S. E. and Hijmans, R. J. 2017. WorldClim 2: new 1-km spatial resolution climate surfaces for global land areas. *International Journal of Climatology* 37(12), 4302–4315.
Fischer, R. A., Rees, D., Sayre, K. D., Lu, Z. -M., Condon, A. G. and Saavedra, A. L. 1998. Wheat yield progress associated with higher stomatal conductance and photosynthetic rate, and cooler canopies. *Crop Science* 38(6), 1467–1475.
Fortescue, J. A., Turner, D. W. and Romero, R. 2011. Evidence that banana (Musa spp.), a tropical monocotyledon, has a facultative long-day response to photoperiod. *Functional Plant Biology* 38(11), 867–878.
Franks, P. J. 2006. Higher rates of leaf gas exchange are associated with higher leaf hydrodynamic pressure gradients. *Plant, Cell and Environment* 29(4), 584–592.
Franks, P. J. and Farquhar, G. D. 2007. The mechanical diversity of stomata and its significance in gas-exchange control. *Plant Physiology* 143(1), 78–87.
Freeman, B. and Turner, D. 1985. The epicuticular waxes on the organs of different varieties of banana (Musa spp.) differ in form, chemistry and concentration. *Australian Journal of Botany* 33(4), 393–408.
Ganapathi, T. R., Higgs, N. S., Balint-Kurti, P. J., Arntzen, C. J., May, G. D. and Van Eck, J. M. 2001. Agrobacterium-mediated transformation of embryogenic cell suspensions of the banana cultivar Rasthali (AAB). *Plant Cell Reports* 20(2), 157–162.
Ganry, J. and Chillet, M. 2008. Methodology to forecast the harvest date of banana bunches. *Fruits* 63(6), 371–373.

Gibbs, D. and Turner, D. 2018. A functional approach to bunch formation in banana. In: Kema, G. and Drenth, A. (Eds) *Achieving Sustainable Cultivation of Bananas*. Cambridge: Burleigh Dodds Science Publishing.

Hamblin, M. T., Buckler, E. S. and Jannink, J. L. 2011. Population genetics of genomics-based crop improvement methods. *Trends in Genetics* 27(3), 98–106.

Harper, A. L., Trick, M., Higgins, J., Fraser, F., Clissold, L., Wells, R., Hattori, C., Werner, P. and Bancroft, I. 2012. Associative transcriptomics of traits in the polyploid crop species Brassica napus. *Nature Biotechnology* 30(8), 798–802.

Hu, G., Koh, J., Yoo, M. J., Pathak, D., Chen, S. and Wendel, J. F. 2014. Proteomics profiling of fiber development and domestication in upland cotton (*Gossypium hirsutum* L.). *Planta* 240(6), 1237–1251.

Huber, A. E., Melcher, P. J., Piñeros, M. A., Setter, T. L. and Bauerle, T. L. 2019. Signal coordination before, during and after stomatal closure in response to drought stress. *New Phytologist* 224(2), 675–688.

Janssens, S. B., Vandelook, F., De Langhe, E., Verstraete, B., Smets, E., Vandenhouwe, I. and Swennen, R. 2016. Evolutionary dynamics and biogeography of Musaceae reveal a correlation between the diversification of the banana family and the geological and climatic history of Southeast Asia. *New Phytologist* 210(4), 1453–1465.

Kaur, N., Alok, A., Shivani, Kaur, N., Pandey, P., Awasthi, P. and Tiwari, S. 2018. CRISPR/Cas9-mediated efficient editing in phytoene desaturase (PDS) demonstrates precise manipulation in banana cv. Rasthali genome. *Functional and Integrative Genomics* 18(1), 89–99.

Khan, A., Belfield, E. J., Harberd, N. P. and Mithani, A. 2016. HANDS2: accurate assignment of homoeallelic base-identity in allopolyploids despite missing data. *Scientific Reports* 6, 29234.

Khanna, H., Becker, D., Kleidon, J. and Dale, J. 2004. Centrifugation assisted Agrobacterium tumefaciens-mediated transformation (CAAT) of embryogenic cell suspensions of banana (Musa spp. Cavendish AAA and Lady finger AAB). *Molecular Breeding* 14(3), 239–252.

Kissel, E., Van Asten, P., Swennen, R., Lorenzen, J. and Carpentier, S. C. 2015. Transpiration efficiency versus growth: exploring the banana biodiversity for drought tolerance. *Scientia Horticulturae* 185, 175–182.

Koh, J., Chen, S., Zhu, N., Yu, F., Soltis, P. S. and Soltis, D. E. 2012. Comparative proteomics of the recently and recurrently formed natural allopolyploid Tragopogon mirus (Asteraceae) and its parents. *New Phytologist* 196(1), 292–305.

Lassoudière, A. 2012. *Le bananier: Un siècle d'innovations techniques*. Paris: Editions Quae.

Machovina, B. and Feeley, K. J. 2013. Climate change driven shifts in the extent and location of areas suitable for export banana production. *Ecological Economics* 95, 83–95.

Martin, G., Baurens, F. C., Droc, G., Rouard, M., Cenci, A., Kilian, A., Hastie, A., Doležel, J., Aury, J. M., Alberti, A., Carreel, F. and D'hont, A. 2016. Improvement of the banana "Musa acuminata" reference sequence using NGS data and semi-automated bioinformatics methods. *BMC Genomics* 17, 243.

Maurel, C., Boursiac, Y., Luu, D. T., Santoni, V., Shahzad, Z. and Verdoucq, L. 2015. Aqusaporins in plants. *Physiological Reviews* 95(4), 1321–1358.

Mboula, L. 2014. Ecophysiology of dwarf plantain hybrids in peri-urban areas of Cameroon. Doctoral dissertation, Université catholique de Louvain.

McAdam, S. A. and Brodribb, T. J. 2015. The evolution of mechanisms driving the stomatal response to vapor pressure deficit. *Plant Physiology* 167(3), 833–843.

McAusland, L., Vialet-Chabrand, S., Davey, P., Baker, N. R., Brendel, O. and Lawson, T. 2016. Effects of kinetics of light-induced stomatal responses on photosynthesis and water-use efficiency. *New Phytologist* 211(4), 1209–1220.

McKown, A. D., Cochard, H. and Sack, L. 2010. Decoding leaf hydraulics with a spatially explicit model: principles of venation architecture and implications for its evolution. *The American Naturalist* 175(4), 447–460.

Mencuccini, M., Mambelli, S. and Comstock, J. 2000. Stomatal responsiveness to leaf water status in common bean (*Phaseolus vulgaris* L.) is a function of time of day. *Plant, Cell and Environment* 23(10), 1109–1118.

Muchow, R. C. and Sinclair, T. R. 1989. Epidermal conductance, stomatal density and stomatal size among genotypes of *Sorghum bicolor* (L.) Moench. *Plant, Cell and Environment* 12(4), 425–431.

Naim, F., Dugdale, B., Kleidon, J., Brinin, A., Shand, K., Waterhouse, P. and Dale, J. 2018. Gene editing the phytoene desaturase alleles of Cavendish banana using CRISPR/Cas9. *Transgenic Research* 27(5), 451–460.

Nyine, M., Uwimana, B., Akech, V., Brown, A., Ortiz, R., Doležel, J., Lorenzen, J. and Swennen, R. 2019. Association genetics of bunch weight and its component traits in East African highland banana (Musa spp. AAA group). *Theoretical and Applied Genetics* 132(12), 3295–3308.

Ohsumi, A., Kanemura, T., Homma, K., Horie, T. and Shiraiwa, T. 2007. Genotypic variation of stomatal conductance in relation to stomatal density and length in rice (*Oryza sativa* L.). *Plant Production Science* 10(3), 322–328.

Page, J. T., Gingle, A. R. and Udall, J. A. 2013. PolyCat: a resource for genome categorization of sequencing reads from allopolyploid organisms. *G3: Genes, Genomes, Genetics* 3(3), 517–525.

Perrier, X., Jenny, C., Bakry, F., Karamura, D., Kitavi, M., Dubois, C., Hervouet, C., Philippson, G. and De Langhe, E. 2018. East African diploid and triploid bananas: a genetic complex transported from South-East Asia. *Annals of Botany* 123(1), 19–36.

Raschke, K. 1970. Leaf hydraulic system: rapid epidermal and stomatal responses to changes in water supply. *Science* 167(3915), 189–191.

Rengel, D., Arribat, S., Maury, P., Martin-Magniette, M. L., Hourlier, T., Laporte, M., Varès, D., Carrère, S., Grieu, P., Balzergue, S., Gouzy, J., Vincourt, P. and Langlade, N. B. 2012. A gene-phenotype network based on genetic variability for drought responses reveals key physiological processes in controlled and natural environments. *PLoS ONE* 7(10), e45249.

Risterucci, A. M., Hippolyte, I., Perrier, X., Xia, L., Caig, V., Evers, M., Huttner, E., Kilian, A. and Glaszmann, J. C. 2009. Development and assessment of Diversity Arrays Technology for high-throughput DNA analyses in Musa. *Theoretical and Applied Genetics* 119(6), 1093–1103.

Robinson, J. C. and Alberts, A. J. 1986. Growth and yield responses of banana (cultivar 'Williams') to drip irrigation under drought and normal rainfall conditions in the subtropics. *Scientia Horticulturae* 30(3), 187–202.

Robinson, J. C. and Bower, J. P. 1987. Transpiration characteristics of banana leaves (cultivar 'Williams') in response to progressive depletion of available soil moisture. *Scientia Horticulturae* 30(4), 289–300.

Robinson, J. C. and Bower, J. P. 1988. Transpiration from banana leaves in the subtropics in response to diurnal and seasonal factors and high evaporative demand. *Scientia Horticulturae* 37(1-2), 129-143.

Robinson, J. C. 1996a. Phenological and physiological responses. In: Atherthon, J. (Ed.) *Bananas and Plantains*. Wallingford: CAB International.

Robinson, J. C. 1996b. Water Requirements and irrigation. In: Atherthon, J. (Ed.) *Bananas and Plantains*. Wallingford: CAB International.

Rouard, M., Droc, G., Martin, G., Sardos, J., Hueber, Y., Guignon, V., Cenci, A., Geigle, B., Hibbins, M. S., Yahiaoui, N., Baurens, F. C., Berry, V., Hahn, M. W., D'hont, A. and Roux, N. 2018. Three new genome assemblies support a rapid radiation in Musa acuminata (wild banana). *Genome Biology and Evolution* 10(12), 3129-3140.

Sack, L. and Holbrook, N. M. 2006. Leaf hydrolics. *Annual Review of Plant Biology* 57, 361-381.

Sági, L., Panis, B., Remy, S., Schoofs, H., De Smet, K., Swennen, R. and Cammue, B. P. 1995. Genetic transformation of banana and plantain (Musa spp.) via particle bombardment. *Bio/Technology* 13(5), 481-485.

Samyn, B., Sergeant, K., Carpentier, S., Debyser, G., Panis, B., Swennen, R. and Van Beeumen, J. 2007. Functional proteome analysis of the banana plant (Musa spp.) using de novo sequence analysis of derivatized peptides. *Journal of Proteome Research* 6(1), 70-80.

Santos, A. S., Amorim, E. P., Ferreira, C. F. and Pirovani, C. P. 2018. Water stress in Musa spp.: a systematic review. *PLoS ONE* 13(12), e0208052.

Sardos, J., Rouard, M., Hueber, Y., Cenci, A., Hyma, K. E., Van Den Houwe, I., Hribova, E., Courtois, B. and Roux, N. 2016. A genome-wide association study on the seedless phenotype in banana (Musa spp.) reveals the potential of a selected panel to detect candidate genes in a vegetatively propagated crop. *PLoS ONE* 11(5), e0154448.

Savvides, A., Fanourakis, D. and Van Ieperen, W. 2011. Co-ordination of hydraulic and stomatal conductances across light qualities in cucumber leaves. *Journal of Experimental Botany* 63(3), 1135-1143.

Siles, P., Bustamante, O., Valdivia, E., Burkhardt, J. and Staver, C. 2011. Photosynthetic performance of banana ('Gros Michel', AAA) under a natural shade gradient. VII International Symposium on Banana: ISHS-ProMusa Symposium on Bananas and Plantains: Towards Sustainable Global Production. *Acta Horticulture* 986, 71-77.

Simmonds, N. W. 1962. *The Evolution of the Bananas*, London: Longmans, Green & Co.

Soltis, D. E., Misra, B. B., Shan, S., Chen, S. and Soltis, P. S. 2016. Polyploidy and the proteome. *Biochimica et Biophysica Acta* 1864(8), 896-907.

Stover, R. H. 1972. *Banana, Plantain and Abaca Diseases*. Kew, UK: Commonwealth Mycological Institute.

Swennen, R., De Langhe, E., Janssen, J. and Decoene, D. 1986. Study of the root development of some Musa cultivars in hydroponics. *Fruits* 41, 515-524.

Tardieu, F. 2016. Too many partners in root-shoot signals. Does hydraulics qualify as the only signal that feeds back over time for reliable stomatal control? *New Phytologist* 212(4), 802-804.

Tardieu, F., Draye, X. and Javaux, M. 2017. Root water uptake and ideotypes of the root system: whole-plant controls matter. *Vadose Zone Journal* 16(9), 1-10.

Thomas, D. S. and Turner, D. W. 1998. Leaf gas exchange of droughted and irrigated banana cv. Williams (Musa spp.) growing in hot, arid conditions. *The Journal of Horticultural Science and Biotechnology* 73(3), 419-429.

Thomas, D. S., Turner, D. W. and Eamus, D. 1998. Independent effects of the environment on the leaf gas exchange of three banana (Musa sp.) cultivars of different genomic constitution. *Scientia Horticulturae* 75(1–2), 41–57.

Tripathi, J. N., Ntui, V. O., Ron, M., Muiruri, S. K., Britt, A. and Tripathi, L. 2019. CRISPR/Cas9 editing of endogenous banana streak virus in the B genome of Musa spp. overcomes a major challenge in banana breeding. *Communications Biology* 2, 1–11.

Tripathi, J. N., Oduor, R. O. and Tripathi, L. 2015. A high-throughput regeneration and transformation platform for production of genetically modified banana. *Frontiers in Plant Science* 6, 1025.

Turner, D. 1990. Modelling demand for nitrogen in the banana. International Symposium on the Culture of Subtropical and Tropical Fruits and Crops. *Acta Horticulture* 275, 497–504.

Turner, D. 2003. Factors affecting the physiology of the banana root system. Montpellier: International Network for the Improvement of Banana and Plantain (INIBAP), pp. 107–113.

Turner, D. and Fortescue, J. A. 2010. Bananas and mangoes at the margins-approaches to flowering for productive management. VI International Symposium on Banana: XXVIII International Horticultural Congress on Science and Horticulture for People. *Acta Horticulturae* 928, 113–120.

Turner, D. and Hunt, N. 1984. Growth, yield and leaf nutrient composition of 30 banana varieties in subtropical New South Wales. *Technical Bulletin* 31, 1–36.

Turner, D. W. and Thomas, D. S. 1998. Measurements of plant and soil water status and their association with leaf gas exchange in banana (Musa spp.): a laticiferous plant. *Scientia Horticulturae* 77(3–4), 177–193.

Turner, D. W., Fortescue, J. A., Ocimati, W. and Blomme, G. 2016. Plantain cultivars (Musa spp. AAB) grown at different altitudes demonstrate cool temperature and photoperiod responses relevant to genetic improvement. *Field Crops Research* 194, 103–111.

Turner, D. W., Fortescue, J. A. and Thomas, D. S. 2007. Environmental physiology of the bananas (Musa spp.). *Brazilian Journal of Plant Physiology* 19(4), 463–484.

Turner, W. 1995. The response of the plant to the environment. In: Gowen, S. (Ed.) *Bananas and Plantains*. London: Chapman and Hall.

Turner, W., Fortescue, J. A. and Thomas, D. S. 2009. Bananas: environment and crop physiology. In: Da Metta, F. (Ed.) *Ecophysiology of Tropical Tree Crops*. New York: Nova Science Publishers.

Vandenhout, H., Ortiz, R., Vuylsteke, D., Swennen, R. and Bai, K. V. 1995. Effect of ploidy on stomatal and other quantitative traits in plantain and banana hybrids. *Euphytica* 83(2), 117–122.

Vandenhouwe, I., Lepoivre, P., Swennen, R., Frison, E. and Sharrock, S. 2003. The world banana heritage conserved in Belgium for the benefit of small-scale farmers in the tropics. *Plant Genetic Resources Newsletter* 19, 18–23.

Vanhove, A. C., Vermaelen, W., Swennen, R. and Carpentier, S. C. 2015. A look behind the screens: characterization of the HSP70 family during osmotic stress in a non-model crop. *Journal of Proteomics* 119, 10–20.

van Wesemael, J. 2019. Mining the Musa biodiversity for drought tolerance: allele discovery *via* integrated phenomics, proteomics and transcriptomics. PhD in Bioscience Engineering, KULeuven.

van Wesemael, J., Hueber, Y., Kissel, E., Campos, N., Swennen, R. and Carpentier, S. 2018. Homeolog expression analysis in an allotriploid non-model crop via integration of transcriptomics and proteomics. *Scientific Reports* 8(1), 1353.

van Wesemael, J., Kissel, E., Eyland, D., Lawson, T., Swennen, R. and Carpentier, S. C. 2019. Using growth and transpiration phenotyping under controlled conditions to select water efficient banana genotypes. *Frontiers in Plant Science* 10, 352.

Varma, V. and Bebber, D. P. 2019. Climate change impacts on banana yields around the world. *Nature Climate Change* 9(10), 752–757.

Wang, G. X., Zhang, J., Liao, J.-X. and Wang, J.-L. 2001. Hydropassive evidence and effective factors in stomatal oscillations of Glycyrrhiza inflata under desert conditions. *Plant Science* 160(5), 1007–1013.

Wang, Z., Miao, H., Liu, J., Xu, B., Yao, X., Xu, C., Zhao, S., Fang, X., Jia, C., Wang, J., Zhang, J., Li, J., Xu, Y., Wang, J., Ma, W., Wu, Z., Yu, L., Yang, Y., Liu, C., Guo, Y., Sun, S., Baurens, F. C., Martin, G., Salmon, F., Garsmeur, O., Yahiaoui, N., Hervouet, C., Rouard, M., Laboureau, N., Habas, R., Ricci, S., Peng, M., Guo, A., Xie, J., Li, Y., Ding, Z., Yan, Y., Tie, W., D'hont, A., Hu, W. and Jin, Z. 2019. Musa balbisiana genome reveals subgenome evolution and functional divergence. *Nature Plants* 5(8), 810–821.

Woo, J. W., Kim, J., Kwon, S. I., Corvalán, C., Cho, S. W., Kim, H., Kim, S. G., Kim, S. T., Choe, S. and Kim, J. S. 2015. DNA-free genome editing in plants with preassembled CRISPR-Cas9 ribonucleoproteins. *Nature Biotechnology* 33(11), 1162–1164.

Yang, H. M., Zhang, J. H. and Zhang, X. Y. 2005. Regulation mechanisms of stomatal oscillation. *Journal of Integrative Plant Biology* 47(10), 1159–1172.

Zait, Y., Shapira, O. and Schwartz, A. 2017. The effect of blue light on stomatal oscillations and leaf turgor pressure in banana leaves. *Plant, Cell and Environment* 40(7), 1143–1152.

Zeiger, E. and Field, C. 1982. Photocontrol of the functional coupling between photosynthesis and stomatal conductance in the intact leaf: blue light and PAR-dependent photosystems in guard cells. *Plant Physiology* 70(2), 370–375.

Zivy, M., Wienkoop, S., Renaut, J., Pinheiro, C., Goulas, E. and Carpentier, S. 2015. The quest for tolerant varieties: the importance of integrating "omics" techniques to phenotyping. *Frontiers in Plant Science* 6, 448.

Index